GUIDE TO THE CHEMICAL INDUSTRY

GUIDE TO THE CHEMICAL INDUSTRY

Technology, R&D, Marketing, and Employment

WILLIAM S. EMERSON
Consultant
Damariscotta, Maine

A Wiley-Interscience Publication
JOHN WILEY & SONS
New York · Chichester · Brisbane · Toronto · Singapore

Library of Congress Cataloging in Publication Data:

Emerson, William S. (William Stevenson), 1913–
Guide to the chemical industry.

"A Wiley-Interscience publication."
Includes bibliographical references and index.
1. Chemistry, Technical. 2. Chemical industries.
I. Title.

TP145.E46 1983 660′.023 83-7035
ISBN 0-471-89040-5

Printed in the United States of America

10 9 8 7 6 5 4 3 2 1

Lovingly dedicated to
Millicent, my dear wife

PREFACE

Chemistry in the Economy [1] has emphasized the need for chemists and chemical engineers, both in industry and academia, to be more broadly familiar with what goes on in the chemical industry. With the encouragement of Professor Robert B. Carlin of Carnegie–Mellon University, I put together a series of seminars that were designed to fill this need for chemistry and chemical engineering students. The late Dr. Byron Riegel very kindly gave me a list of the seminars he was giving at that time at Northwestern University along these same lines.

My original plan was to adapt the material I had helped present to a group of research directors in Egypt, which Dr. Lawrence W. Bass, the director of that course, has since published [2]. However, it became apparent very early on that the needs of Egyptian research directors and those of American chemists and chemical engineers bore little resemblance to each other. The result has been a series of quite different seminars, one or more of which have been delivered at some 23 colleges and universities throughout the country. I have received feedback from students who have attended some of these seminars that their transition to industrial life has been less traumatic than otherwise would have been the case. I am particularly grateful to Professors C. S. Marvel, Henry K. Hall, Jr., and Lee B. Jones of the University of Arizona and to Professor Therald Moeller of Arizona State University for their enthusiastic support during the early days of this venture.

Once the nature and content of the seminars had become reasonably established, it appeared that it would be helpful if they were available in book form—hence the present volume. This book will give chemistry and chemical engineering students an understanding of what the chemical industry is like. It also will allow chemists and chemical engineers, in both industry and academia, an opportunity to refresh their minds with regard to the total industrial chemical picture. The aim is understanding, not knowledge. For instance, there are excellent treatises that cover industrial chemistry, polymer chemistry, and chemical engineering in far more detail and are presently being used as the texts for full-year courses at many universities.

This book can easily be read by the chemist or chemical engineer in a brief period of time. In the case of students it can be used as the basis for a one-hour-a-week, one-semester seminar series that would not disrupt the full-time curriculum of required courses.

Since each chapter constitutes a single seminar, there is some repetition among chapters. However, every effort has been made to eliminate redundancy and leave only that repetition which is desirable for emphasis. It also should be recognized that most of the material can be found in various places in the literature, both in journals and in parts of books, but nowhere else has it been assembled in one brief volume. The book also reflects some of my personal biases, which have been acquired over the years in the chemical industry.

I am grateful to Professor Charles G. Overberger of the University of Michigan and to Professor Cheves Walling of the University of Utah for very helpful general suggestions about the content of the book, arrangement, and emphasis. The seminars on industrial chemistry and petrochemistry were prepared at the request of Professor Louis D. Quin of Duke University in connection with a series of talks I gave at that school. I also am very grateful to Professor Quin for urging me to include a chapter on polymer chemistry and one on chemicals from petroleum, for urging me to emphasize chemical engineering, and for his encouragement and support of the entire idea.

Furthermore I wish to express my particular gratitude to a number of people for their encouragement and helpfulness:

To Professor George B. Butler of the University of Florida for a very helpful copy of the chapter on polymer chemistry in his textbook [3].

To Professor Norman H. Cromwell of the University of Nebraska for reviewing the chapters on research, on development, and on industrial chemistry.

To Professor Dana W. Mayo of Bowdoin College for comments on these same chapters.

To Professor C. S. Marvel of the University of Arizona for reviewing the chapter on polymer chemistry.

To Professor Hal G. Johnson of Northern Illinois University, formerly director of Monsanto Company's general development department, for reviewing the chapter on marketing.

To Frederick M. Murdock, Esq., formerly associate director of Monsanto Company's patent department, for reviewing the material on U.S. patents and for providing background information for the section on foreign patents.

To Michael N. Meller, Esq., of Handal, Meller, and Morofsky for providing background information for the section on foreign patents and for a most helpful telephone interview that cleared up several important points in this connection.

To Mr. Monte C. Throdahl, group vice-president, Monsanto Company, and his colleagues for reviewing the entire manuscript.

To Dr. John Mahan and his colleagues at the Phillips Petroleum Company for reviewing the chapter on chemicals from petroleum.

To Mr. Paul Sanderson and Mr. Charles Jenest of Arthur D. Little, Inc., for many of the marketing and production figures in Chapters 3 and 5.

To Professor Norman Rabjohn of the University of Missouri and to Ms. Norma B. Mar of the Chemists Club (New York) Library for assistance with references.

To my wife, Millicent, and my daughter, Linda Rae, for helpful comments here and there in the manuscript.

To Mr. and Mrs. Kritz Meserve of Victoria Print Shop, Damariscotta, Maine, for preparing the printed copies of some of the more complicated formulas.

I also wish to express my appreciation to my secretary, Mrs. Mary N. Orrick of Pemaquid Point, Maine, for the many hours spent typing the material, much of it in a totally unfamiliar language.

REFERENCES

1. *Chemistry in the Economy,* American Chemical Society, Washington, D.C., 1973, Chapter 24.
2. Lawrence W. Bass, *The Management of Technical Programs,* Praeger, New York, 1965.
3. G. B. Butler and K. D. Berlin, *Fundamentals of Organic Chemistry,* Ronald Press, New York, 1972, Chapter 34.

WILLIAM S. EMERSON

Damariscotta, Maine
July 1983

CONTENTS

PART ONE CHEMISTRY AND PROCESSING

PART THREE CAREERS

GUIDE TO THE CHEMICAL INDUSTRY

PART ONE

CHEMISTRY AND PROCESSING

1

THE CHEMICAL INDUSTRY

I. INTRODUCTION

The object of this chapter is to give the beginning chemist some idea of the scope of the chemical industry and therefore of the career opportunities available to him or her as a member of the chemical or chemical engineering profession. An attempt has been made to indicate as many as possible of the industrial areas in which chemistry finds use. Obviously such a survey cannot be complete in one short chapter. Statistics have been omitted deliberately with the hope that the vastness of at least some of the operations is obvious to the reader from everyday experience.

For the purposes of this chapter the chemical industry is defined as that segment of the economy which isolates very simple, naturally occurring raw materials, transforms them through chemical reactions into commercially useful products, and participates industrially in the use of these products. Any segment of the industry or any individual company may only participate in one of these three activities. This definition extends the scope of the chemical industry far beyond what are known as strictly chemical companies. However, it does rule out such industries as coal mining, since coal is pretty much sold and used as fuel in the form in which it is mined. As the sources of petroleum and natural gas become depleted, so not only liquid and gaseous fuels, but also key chemical raw materials must be manufactured from coal by chemical processes, coal mining along with its initial conversions may well become part of the chemical industry. One present exception to initial processing being part of the chemical industry will be mentioned shortly, but this exception is conducted by chemical companies, not by coal companies. Likewise the use of ethical pharmaceuticals is not part of the chemical industry, since it is supervised by physicians.

To emphasize the simplicity of the basic chemical raw materials, a minute will be taken here to review the preparation of some of the key ones.

This chapter is based on material found in *Chemistry in the Economy,* American Chemical Society, Washington, D.C., 1973, and R. Norris Shreve, *Chemical Process Industries,* Third Edition, McGraw-Hill, New York, 1967.

Chlorine and caustic soda are very important. They are prepared along with hydrogen by the electrolysis of aqueous sodium chloride:

$$2NaCl + 2H_2O \xrightarrow{\text{elec}} Cl_2 + 2NaOH + H_2 \tag{1}$$

Most of the salt comes either from a mine or from a brine well.

Both nitrogen and oxygen are used as such. They are obtained by the fractional distillation of air. Most of the nitrogen is first converted to ammonia by hydrogenation and then to a whole host of nitrogen-containing chemical compounds.

$$N_2 + 3H_2 \xrightarrow[\text{500°C, 150–300 atm}]{Fe \cdot K_2O \cdot Al_2O_3} 2NH_3 \tag{2}$$

Probably the most important mineral acid is sulfuric. Most of it is prepared by the stepwise oxidation of sulfur:

$$S + O_2 \longrightarrow SO_2 \xrightarrow[\text{575°C}]{O_2,\ V_2O_5} SO_3 \xrightarrow{H_2O} H_2SO_4 \tag{3}$$

The sulfur is mined as such from underground deposits (Frasch process), or obtained as sulfur dioxide either from roasting sulfide ores such as iron pyrites or from scrubbing furnace off gases.

Phosphoric acid is the basis for phosphorus chemistry. Phosphate rock is mined as such, and much of it is treated with sulfuric acid to give phosphoric acid or various phosphates:

$$M_3PO_4 + H_2SO_4 \longrightarrow H_3PO_4 \tag{4}$$

Calcium oxide is an important chemical by itself and as the corresponding hydroxide after hydration. Limestone (calcium carbonate) is mined as such and converted to calcium oxide by heat:

$$CaCO_3 \xrightarrow{\Delta} CaO + CO_2 \xrightarrow{H_2O} Ca(OH)_2 \tag{5}$$

Hydrogen has already been mentioned. Most of it is made by treating methane or naphtha with steam at high temperatures:

$$C_xH_y + H_2O \xrightarrow{Ni \cdot Al_2O_3 \cdot SiO_2} H_2 + CO \tag{6}$$

It also can be made from coke:

$$C + H_2O \longrightarrow H_2 + CO \tag{7}$$

Although this latter process is uneconomical at present, it could become important as the world's supplies of petroleum and natural gas become depleted.

Benzene is the raw material for many aromatic chemicals. For a long time it was obtained by the fractional distillation of coal tar. Now it is manufactured by cracking and reforming petroleum.

Ethylene and propylene are the raw materials for a wide variety of aliphatic chemicals. They are obtained by cracking petroleum.

At one time acetylene was the raw material for many aliphatic chemicals, although ethylene is now used instead, as petroleum has superseded coal as the source of carbon compounds. The original acetylene process was based on calcium carbide:

$$3C + CaO \xrightarrow{2000°C} CO + CaC_2 \xrightarrow{H_2O} HC{\equiv}CH \tag{8}$$

Acetylene now is obtained by cracking methane at very high temperatures.

This very brief review shows how very simple are the raw materials upon which the chemical industry is based. The conversion of these raw materials to a myriad of other useful chemicals has been the subject of the chemist's and chemical engineer's education. Now how are some of these many chemicals used in a large number of fields?

Following are two specific examples that illustrate the subject. The largest use for ethylene glycol is as a permanent automotive radiator coolant, permanent because the radiator operates year round above the boiling point of water. Although the system is under some pressure, this pressure is minimized by using ethylene glycol–water mixtures as the fluid. The preparation of this coolant does not involve simply mixing the glycol with water in the desired proportions. Ethylene glycol decomposes and oxidizes when subjected to continuous heating. This oxidation and decomposition has been studied in detail so that suitable antioxidants and stabilizers can be added to the coolant. Water corrodes metals, particularly at elevated temperatures. Therefore, the rate of corrosion of the various metals used in automobile radiator construction has been studied so that appropriate corrosion inhibitors can be added to the coolant.

Tetraethyl lead is added to gasoline to improve the octane rating. The amount added depends on the octane rating of the unleaded gasoline, which in turn depends on the structures of the various hydrocarbons that constitute it. Thus the background for effective gasoline formulation is a detailed study of hydrocarbon oxidation. Also, if left to its own devices, lead builds up in the cylinders of the automobile. This has resulted in the use of ethylene dibromide as a lead scavenger, since lead bromide is sufficiently volatile to escape in the exhaust. All of this chemistry is in addition to developing the process for manufacturing tetraethyl lead:

$$CH_2{=}CH_2 + HCl \longrightarrow CH_3CH_2Cl \xrightarrow[40\text{–}60°C]{Na \cdot Pb} Pb(CH_2CH_3)_2 \tag{9}$$

Now specific fields in which chemistry is a major, although not always a totally obvious ingredient, will be examined. Petroleum refining and pharmaceuticals are not included, since most chemists consider oil companies and drug manufacturers to be in the chemical business.

II. FERTILIZERS

In order to grow, plants need various nutritive elements. Not all these elements occur in sufficient quantity in every soil, and they may become depleted and have to be replaced when land is farmed season after season. The necessary additions and replacements are made by means of mineral fertilizers.

The major elements that plants require are nitrogen, phosphorus, and potassium. Secondary requirements are for calcium, magnesium, and sulfur. Trace elements that are needed include boron, copper, iron, manganese, molybdenum, zinc, and possibly chlorine, silicon, and sodium. This knowledge is based on extensive studies of plant nutrition, metabolism, and biochemistry. Obviously soils must be analyzed to determine which of the above fertilizer components are needed and in what quantities.

Of the principal elements, phosphorus occurs naturally largely as fluorapatite, $CaF_2 \cdot 3Ca_3(PO_4)_2$, which is useless as a fertilizer ingredient because of its extreme water insolubility. The simplest treatment of this ore for fertilizer use is the preparation of normal superphosphate by acidification of the fluorapatite with sulfuric acid. (Superphosphate is a mixture of calcium phosphates and calcium sulfate.)

$$CaF_2 \cdot 3Ca_3(PO_4)_2 + 7H_2SO_4 + 3H_2O \longrightarrow 3CaH_4(PO_4)_2 \cdot H_2O + 7CaSO_4 + 2HF\uparrow \quad (10)$$

In some processes the acidification is carried out with nitric acid, which gives calcium nitrate as the co-product and thereby adds nitrogen values to the superphosphate.

When larger amounts of sulfuric acid are used in the acidification, the products are phosphoric acid and gypsum. The solid gypsum is separated from the phosphoric acid by filtration.

$$CaF_2 \cdot 3Ca_3(PO_4)_2 + 10H_2SO_4 + 20H_2O \longrightarrow 6H_3PO_4 + 10CaSO_4 \cdot 2H_2O + 2HF\uparrow \quad (11)$$

The resulting wet-process phosphoric acid can be used to acidify the fluorapatite to give triple superphosphate, which obviously is much higher in phosphorus than normal superphosphate:

$$CaF_2 \cdot 3Ca_3(PO_4)_2 + 14H_3PO_4 \longrightarrow 10Ca(H_2PO_4)_2 + 2HF\uparrow \quad (12)$$

Although these equations appear to be reasonably simple, the chemistry of the various phosphates and polyphosphates is very complex. Phosphate chemistry is a large and still-growing field of inorganic chemistry.

The second important fertilizer element is nitrogen. As every chemist knows, the ammonia produced by nitrogen fixation can be oxidized to nitric acid. In soil fertilization the ammonia can be injected into the soil as such, or it can be sprayed on the land in water solution as ammonium hydroxide. Some ammonia is used to neutralize phosphoric acid to produce diammonium hydrogen phosphate, a standard fertilizer ingredient. Urea also is used extensively as a nitrogen source in fertilizer formulations:

$$2NH_3 + CO_2 \xrightarrow[\text{pressure}]{\Delta} H_2NCONH_2 + H_2O \quad (13)$$

Mention has been made of the use of nitric acid to acidify phosphate rock. Nitric acid also can be neutralized with ammonia to give ammonium nitrate, a constituent of many high-nitrogen fertilizers.

The third important fertilizer element is potassium. Much of it occurs naturally as the chloride (the salt that is used in fertilizer formulations) in sedimentary deposits in New Mexico and in Saskatchewan. It is mined and separated from the sodium chloride and magnesium salts, which occur with it, by differential water solubility or froth flotation. Some of the very deep deposits are solution- rather than shaft-mined.

A large source of potassium chloride is Searles Lake in California. This is a dry lake on the surface, the depths of which are saturated with a brine containing a variety of salts. These salts are separated by a process involving the very skillful application of Gibbs' phase rule to produce potassium chloride, sodium sulfate, sodium carbonate, and borax. Minor quantities of lithium salts and of bromides also are obtained. Much of the sodium carbonate is converted to the bicarbonate by carbonation with flue gas; otherwise the carbon dioxide in the flue gas would be wasted. The potassium chloride (potash) actually is crystallized from a supersaturated solution of the more slowly crystallizing borax.

Most fertilizers are granular mixtures of such salts as diammonium hydrogen phosphate, potassium chloride, and ammonium nitrate. However, the present trend for large-scale use is toward solutions, since they are easier to handle.

The chemistry of fertilizers is inorganic chemistry plus a great deal of highly sophisticated physical chemistry and chemical engineering.

III. PESTICIDES

Obviously the pesticide business is part of the chemical industry. However, the average chemist or chemical engineer probably is not aware of its tremendous scope.

What are the needs of those who have to fight pests, particularly agricultural pests? The farmer has to combat insects, fungi, nematodes, weeds, and rodents. Large portions of the world's population have to fight insect carriers of such deadly diseases as malaria (mosquitos), yellow fever (mosquitos), typhus (lice), bubonic plague (fleas), sleeping sickness (tsetse fly), typhoid fever and amoebic dysentary (house fly), and dengue and encephalitis (mosquitos). Foresters have insect, fungus, and weed problems. Those who store grain must fight rodents and other causers of spoilage. Those who own houses and other wooden facilities that touch the ground, such as utility poles, railroad ties, and fence posts, must combat termites and rot.

All these pests are combated with chemicals. Insecticides will be considered first. Some of these, such as benzene hexachloride, DDT, methoxychlor, and lead arsenate, are stomach poisons that kill the insect when it eats the plant that has been treated with the compound. Other insecticides (including some of the above), such as benzene hexachloride, DDT, chlordane (and similar chlorinated cyclic compounds), Sevin, parathion, malathion, pyrethrins, and rotenone, kill the insect on contact. Still others, such as hydrogen cyanide and carbon disulfide, kill by attacking the insect's respiratory tract and are used primarily as fumigants. More recently, insecticides have been developed that are absorbed unchanged by the plant and become systemic. They are particularly effective against sucking insects such as aphids and mites, which eat the plant or suck its juices.

An inorganic insecticide that still is used extensively against the potato beetle is lead arsenate. Usually it is made from lead nitrate.

$$Pb(NO_3)_2 + H_3AsO_4 \longrightarrow PbHAsO_4\downarrow + 2HNO_3 \qquad (14)$$

Although highly toxic to man, it offers no residue problems, since the crop it protects, the potato, is in the ground and is not touched by the insecticide as applied to the foliage.

Of the synthetic organic insecticides, the three most important classes are the chlorinated hydrocarbons, the organic phosphates, and the carbamates. DDT is the best known chlorinated hydrocarbon. It is prepared as follows:

$$2C_6H_5Cl + CCl_3CHO \xrightarrow{H_2SO_4} Cl\text{-}C_6H_4\text{-}CH(CCl_3)\text{-}C_6H_4\text{-}Cl \qquad (15)$$

It has been of tremendous importance as a broad-spectrum insecticide (it kills almost any bug), particularly in combating the mosquitos that carry malaria and yellow fever. An analogue, methoxychlor, also finds wide usage:

$$2C_6H_5OCH_3 + CCl_3CHO \xrightarrow{H_2SO_4} CH_3O\text{-}C_6H_4\text{-}CH(CCl_3)\text{-}C_6H_4\text{-}OCH_3 \qquad (16)$$

Other chlorinated hydrocarbon insecticides include lindane (the most active of the benzene hexachloride isomers), and a large number of chlorinated terpene types such as chlordane, aldrin, dieldrin, and so forth.

These broad-spectrum insecticides, the chlorinated hydrocarbons, are very stable and tend to build up in food chains, so the predator at the end of the chain can suffer appreciably. As a result, insecticides of this type are coming into more and more disfavor and are being replaced by others that do not have the disadvantage of this residual toxicity.

The second very important class of synthetic organic insecticides, which also is broad-spectrum, is the organic phosphates. Most of these relatives of the nerve gases are very toxic to man, but they hydrolyze rather quickly in the atmosphere, so they only stay around for a couple of weeks and therefore offer no residual toxicity problems. In their manufacture all kinds of precautions are taken against worker exposure, and the plant operators usually are required to take blood tests once a month. Needless to say, good chemical engineering has to go into the design and construction of these manufacturing plants. Typical examples of this class of insecticides are parathion and the surprisingly less toxic malathion:

$$O_2N-C_6H_4-OP(=S)(OC_2H_5)-OC_2H_5, \quad C_2H_5OCOCH(CH_2COOC_2H_5)SP(=S)(OCH_3)-OCH_3 \qquad [1]$$

The third important group of synthetic organic insecticides is the carbamates, esters of nitrogen-substituted carbamic acids, of which Sevin is perhaps the best known.

$$C_{10}H_7-OCONHCH_3 \qquad [2]$$

Next to insecticides, herbicides probably are the pesticides most familiar to chemists. Two of the earliest were 2,4-D and 2,4,5-T, which are toxic to broad-leaved plants but do not injure grasses.

$$C_6H_5OH + Cl_2 \rightarrow Cl-C_6H_3(Cl)-OH \xrightarrow[NaOH]{ClCH_2CO_2H} Cl-C_6H_3(Cl)-OCH_2CO_2H \qquad (17)$$

$$\text{1,2,4,5-}C_6H_2Cl_4 \xrightarrow[OH^-]{H_2O} \text{2,4,5-}Cl_3C_6H_2\text{—}OH \xrightarrow[NaOH]{ClCH_2CO_2H} \text{2,4,5-}Cl_3C_6H_2\text{—}OCH_2CO_2H \quad (18)$$

They are used extensively in a general way against broad-leaved weeds, brush on rangeland, and brush and weeds along highways and railroad rights of way. The herbicide 2,4,-D is a big help in lawns against plantains and dandelions.

Since these two early examples that were in commercial use shortly after World War II, a wide variety of other structures such as ureas, amides, triazines, and dinitroanilines have come into use. The trend is toward selectivity. Many of these newer herbicides are directed toward controlling one or two specific weeds that plague the growers of such large-scale crops as corn, wheat, and soy beans.

Fungi and rot also are problems. For their control on crops, sulfur, copper compounds, and dithiocarbamates are typical of the materials that are employed. Fortunately, some of the dithiocarbamates are systemic. For protecting fence posts, utility poles, and railroad ties, both creosote oil and pentachlorphenol are used.

Nematodes are tiny worms that are particularly deleterious to root crops. They deform the root and cripple the plant. They are controlled by such soil fumigants as dichloropropene, dichloropropane, ethylene dibromide, methyl bromide, and chloropicrin.

Rodents are another group of pests. Rats, in particular, can be very destructive to stored grain. Warfarin, which functions by causing internal hemorrhages, is the leading rat killer.

$$\text{4-hydroxycoumarin—}C(OH){=}C\text{—}CH(C_6H_5)CH_2COCH_3 \quad [3]$$

This wide variety of chemicals is the result of a tremendous program of organic synthesis coupled with very extensive testing in the greenhouse and then in the field. A great deal of biochemistry is involved in plant, insect, and animal metabolism studies (often with tagged compounds), and very detailed and highly sophisticated analytical chemistry is involved in studies of metabolites and residues. Likewise there is a great deal of formulation work in this field, since formulation can have a tremendous effect on activity. Obviously a wide variety of chemical manufacturing methods is employed in making so diverse a group of compounds.

The present trend in the pesticide business is toward selectivity and systemics. Selectivity is particularly valuable in the case of insecticides, since broad-spectrum compounds not only attack the pest, but also kill its natural insect predators as well as other useful insects such as bees. Likewise the effect of pesticides on the environment is becoming increasingly recognized. A new area that is becoming of interest is that of plant growth regulators, of which a good example is α-naphthaleneacetic acid. It is used to prevent the premature drop-off of apples.

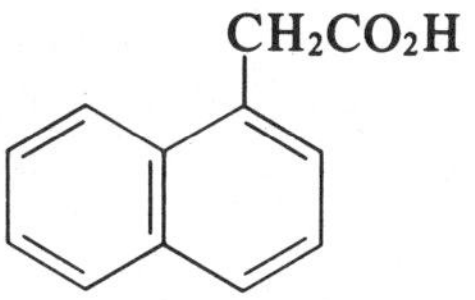

CH_2CO_2H

[4]

IV. FOOD PROCESSING

At first glance most people do not think of food processing as a branch of the chemical industry. It most certainly is. Modern methods for preserving foods—canning, freezing, and drying—although improvements over techniques that had been in use for centuries, have been extended tremendously and now are based on sound biochemistry and chemical engineering. The enormous variety of foods that are available year round making a balanced diet possible simply did not exist a hundred years ago, and still does not exist in many developing countries.

The fundamental nutritional and biochemical studies will be examined first. Proteins, fats, carbohydrates, vitamins, and minerals have been identified as essential dietary constituents. The amino acids that make up the proteins have been isolated and identified. This amino acid identification was most important, since some plant proteins are deficient in certain amino acids that the human body itself cannot synthesize, but which nonetheless are essential to human existence.

Wheat, rice, corn, grain sorghum, and most cereals are fortified with lysine, an amino acid the human body cannot synthesize. Methionine is added to cereals, and threonine might well be added to rice, and tryptophan to corn and grain sorghum. There are commercially feasible syntheses for lysine and methionine, but good processes still are needed for the synthesis of threonine and tryptophan.

Today many foods are fortified with vitamins and minerals at the level which a person needs daily. Vitamin C, which prevents scurvy, is added to many beverages, particularly soft drinks. Vitamin D, which prevents rickets, is added to milk, butter, and margarine. In the mineral area, iodine, which prevents goiter, is added to table salt. These are just a few examples.

In the area of food preservation, Pasteur showed that bacteria destroyed foods. Therefore, it obviously was necessary to cook such foods long enough to destroy the bacteria contained therein. Although it had been known from ancient times that sulfite inhibited the browning of dried foods, the elucidation of the mechanism of browning involved a great deal of complex sugar chemistry. Enzymes are deleterious to frozen foods. Therefore, frozen foods are blanched before freezing in order to destroy these enzymes.

These studies involved highly sophisticated biochemistry and organic chemistry. Obviously their application involves equally sophisticated chemical engineering, particularly in the case of the freezing, storing, and transportation of frozen foods.

Since fats tend to oxidize and turn rancid, such antioxidants as *n*-propyl gallate, butylated hydroxyanisole, and butylated hydroxytoluene are added to preserve them.

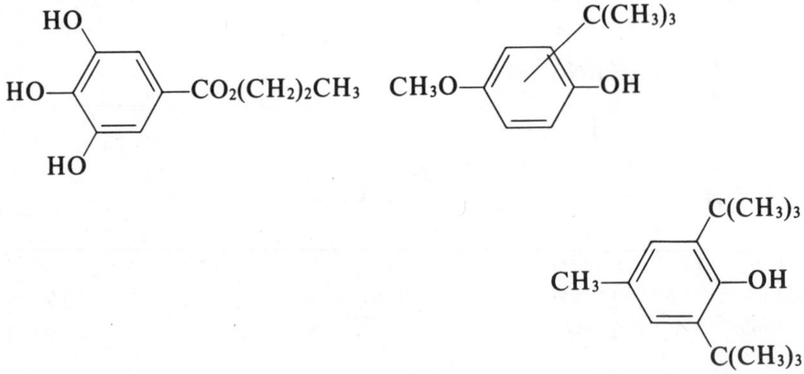

HO, HO, HO — $CO_2(CH_2)_2CH_3$ $\qquad$ CH_3O — OH, $C(CH_3)_3$

CH_3 — OH, $C(CH_3)_3$, $C(CH_3)_3$ $\qquad$ [5]

Benzoic acid and its sodium salt and sorbic acid and its potassium salt preserve acid foods against yeasts and bacteria. The alkyl esters of *p*-hydroxybenzoic acid exhibit antimicrobial activity at pHs higher than those at which benzoic acid and sodium benzoate are useful. Sulfurous acid is rather broad-spectrum and protects many foods against yeasts, molds, and bacteria.

Actually foods themselves are modified chemically to improve their properties. Starch granules are toughened by crosslinking with small amounts of such chemicals as phosphorus oxychloride, epichlorhydrin, and acrolein so that they swell but do not disintegrate when cooked. Monoglyceride shortenings, which really are emulsifying agents, are prepared by transesterifying triglycerides with glycerol. They have helped to mechanize the baking industry by providing reliable and reproducible shortenings. Margarine is prepared by deodorizing and partially hydrogenating soybean or cottonseed oil. In order to make it taste more like butter, biacetyl and acetoin are added.

$$CH_3COCOCH_3 \qquad CH_3CHOHCOCH_3 \qquad [6]$$

These have been shown to be the major flavoring ingredients in butter.

Some foods are modified even before isolation in edible form. The Swift Company invented the process of injecting papain (an enzyme) into animals that are about to be slaughtered. This tenderizes the meat (meat often is bought on the basis of tenderness) and obviates the necessity of aging. Unaged meat is bright red and therefore attractive to the consumer, whereas aged meat is dull brown.

The above examples illustrate the tremendous amount of biochemistry that is involved in food technology. All the additives have to be manufactured, and the chemical modifications must be effected by economical and commercially attractive processes. A great deal of chemical engineering is involved, and expensive and sophisticated equipment is used, such as stainless steel in milk pasteurization. Plant sanitation chemicals are used widely, and quality control must be as good as it is in the pharmaceutical industry.

V. SOAPS AND DETERGENTS

The soap and detergent business is part of the chemical industry and is a huge business in its own right. Interestingly, it has followed much the same pattern as the pharmaceutical industry. The big soap companies have become primarily formulators and packagers. They buy most of their ingredients from the large chemical companies.

Most people think of soap in terms of personal use and think of various detergent formulations for washing clothes and home furnishings. Although the domestic market is still the largest surfactant market, large quantities of detergents are used to clean buildings, public places, buses, aircraft, and railway cars; to clean metal surfaces before plating, enameling, or painting; in the many cleaning and scouring operations involved in textile manufacture; in ore flotation and separation; in emulsion polymerization to make synthetic rubber; and to clean fruits, vegetables, and eggs before packaging.

Historically, the original surfactant was soap, with whose manufacture and properties all chemists are familiar. Soap is still used as the primary ingredient in almost all detergent bars.

In order to better meet most of the cleaning needs listed, chemists have developed a wide variety of synthetic detergents. Although all synthetic detergents have an oil-soluble part of the molecule, the water-soluble part can be anionic, cationic, amphoteric, or nonionic.

Sodium dodecylbenzenesulfonate is by far the largest-volume synthetic surfactant. It is cheap, and its calcium salt is water soluble, so that it is effective in hard water. Originally it was prepared by alkylating benzene with propylene tetramer followed by sulfonation and neutralization:

$$C_6H_6 + CH_3\underset{\substack{|\\CH_3}}{C}HCH_2\underset{\substack{|\\CH_3}}{C}HCH{=}\underset{\substack{|\\CH_3}}{C}(CH_2)_2CH_3 \longrightarrow$$

$$C_6H_5{-}C(CH_3)[(CH_2)_2CH_3][CH_2\underset{\substack{|\\CH_3}}{C}HCH_2CH(CH_3)_2] \xrightarrow{H_2SO_4} C_{12}H_{25}{-}C_6H_4{-}SO_3H$$

$$\downarrow NaOH$$

$$C_{12}H_{25}{-}C_6H_4{-}SO_3Na \qquad (19)$$

In an alkybenzene, the most vulnerable point for degradative attack is the hydrogen atoms on the carbon atom attached to the benzene ring. In the above dodecylbenzene there are no such hydrogen atoms, so biological degradation is slow. When biodegradability became a problem in sewage disposal plants, in streams, and even in wells, the dodecylbenzene manufacturers shifted to a different structure, which is much more readily degraded biologically. Here is a typical example:

$$CH_2{=}CH_2 \xrightarrow{\text{polymerize}} CH_3(CH_2)_9CH{=}CH_2 \qquad (20)$$

$$CH_3(CH_2)_{10}CH_3 \xrightarrow{\text{crack}} CH_3(CH_2)_9CH{=}CH_2 \qquad (21)$$

$$CH_3(CH_2)_{10}CH_3 \xrightarrow{Cl_2} CH_3(CH_2)_9CHClCH_3 \longrightarrow CH_3(CH_2)_9CH{=}CH_2 \qquad (22)$$

$$CH_3(CH_2)_9CH{=}CH_2 \xrightarrow{C_6H_6} C_6H_5{-}CH(CH_3)(CH_2)_9CH_3 \qquad (23)$$

Nonionics are important when foaming is undesirable, as in automtic home washing machines and in some industrial processes. The preparation of a typical industrial nonionic detergent is shown in Equation 24.

$$C_6H_5OH + CH_3\underset{\substack{|\\CH_3}}{C}HCH_2\underset{\substack{|\\CH_3}}{C}HCH{=}\underset{\substack{|\\CH_3}}{C}H \longrightarrow$$

$$(CH_3)_2CHCH_2\underset{\substack{|\\CH_3}}{C}HCH_2\underset{\substack{|\\CH_3}}{C}H{-}C_6H_4{-}OH \xrightarrow[OH^-]{CH_2{-}CH_2\text{ (O-bridged)}}$$

$$(CH_3)_2CHCH_2\underset{\substack{|\\CH_3}}{C}HCH_2\underset{\substack{|\\CH_3}}{C}H{-}C_6H_4{-}O{-}\left[(CH_2)_2O\right]_n{-}H \qquad (24)$$

Cationics, in which the polar group is a quaternary ammonium salt, are disinfectants. Since these compounds are not outstanding surfactants in themselves, they usually are used in conjunction with nonionics.

Commercial detergents are formulations in which the surfactant is only one ingredient, albeit the most important one. Of almost equal importance are builders, which soften the water and enhance the performance of the surfactant. The earlier builders were sodium carbonate (washing soda) and sodium sulfate. More recently, condensed phosphates have been found to be extremely efficient. Sodium tripolyphosphate is the most important example:

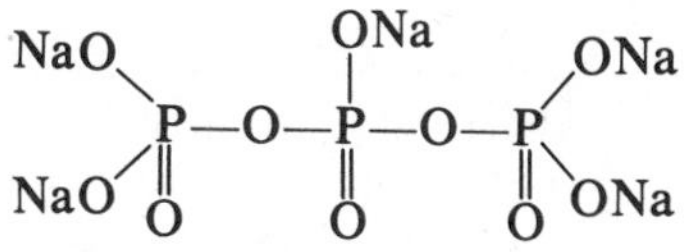

$$(NaO)_2P(=O){-}O{-}P(=O)(ONa){-}O{-}P(=O)(ONa)_2 \qquad [7]$$

Laundry formulations contain antiredeposition agents, of which carboxymethylcellulose is the most widely used. There is no sense in removing dirt from one part of a garment if the dirt is going to redeposit somewhere else. Laundry formulations also contain fluorescent compounds, which serve as brighteners, enzymes that help remove certain stains, and anticorrosion agents that help protect the metal parts of the washing machine.

This brief summary covers some of the high points in the soap and detergent industry. Obviously there are many other types of compounds and formulations in commercial use. All these different structures and formulations are the result of extensive organic synthesis programs, detailed physical chemical studies, and some biochemistry (enzymes). It can be judged from the very size of the industry alone that a great deal of chemical engineering has gone into process development and plant construction.

VI. SURFACE COATINGS

The surface coatings (paint, etc.) industry divides itself into two business categories:

1. *Trade Sales.* These are pretty much off-the-shelf items, which are used to paint new houses, repaint old ones, and refinish automobiles, vehicles, equipment, and furniture.
2. *Industrial Finishes.* These comprise factory-applied coatings, which are applied to vehicles, machinery, appliances, furniture, and containers before they are sold.

Both types of coatings have the same components:

1. A binder comprising resins, drying oils, and latices. This area has provided many opportunities for the polymer chemist, and most of the rest

of this section describes the various types that make up the myriad of polymers available to the coatings formulator.

2. A volatile solvent, which may be an organic chemical (usually a hydrocarbon) or water.
3. A pigment, usually inorganic in nature. Varnishes, of course, are unpigmented.
4. Such other items are stabilizers, emulsifiers, defoamers, antiskinning agents, antisettling agents, driers, ultraviolet-radiation absorbers, mildewcides, fungicides, thickeners, and flatting agents.

It can be seen at once that surface coatings are complex formulations, capable of almost infinite variation, depending on the use to which they are to be put.

The original binders were polyunsaturated vegetable oils, like linseed, which dry (polymerize) very slowly. A tremendous advance was made in the binder field when these vegetable oils were replaced by nitrocellulose in an organic solvent. This changed paint-drying times from days to hours and facilitated the mass production of automobiles. Since this first step, other surface-coating trends have been to longer-lasting coatings, to cheaper coatings, and to formulations containing more solids at application viscosities, so the products require fewer coats.

The trend in solvents has been to go to aqueous systems, and, more recently, to no solvent at all. The big step in pigment development was to replace white lead by titanium dioxide. Titanium dioxide coatings weather by slow erosion rather than by checking and cracking, so they are easy to keep clean and to refinish. They are less poisonous than lead-based coatings.

The original method of surface-coating application was by brushing. For industrial finishes this was soon replaced by spraying. Now electrodeposition from water has entered the picture. This process eliminates hydrocarbon emissions and the related fire hazard. It coats the interiors (inside corners) of boxes, and there is no waste, as there is in spraying.

As mentioned previously, nitrocellulose was the first synthetic coating resin. The step that followed was to make phenol-formaldehyde resins oil soluble. In the case of phenol itself, the resins prepared therefrom are highly crosslinked and therefore oil insoluble.

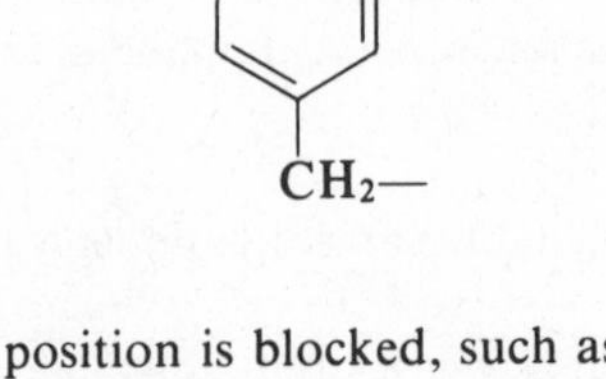

OH

$—CH_2$ $CH_2—$

[8]

$CH_2—$

However, when the para position is blocked, such as by a methyl, *t*-butyl, or phenyl group, the resulting polymer has a straight uncrosslinked chain, and therefore is oil soluble.

$$-CH_2-C_6H_2(OH)(CH_3)-CH_2-C_6H_2(OH)(CH_3)-CH_2- \quad [9]$$

The next group of resins to be developed was the alkyds. It still constitutes the most important type of resin in the U.S. coatings industry. Alkyds are made by condensing a polyol, such as glycerol or pentaerythritol, with a dibasic acid (often as the anhydride), such as phthalic anhydride or maleic anhydride, and a drying oil, such as linseed or many others. Following is a typical structure.

$$C_6H_4(CO_2CH_2CH(O-C(=O)-CH{=}CHC(=O)-)CH_2OCO(CH_2)_7CH{=}CHCH_2CH{=}CH(CH_2)_4CH_3)(C(=O)-) \quad [10]$$

Obviously, by varying the ingredients and their ratios one to another, the amount of unsaturation and crosslinking can be varied enormously, and therefore the oil solubility, compatibility with other resins, curing rate, and final film properties. Also there are enough reactive points in the molecule so that alkyds can be condensed with other monomers and polymers, such as styrene, phenol-formaldehyde resins, acrylics, silicones, and so forth.

Another development that followed the development of the alkyds was the use of polymer emulsions in water-based paints. The first commercially successful film former in this category was butadiene-styrene latex, but other polymers such as polyvinyl acetate soon followed.

The most successful of these latices have been the acrylics, because of the superior properties possessed by the final films. These latices are based on various esters of acrylic and methacrylic acids and on acrylamide (which can be crosslinked with formaldehyde). They can be produced not only as emulsions, but also in water-soluble form, by leaving a few of the carboxyl groups unesterified and forming water-soluble salts with these free carboxyl groups. The key to the success of acrylics in coatings was the E. I. DuPont de Nemours and Company's finding that a polymer molecular weight of close to 100,000 (not much above and not much below) gives the desired properties. That company developed a process to produce such a polymer, and today most mass-produced U.S. automobiles are finished with acrylics.

Amino polymers also have found their way into the coatings industry. Both urea-formaldehyde and melamine-formaldehyde resins are used, primarily to improve the properties of alkyds.

Epoxies also are important. Many of them are based on the condensation product of bisphenol A and epichlorohydrin:

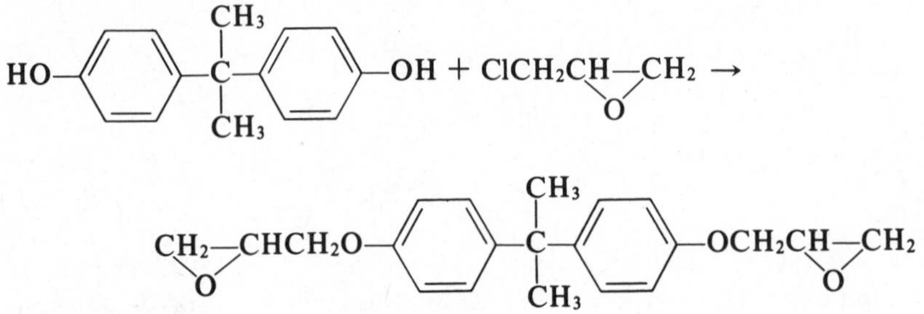

$$HO-C_6H_4-C(CH_3)_2-C_6H_4-OH + ClCH_2\underset{\diagdown O \diagup}{CH-CH_2} \rightarrow$$

$$\underset{\diagdown O \diagup}{CH_2-CH}CH_2O-C_6H_4-C(CH_3)_2-C_6H_4-OCH_2\underset{\diagdown O \diagup}{CH-CH_2} \qquad (25)$$

Obviously the diepoxide so produced can be, and is, cured with almost anything.

Many other polymers such as unsaturated polyesters and fluorocarbons also are used in coatings. The whole field of pigments has not been touched upon here—their compatibility with the rest of the coating system, their preparation, and their durability, including light fastness.

Although probably not familiar to the average chemist or chemical engineer, this whole field of surface coatings is based on chemistry, with the emphasis on complex polymer chemistry and formulation.

VII. GLASS

Glass has been known for at least 8000 years. Most of it (95%) is still made by the ancient process of melting soda ash (Na_2CO_3), limestone ($CaCO_3$), and sand (SiO_2). The carbon dioxide and any nitrates (as nitrogen oxides) and any sulfates (as SO_3) are driven off, and the melt is cooled. All glass is annealed (held at certain temperatures) and cooled slowly to relieve internal stresses.

Because glass is hard, transparent, chemically inert, and readily formable into a variety of shapes, it finds many uses. About half of its production is for containers, and everyone certainly is familiar with many other uses.

Much of glass technology is based on mechanical engineering. The manufacture of bottles, light bulbs, and sheet glass is all automatic. However, many improvements in glass properties, which lead to specialty uses, are based on sophisticated inorganic and physical chemistry. Here are a few examples.

Over a period of time, water dissolves traces of soda-lime-silica glass. However, if the surface of the glass is given a fine coat of sulfur during the annealing process, this dissolving does not take place. As a result, containers for medicines can be made from ordinary glass, instead of from the much more expensive borosilicate glasses, which are not leached by water.

Increasing the alkaline earth content of the glass at the expense of the alkali metal content reduces the glass viscosity at the melting temperature and increases it at the forming temperature, thus speeding up the time of bottle formation by reducing the time in the mold.

The addition of borax to the glass formulation markedly increases the resistance of the final product to thermal shock. Although the formulation originally was developed to prevent the fracturing of hot railroad lantern globes by rain or snow, it has found extensive use in cooking utensils and laboratory glassware. We are all familiar with Pyrex. This glass also resists attack by acids, thereby doubling the life of battery jars. Because of its excellent electrical resistance it is used in electric-power insulators.

A glass made entirely from titanium dioxide and silicon dioxide has an ultra low coefficient of expansion and therefore finds use in telescope mirror blanks.

Ordinary soda-lime-silica glass can be colored or tinted by the inclusion of certain metallic oxides in the formula. Tinted glass is used in windows (also in automobile windshields) to reduce glare and to absorb solar heat. The same result is accomplished by a very thin metallic coating on the glass.

Automobile safety has been increased enormously by the invention of safety glass, in which a sheet of polyvinyl butyral is laminated between two thin sheets of plate glass. Since the butyraldehydridazation is random, there also are hydroxyl groups along the polymer chain. This transparent polymer adheres to the glass so strongly that the glass does not shatter. On impact, the glass fragments remain attached to the polymer film and do not shower on and lacerate the vehicle occupants.

$$\left[\begin{array}{c} -\mathrm{CHCH_2CHCH_2}- \\ \mathrm{O{-}CH{-}O} \\ | \\ \mathrm{(CH_2)_2CH_3} \end{array}\right]_n \qquad [11]$$

Certain glass formulation can be so heat treated that they actually do crystallize. The resulting glass ceramics are called Pyroceram.

Protection against x-rays is important. Strontium oxides are used in the formulations of the glass for color television tubes to protect the viewers from these rays.

It has been found that the addition of rare earth oxides, particularly that of lanthanum, to the glass formulation enhances the optical properties of the final product. This has given improved resolution and aperture to aerial-observation instruments and photographic cameras.

These are just a few of the many chemical modifications of the ancient soda-lime-silica glass. As any chemist or engineer can readily see, glass manufacture is a very real part of the chemical industry, based on extensive contributions from inorganic chemistry backed up by very fundamental physical chemical and engineering studies.

VIII. METALLURGY

Although a great deal of metallurgy is based on mechanical engineering, as in the fabrication of steel, the processes for winning the pure metals are essentially all chemical. In this section the discussion will be confined for reasons of space to only two metals, iron and aluminum.

The fundamental process for making steel has not changed over the years, although chemistry has contributed tremendously to many improvements in it. Basically, a mixture of iron ore, coke, and limestone is treated with an air blast. This converts the coke to carbon monoxide, thereby supplying the heat to keep the iron molten. Any water vapor in the air also reacts with the coke (endothermally) to give carbon monoxide and hydrogen (the water gas reaction) (See p. 42). The carbon monoxide and hydrogen reduce the iron oxides to the metal, pig iron. This pig iron, in the molten state, is blown with oxygen to remove such impurities as phosphorus, silicon, manganese, and sulfur, which either leave the furnace as gases or go into the slag. The resulting product, a mixture of several physical forms of iron and small amounts of iron carbides, is steel. Its properties are varied and controlled by the amount of carbon present and by the addition of small amounts of other metallic elements such as manganese, cobalt, and nickel.

What has chemistry contributed to this process? Most contributions have been through physical chemistry. From them a thorough understanding of the fundamental chemistry of steel manufacture has been developed. The equilibria and rates of various reactions in metals have been determined, and this has led among other things to the discovery of new alloys. Phase rule studies have been of extreme importance. All steel manufacture is based on a thorough understanding of the phase relationships among the various forms of iron and of its carbides. Phase diagrams of such systems as $CaO \cdot Al_2O_3 \cdot SiO_2$, $CaO \cdot MgO \cdot SiO_2$, $CaO \cdot MgO \cdot Al_2O_3$, and $Al_2O_3 \cdot MgO \cdot SiO_2$ have led to a much better understanding of slag behavior and of refractory-furnace construction.

Recently there has been a tremendous advance in analytical methodology in which instrumentation has replaced the old methods of wet chemistry. The use of such modern techniques as neutron activation, spectroscopy, x-ray fluorescence, and so forth, has enabled many routine process-control determinations that used to take hours to be made in a matter of minutes.

What about actual process improvements, of which there have been many? Following are a few examples.

Since the reaction of coke with steam is endothermic, variations in the moisture content of the air blast used to lead to uneven furnace performance. Now some steam is added to the air, and the total amount of steam is kept uniform. This steam addition also provides more hydrogen for ore reduction.

Some of the coke is now replaced by oil, which also supplies more hydrogen to the mixture in the furnace. The amount added depends on the cost of the auxiliary fuel versus that of coke at the location in question.

Flaking of steel is caused by dissolved hydrogen. Molten steel is now dehydrogenated by casting the ingot *in vacuo*.

Steel is pickled by dipping in acid to remove oxides. The pickling process has been made continuous, and hydrochloric acid has replaced much of the sulfuric acid that formerly was used exclusively. Hydrochloric acid is cheaper and is more easily disposed of.

Tin coating is now effected by a high-speed electrotinning process using an alkaline sodium stannate (Na_2SnO_3) or similar bath. When the steel is to be coated with a lacquer or plastic, as for use in beer or soft drink cans, a chromium plate is applied electrolytically.

The production of aluminum is strictly chemical. Alumina is electrolyzed in a bath of fused cryolite (Na_3AlF_6), which contains some additional aluminum fluoride (AlF_3) and a little fluorspar (CaF_2):

$$2Al_2O_3 \xrightarrow{\text{elec}} 4Al + 3O_2 \tag{26}$$

Considerable chemical engineering is involved in aluminum production, since the reaction takes place at 940–980°C. The melt is held in a carbon-lined steel cell, and the anode also is made of carbon. This anode must be replaced periodically, since it is attacked by the evolved oxygen. Likewise the melt is sufficiently corrosive so that the cell lining has to be replaced every two to four years. The molten aluminum itself serves as the cathode.

IX. PHOTOGRAPHY AND COPYING

Although photography and copying are not visibly a part of the chemical industry, all the basic processes in photography and copying depend entirely on chemistry. Obviously physics and optics enter into the picture as well. There is a tremendous amount of chemical engineering in film manufacture and development. In this field we are dealing with a system that embraces nonchemical disciplines. There is a great deal of complementary mechanical engineering in the design and construction of cameras and copiers.

The polymeric film or paper support involves cellulose or other polymer chemistry.

In all the processes involved, there is some kind of a layer that is sensitive to image recording. In the case of photography, this process embraces the chemistry of gelatin, of silver salts, and of a plethora of other chemical compounds.

The exposure of the photosensitive material, whatever it is, is based on photochemistry.

Processing (developing) of the final visible image is in all cases chemical.

Now let's look at specifics. In this limited space, specifics will have to be confined to principles. Furthermore, many of the details are proprietary and kept secret by the various companies involved.

Black and white photography is based on light's striking silver halide crystals that are suspended colloidally in gelatin. The light changes the structure of the crystals it strikes so that in the light-exposed areas the developing agent (usually hydroquinone) reduces the silver halide to metallic silver. The unexposed silver halide is then washed out (usually with sodium thiosulfate) to leave a negative image. The gelatin layer also contains dyes and other sensitizing agents such as allyl thiocarbamide:

$$CH_2{=}CHCH_2NHCSNH_2 \quad [12]$$

The earliest plastic supporting film was cellulose nitrate, which unfortunately is inflammable and discolors with age. For these reasons cellulose acetate was used instead. More recently polyester (polyethylene terephthalate) has come into widespread use.

In color photography, photosensitive dyes are used in conjunction with silver halides. There are many variations of the basic scheme. However, three thin, light-sensitive filtering layers are used in contact with one another, and each one has to be coated. The dyes, which are different in each layer, must not migrate. Nontoxic developing agents and couplers had to be discovered.

Recently, diffusion transfer has come into widespread use in photography. Here the amount of exposure governs the diffusion from the photosensitive layer to the image-receiving layer to give a black and white image. Diffusion transfer is based on soluble silver complexes. It is the basis for the Polaroid process.

Another field is copying—non-silver halide reprography.

In thermography an image is recorded as a heat pattern. The heat-sensitive layer contains a heavy metal soap, such as ferric stearate, and a component, such as gallic acid, that reacts with the soap to give a visible image upon heating. When the film and the original to be copied are moved past an infrared source, the dark printed areas absorb heat and produce an image on the film.

Other forms of thermography involve the transfer under heat of a dye or a dye precursor from an intermediate to a receptor sheet. Variations of this principle include Minnesota Mining and Manufacturing Company's Thermofax and Eastman Kodak Company's Verifax.

In electron beam reprography, an electron beam, focused *in vacuo,* impacts on a material sensitive to electrons to produce a visible image after processing. Plenty of chemistry was involved in the development of the electron-sensitive material. This kind of reprography finds use in computers, since by this technique 100,000 characters per second can be recorded.

In electroconductive electrophotography, certain photoconductors maintain a latent image as states of increased electrical conductivity. This latent image can be developed into a visible image.

In electrostatic reprography, an electrostatic image is formed on an insulating surface that is photoconductive. The image is developed with finely divided pigment particles, which are then transferred to a permanent support, such as paper, and are fixed there. This is the basis of the familiar Xerox process.

Modern engineering and construction work is based on a very valuable set of master drawings, of which many copies are needed by contractors, engineers, foremen, and so forth. Traditionally these copies have been made by the blueprint process, in which ferric ammonium citrate is reduced to the ferrous state by light in the presence of the organic material in which it is incorporated. The ferrous ion reacts with potassium ferricyanide to give Prussian blue, and thereby a white-on-blue copy.

Recently the diazotype process has pretty well displaced blueprints. This process depends on the coupling of diazonium salts in alkaline medium with phenols or enols to produce a colored compound and the decomposition of unreacted diazonium salt by light. The image thus produced is positive.

In photoprinting, a polymer film is placed on a metal backing. A negative is held against the film, and the unshielded polymer areas are crosslinked by light. After the uncrosslinked polymer is washed away, the positive image remains as a raised surface that can be inked and used to print. This is an excellent way to reproduce a photograph, and for text the photoprinting method is much more convenient than the older letterpress method using lead type.

X. CONCLUSION

This chapter should give the reader some idea of the scope of the chemical industry. Even so, there has been no mention of textiles, rubber, personal-care products, nuclear energy, pulp and paper, ceramics, cement, electronic equipment, electrical equipment, industrial gases, explosives, industrial carbon, leather, synthetic fibers, fermentation, and many others.

2

CHEMICALS FROM PETROLEUM

I. INTRODUCTION

In the early days of organic chemistry, the building blocks for many organic chemicals came from the fractional distillation of coal tar, the liquid driven from coal when it is pyrolyzed to produce coke. Since most of the chemicals thus obtained were aromatic, this lead to the more extensive development of aromatic chemistry compared with aliphatic at that time.

More recently, petroleum, natural gas, and natural gas liquids have superseded coal tar as the primary source of organic chemicals. In the future, shale oil and tar sands may take the place of petroleum and natural gas.

Natural gas is composed mostly of methane, although a mixture of higher hydrocarbons (up to about C_6 casinghead gasoline) is separated from it. Shale oil (from the pyrolysis of oil shale) and the liquid from tar sands, separated by hot water flotation, are somewhat similar to petroleum in composition. All these raw materials can be converted to useful organic chemicals by procedures similar to those used in the refining of petroleum. Therefore, this chapter will be limited to a discussion of petroleum refining.

II. COMPOSITION OF PETROLEUM

Petroleum is a mixture of a tremendous number of organic compounds of different basic structures and varying molecular weights. Most of them are hydrocarbons. These comprise normal and branched-chain paraffins, naphthenes, and aromatics. Higher cyclics such as polyalkylindanes, and higher aromatics

---

This chapter is based mostly on material contained in W. L. Nelson, *Petroleum Refinery Engineering,* Fourth Edition, McGraw-Hill, New York, 1968; R. N. Shreve, *Chemical Process Industries,* Third Edition, McGraw-Hill, New York, 1967; and B. G. Reuben and M. L. Burstall, *The Chemical Economy,* Longman, London, 1973.

such as polyalkylanthracenes and phenanthrenes as a minimum are believed to occur as well, together with other even more highly condensed and much higher molecular weight compounds.

Petroleum also contains sulfur and its compounds in varying amounts. A "sour" crude contains dissolved hydrogen sulfide. Sulfur itself may be present in elementary form. Of the organic sulfur compounds, probably the commonest types are mercaptans, disulfides, sulfides, and thiophenes.

Nitrogen compounds also occur in petroleum. In general these comprise more highly condensed structures, such as the alkylquinolines. Oxygen also can occur in similar condensed structures and in such compounds as the naphthenic acids. In addition there are traces (usually parts per million) of metalorganic compounds that contain such elements as iron, vanadium, nickel, and arsenic.

These, then, are the general structures of the types of compounds present in petroleum. Within each structural type, molecular weights vary tremendously. When petroleum is distilled, first at atmospheric pressure and then under vacuum, a series of compound mixtures is obtained. Most volatile are the dissolved gases, which are low-molecular-weight hydrocarbons (up to about C_5). Next in boiling point are gasoline and solvent naphtha, followed by kerosene. Then come light and heavy fuel oils and diesel oils. The last to distil are the heavy mineral oils, lubricating oils, and waxes. The pot residue contains even heavier fuel and lubricating oils, petrolatum, road oils, asphalt, and coke. Each fraction is a mixture of a tremendous number of different compounds, which have the basic structures indicated above.

III. ANALYSIS OF PETROLEUM

It should be obvious to any well-trained chemist that most of the customary analytical methods are of very limited value in characterizing petroleum. For instance, a molecular-weight determination on the entire crude would be irrelevant. However, it is useful in the case of lower-boiling fractions. Therefore, petroleum chemists have developed a number of empirical tests upon which they can base the procedures for handling the crude and operating the refinery. The exact conditions for conducting each test, such as temperature, time for carrying out each manipulation, apparatus used, and so forth, usually are specified exactly. These are standard, numbered tests described in the compendia and the annual publications of test methods of the American Society for Testing and Materials as well as in similar publications of the Institute of Petroleum (U.K.).

Of course, petroleum analysts rely very heavily on such modern techniques as x-ray and electron diffraction; electron micrography; and mass, emission, ultraviolet absorption, Raman, and infrared spectroscopy.

Here are a few examples of the techniques of petroleum analysts. There are many, many others. Most are used both on crudes and on finished products.

Boiling Range. A chromatographic technique, or GLC simulated distillation, is used that generates boiling-point data of a crude oil or a crude-oil fraction up to about 600 or 650°C.

Analysis of a Crude. In the analysis of a crude-oil fraction, a combination of nuclear magnetic resonance–liquid chromatography, GLC, and mass spectroscopy is employed. The latter can be used to determine hydrocarbon types in petroleum fractions up to a boiling point of 1100°F. It gives a breakdown of paraffins, naphthenes, and aromatics (up to five rings). Nonaromatic olefins can be determined by bromine number (usually on a distillate). Aromatics also can be measured by the aniline point, which is the minimum equilibrium solution temperature for equal volumes of aniline and sample.

Specific Gravity. This is determined at 60°F. There are specific tests and procedures depending on whether a hydrometer, a pycnometer, or some other measuring device is used. Petroleum products are sold on the basis of volume delivered, corresponding volume at 60°F as calculated from standard tables. Weight is important in determining freight rates, tankers' cargoes, and the power necessary to pump the material. Also important is whether the material will sink or float in water or even separate from it.

Viscosity. Both absolute viscosity and relative (to water) viscosity at 68°F are determined in centipoises. The kinematic viscosity also is of interest:

$$\text{Kinematic viscosity} = \frac{\text{viscosity}}{\text{specific gravity}}$$

The fluidity is the reciprocal of the viscosity. The common viscosimeters do not measure in centipoises. They have their own tables, which are used by analysts. All viscosity measurements are interconvertible by the use of appropriate tables. Viscosity bears on structure as well as necessary pumping power and the movement of the crude in the plant. For oils, the viscosity index, the change of viscosity with temperature, is very important.

Characterization Factor. Various factors are calculated for comparison purposes. For instance, the characterization factor is based on boiling point and specific gravity:

$$K = \frac{\sqrt[3]{T_B}}{S}$$

In this equation T_B is the average molal boiling point in degrees Fahrenheit absolute (°F + 460) and S is the specific gravity at 60°F. The characterization factor has been found to correlate with viscosity, aniline point, molecular weight, critical temperature, viscosity index, and percentage of hydrogen. Thus it can be estimated from a wide variety of laboratory data. The correlation with hydrogen content can be used as a measure of paraffinicity.

Vapor Pressure. This is determined at 100°F. An indication of the pressure the material will generate in a closed container is important for safety in transportation, vapor lock in gasoline, the types of tanks that will be necessary in which to store the material, and its starting characteristics as a motor fuel. The vapor pressure obviously is based on the low-molecular-weight-hydrocarbon content of the sample in question.

Flash and Fire Points. The flash point is the temperature at which vapors above the liquid explode in the presence of a flame. The fire point is the temperature at which these vapors burn continuously. Again these values have a bearing on the transportation of the crude and its handling and that of its fractions in the refinery.

Cloud and Pour Points. The cloud point is the temperature at which the cooled liquid becomes cloudy and precipitation begins. The pour point is that temperature at which the cooled liquid no longer will pour. Both of these values indicate how cold you can handle the material before real trouble ensues.

Elemental Analysis. Most important is that for sulfur, since sulfur compounds cause corrosion, catalyst poisoning, and other problems in refining and transportation. The sulfur analysis indicates which products probably will need further purification before they are suitable for commerce. In burnable products (gasoline and oils), the sample is burned and the gas so produced is passed into a sodium carbonate solution. Sulfur is determined by back-titration. For higher-molecular-weight materials, the standard oxygen bomb method is employed. Corrosive properties are estimated by the effect of the sample on a strip of polished copper. Analysis for trace metals also is important since these elements also can poison refinery catalysts.

IV. REFINERY OPERATION

Now that each shipment of crude has been fairly well characterized, what happens to it in the refinery? Basically four things happen.

1. The more volatile constituents are separated by distillation.
2. Each fraction, even to the lowest boiling, may be broken down into smaller molecules. In recent years much emphasis has been placed on breaking down the higher-boiling fractions (including much of the residue) into smaller, more-valuable species.
3. The small, low-boiling molecules are converted by various chemical reactions into compounds for which there is the most-profitable market demand.
4. These products (such as gasoline or fuel oil) may be given some form of final purification, such as to remove sulfur compounds, improve color, and so forth.

From this discussion it can be seen that most refineries are broadly similar in construction. Equally obviously, no two are run in exactly the same fashion, since every batch of crude is different, and therefore each batch requires slightly different processing. Furthermore, each refinery sells to a different market, which also probably varies from month to month for the individual refinery in question. For instance, in the United States there is a big demand for motor fuel in the summer and for heating oil in the winter, and the refineries must respond in appropriate fashion. Likewise, in contrast to others, a given refinery may have a big petrochemical-raw-material market, such as for some compound such as ethylene, propylene, benzene, or butadiene. If the petroleum company in question manufactures polyethylene, polypropylene, or polystyrene, as some of them do, then the particular refinery in question has an internal, built-in chemical market.

All refineries are run continuously. This means that many processes are conducted simultaneously. Mixtures are fractionated, molecules are broken down and rebuilt, and product streams are separated and further purified all at once. Changes in product mix are brought about by greater raw-material flows to a given reaction, perhaps even by a change in the number of reactors performing a given chemical reaction, and by changes in processing conditions, such as different catalysts, different reaction temperatures, and different pressures.

The idea that many different operations go on simultaneously and continuously in one plant, many of them dependent on each other, and that a changing variety of products is constantly produced is foreign to the thinking of many chemists. They tend to think of a single synthesis, which is conducted step by step in striking contrast to refinery operation. Thus when they make sulfanilic acid, they nitrate benzene and purify the single product (nitrobenzene), reduce it to aniline, purify the aniline thus produced, and then bake the aniline sulfate to give the desired sulfanilic acid.

$$C_6H_6 \xrightarrow[H_2SO_4]{HNO_3} C_6H_5NO_2 \xrightarrow[\text{catalyst}]{H_2} C_6H_5NH_2 \xrightarrow[\text{then } \Delta]{H_2SO_4} H_2N{-}C_6H_4{-}SO_3H \quad (1)$$

In general, in a refinery the crude is first at least topped, by feeding it into a still, which is tapped at different places to take off fractions with different boiling points. The residue from this operation may be vacuum or steam distilled.

It should be emphasized again that all refineries are operated so as to produce the most desirable products. The desirability of these products depends on their performance and the markets for them.

V. REFINERY CHEMISTRY

Molecular breakdown of a crude in a refinery is effected by cracking. Thermal cracking is nothing but straight pyrolysis. Catalytic cracking is pyrolysis in the presence of a catalyst such as silica-alumina, processed clay, or a combination of silica-alumina and a zeolite. Because of the irregularities in its crystal structure, a silica-alumina catalyst is strongly acidic.

Catalytic cracking has the advantage over thermal cracking in that it is easier to control and takes place at a lower temperature and thereby requires less energy. A moving or fluid bed reactor (the details of which are discussed later in the book) usually is used. Steam often is added to a cracking operation to lower the partial pressure of the reactant and achieve a high flow rate. Recently, cracking in the presence of hydrogen has been developed as a means of obtaining a greater yield of low-boiling materials from the very high-boiling fractions. Hydrocracking is one of the more versatile processes in a refinery, since it can be utilized to convert high-boiling fractions of crude oil into gasoline, high-quality distillate, or LPG (liquefied petroleum gas), depending on catalysts and conditions. Following are two typical chemical reactions that may take place during the cracking process:

$$n-C_{30}H_{62} \longrightarrow CH_3(CH_2)_5CH{=}CH_2 + CH_3(CH_2)_{10}CH_3 + CH_3(CH_2)_7CH{=}CH_2 \quad (2)$$

$$\text{(cyclohexyl)}-CH_2CH_2R \rightarrow \text{(benzene)} + RCH{=}CH_2 + 4H_2 \quad (3)$$

The second reaction (Equation 3) probably proceeds stepwise, so that partially dehydrogenated products may be obtained before final scission. Their nature and amount depend on cracking temperature and residence time in the cracker. Obviously molecular breakdown can continue if the products are not removed quickly from the reactor. In gasoline production, molecules in the C_8 range are separated as crude product, and higher-molecular-weight molecules are recycled to the cracking operation. The lower-molecular-weight molecules also are separated and then subjected to some rebuilding operation such as one of those described later in this chapter.

Perhaps the best way to visualize the variety of molecular breakdowns that can occur during cracking is to examine the thermal decomposition of *n*-butane [1]:

$$CH_3(CH_2)_2CH_3 \rightleftharpoons CH_3CH_2CH{=}CH_2 \text{ or } CH_3CH{=}CHCH_3 \quad (4)$$

$$CH_3(CH_2)_2CH_3 \longrightarrow CH_3CH{=}CH_2 + CH_4 \quad (5)$$

$$CH_3(CH_2)_2CH_3 \longrightarrow CH_2{=}CH_2 + CH_3CH_3 \quad (6)$$

$$2CH_3(CH_2)_2CH_3 \longrightarrow n-C_4H_8 + CH_3CH{=}CH_2 + CH_4 \quad (7)$$

$$2CH_3(CH_2)_2CH_3 \longrightarrow n-C_4H_8 + 2CH_3CH_3 \quad (8)$$

Thermal cracking leads to high yields of ethylene and therefore is particularly valuable to the petrochemical industry. Catalytic cracking produces more branched-chain compounds and therefore is of more value in gasoline manufacture. This difference is because thermal cracking is a free-radical chain reaction, and catalytic cracking involves carbonium-ion intermediates. Under cracking conditions the free-radical intermediates fragment about as fast as they rearrange, whereas the carbonium ions rearrange much faster than they fragment.

Cracking is a capital-intensive operation, particularly when the desired products are low-molecular-weight materials. The cracking must be conducted at high temperatures (~1475°F) and the evolved gases quenched and fractionally distilled under pressure. This means there are real economies in building a very large plant in which centrifugal compressors may be used.

In the molecular rebuilding reactions, certain changes in raw-material structure are desirable. Branched-chain compounds are more useful in product synthesis than are their straight-chain isomers. They are prepared by isomerization:

$$CH_3(CH_2)_2CH_3 \xrightarrow{AlCl_3} CH_3\overset{\overset{\displaystyle CH_3}{|}}{C}HCH_3 \qquad (9)$$

Olefins are far more reactive chemically than are paraffins. Therefore, dehydrogenation of a hydrocarbon stream may be used prior to other transformations:

$$CH_3(CH_2)_2CH_3 \xrightarrow{Cr_2O_3 \cdot Al_2O_3} CH_3CH_2CH{=}CH_2 + H_2 \qquad (10)$$

Aromatization is a form of dehydrogenation. In petroleum refining, molecular rearrangement often occurs along with aromatization. The process is broadly called reforming, of which there are several variations. Perhaps the commonest catalyst in reforming use is platinum on alumina. Following are some typical reactions that occur during reforming:

$$\text{(cyclohexane ring)}\text{—}CH_3 \longrightarrow \text{(benzene ring)}\text{—}CH_3 + 3H_2 \qquad (11)$$

$$\text{(cyclopentane ring)}\text{—}CH_3 \longrightarrow \text{(benzene ring)} + 3H_2 \qquad (12)$$

$$CH_3(CH_2)_5CH_3 \longrightarrow \text{(benzene ring)}\text{—}CH_3 + 4H_2 \qquad (13)$$

Some paraffin isomerization, some olefin hydrogenation, and some cracking also occur during reforming. Likewise, during the process sulfur may be removed from the stock, of which one of the simplest examples is shown in Equation 14.

$$C_4H_4S \text{ (thiophene ring)} + 4H_2 \rightarrow C_4H_{10} + H_2S \qquad (14)$$

It generally is desirable to remove sulfur compounds to a very low level (<1 ppm) prior to catalytic reforming in order to reduce poisoning when noble metal catalysts are used.

Olefin polymerization is a means of rebuilding small molecules into larger ones. The process may be entirely thermal or may be catalyzed by such compounds as phosphoric acid on an inert support. Following is a typical reaction:

$$2CH_3\overset{CH_3}{\overset{|}{C}}{=}CH_2 \xrightarrow[H_3PO_4]{\Delta \text{ or}} CH_3\underset{\underset{CH_3}{|}}{\overset{\overset{CH_3}{|}}{C}}CH{=}\overset{\overset{CH_3}{|}}{C}CH_3 + CH_3\underset{\underset{CH_3}{|}}{\overset{\overset{CH_3}{|}}{C}}CH_2\overset{\overset{CH_3}{|}}{C}{=}CH_2 \qquad (15)$$

Diisobutylene may be used to alkylate phenol and thus to give *p*-octylphenol, a valuable detergent intermediate.

Tertiary paraffins can be alkylated in the presence of such catalysts as sulfuric acid, hydrofluoric acid, or aluminum chloride to give even more valuable, highly branched compounds. Following is again a typical reaction:

$$CH_2{=}\overset{\overset{CH_3}{|}}{C}CH_3 + CH_3\overset{\overset{CH_3}{|}}{C}HCH_3 \xrightarrow{H_2SO_4} CH_3\underset{\underset{CH_3}{|}}{\overset{\overset{CH_3}{|}}{C}}CH_2\overset{\overset{CH_3}{|}}{C}HCH_3 \qquad (16)$$

Aromatics also can be alkylated. Of special interest is the production of cumene, which not only is the raw material for the manufacture of phenol, but which also is a valuable constituent of high-octane gasoline:

$$C_6H_6 + CH_3CH{=}CH_2 \xrightarrow{H_3PO_4} C_6H_5CH(CH_3)_2 \qquad (17)$$

VI. PRODUCT SEPARATION AND ISOLATION

From Section V it can be seen that a tremendous variety of chemicals is produced in a refinery. These chemicals have to be separated into useful products, either as individual compounds, or, more likely, as mixtures for which there may be a very large commercial use, such as gasoline or lubricating oil. Since essentially all these chemicals are liquids or gases, initial separation and purification usually is by distillation. The gases can be distilled under pressure, and, if necessary, the high-boiling oils under vacuum. However, in specific cases other methods may be more economical.

For instance, low-boiling hydrocarbons such as propane and the butanes may be separated from methane by absorption in a heavier oil. They can be recovered by steam stripping. The mixture thus separated is sold commercially as LPG ("bottled gas") for household use, or it can be used as a refinery raw material as shown in the previous section. This absorption process may be used to remove these same hydrocarbons from natural gas.

Standards for the safe handling of LPG are set by the National Board of Fire Underwriters.

The trend now in natural gasoline plants is to refrigerate the methane by means of an expansion engine. Such a plant is called an expander plant. This refrigeration allows low-temperature fractionation of the ethane and other heavier components from the methane. These heavier components subsequently may be processed further.

The next fraction is gasoline, which boils up to about 375°F. There are many gasolines whose properties depend on where and when they are to be used. Specifications depend on such criteria as vapor pressure and percentage evaporated at 140°F. The most important characteristics are freedom from water, gum, and corrosive sulfur.

Following gasoline is kerosene, b.p. <370–500°F. There is always a little gasoline in kerosene to improve inflammability. The properties of kerosene are determined by its behavior in a lamp. Since smoking is undesirable, kerosene should contain a minimum of aromatics.

Naphthas and solvents overlap both gasoline and kerosene with boiling-point ranges somewhere in the broad spectrum of 95–455°F. The chemical structure of solvents varies enormously depending upon what the specific solvent is supposed to dissolve.

Jet fuel is kerosenelike in nature, with the added specification of such freezing points as −40°F or −76°F.

Above kerosene come the distillate fuels such as tractor, stove, and furnace oils. The flash point for domestic fuels is governed by safety considerations.

Color bodies and color formers always are a problem in petroleum products. In this connection sulfur dioxide is used to extract olefins, aromatics, and sulfur compounds from kerosene. Furfural is used to extract color bodies and sulfur and oxygen compounds from lubricating oils. Sulfuric acid is used very widely to remove color bodies and highly reactive gum formers from a number of fractions. The strength of the sulfuric acid, the amount of it used, and the temperature of its treatment vary with the job to be done. Adsorption of the colored impurities in lubricating oil on charcoal also is used. After washing with sulfuric acid, paraffin wax is given a final cleanup by percolation through a bed of attapulgus clay.

VII. CHEMICAL TREATMENTS

The treatments described in Section VI are largely physical in nature. Every so often a purely chemical treatment is necessary. Mercaptans are particularly undesirable constituents of petroleum products, both because they are sulfur compounds and because their aroma can hardly be described as pleasant. Two chemical treatments may be used to convert them to the less-objectionable disulfides. The so-called doctor solution is a mixture of sodium plumbite and sulfur:

$$2RSH + Na_2PbO_2 \longrightarrow (RS)_2Pb + 2NaOH \tag{18}$$

$$(RS)_2Pb + S \longrightarrow R_2S_2 + PbS \tag{19}$$

Cupric chloride also can be used:

$$4RSH + 4CuCl_2 \longrightarrow 2R_2S_2 + 4HCl + 4CuCl \tag{20}$$

$$4CuCl + 4HCl + O_2 \longrightarrow 4CuCl_2 + 2H_2O \tag{21}$$

In addition, about 40% of the crude oil entering a refinery undergoes a catalytic hydrodesulfurization at some stage of its processing, and this operation has become more important, since crude oils are now higher in sulfur content. Hydrodesulfurization converts the sulfur-containing compounds previously mentioned into hydrogen sulfide and lower-boiling hydrocarbons.

VIII. KEY PETROCHEMICAL BUILDING BLOCKS

At this point perhaps the best way to bring things into focus is to outline how certain key chemical building blocks are obtained from refineries as pure chemicals. This will summarize and reemphasize much of the chemistry that has just been discussed.

Many books on industrial chemistry simply state that ethylene is obtained by cracking higher hydrocarbons. This certainly is true. However, for many years it was prepared specifically by the cracking of propane:

$$CH_3CH_2CH_3 \longrightarrow CH_2{=}CH_2 + CH_4 \tag{22}$$

In a refinery cracking operation, higher paraffins also yield ethylene by much the same mechanism. More recently the yields by the dehydrogenation of ethane have been raised to the point where this process also is competitive:

$$CH_3CH_3 \longrightarrow CH_2{=}CH_2 + H_2 \tag{23}$$

The ethylene and unreacted ethane can be separated by fractional distillation under pressure or by adsorption of the ethylene on charcoal.

In 1979 ethylene uses were as follows [2]: polyethylene, 44%; vinyl chloride, 13%; ethylbenzene, 8%; ethylene oxide, 16%; and ethanol, aliphatic alcohols, α-olefins and vinyl acetate, most of the rest.

Propylene is prepared in basically the same ways as ethylene is:

$$CH_3(CH_2)_2CH_3 \longrightarrow CH_3CH{=}CH_2 + CH_4 \qquad (24)$$

$$CH_3CH_2CH_3 \longrightarrow CH_3CH{=}CH_2 + H_2 \qquad (25)$$

The preparation of butadiene has been developed in detail because of the importance of this chemical as a raw material for the manufacture of synthetic rubber. The feed to the cracker is a butane-butylene stream. Isobutylene is removed from this stream by polymerization with sulfuric acid. After the first cracking step over a magnesium oxide or chromia-alumina catalyst, the unreacted butane is separated from the product butylenes by extractive distillation, originally with aqueous acetone, but more recently with acetonitrile. Furfural also has been used in this step. In extractive distillation a solvent of low volatility is used to increase the boiling-point spread and therefore facilitate the separation of two more closely boiling chemicals.

The second step in the preparation of butadiene has been the oxydehydrogenation of the *n*-butene mixture over either a tin-lithium-phosphorus or ferrite catalyst under more-extreme conditions than those used in the previous operation. The butadiene so produced is extracted from the unreacted butenes by means of a selective solvent such as furfural, acetonitrile, dimethyl formamide, or *N*-methylpyrrolidone.

However, it should be pointed out that by-product butadiene from cracking to produce ethylene probably will soon displace all butene-to-butadiene dehydrogenation plants.

Isoprene is prepared by the catalytic dehydrogenation of a mixture of 2-methyl-1-butene and 2-methyl-2-butene. These petroleum cracking products can be extracted from refinery gases with 65% sulfuric acid. By starting with the two-olefin mixture rather than with isopentane, the formation of by-product piperylene is avoided, thus eliminating its attendant expensive separation step.

Isoprene also can be made from propylene, as shown in equation form:

$$2CH_3CH{=}CH_2 \rightarrow CH_2{=}\underset{}{\overset{\displaystyle CH_3}{\overset{|}{C}}}(CH_2)_2CH_3 \rightarrow CH_3\overset{\displaystyle CH_3}{\overset{|}{C}}{=}CHCH_2CH_3 \xrightarrow{\Delta}$$

$$CH_2{=}\overset{\displaystyle CH_3}{\overset{|}{C}}CH{=}CH_2 + CH_4 \qquad (26)$$

The Prins reaction also may be used. It is under continuing study today, probably because the isoprene produced is readily obtained in very pure form, ideal for polymerization.

$$CH_3C(CH_3){=}CH_2 + 2CH_2O \xrightarrow{H^+} (H_3C)_2C\begin{matrix}CH_2CH_2 \\ O{-}CH_2\end{matrix}\!\!>O \rightarrow CH_2{=}C(CH_3)CH{=}CH_2 + CH_2O + H_2O \quad (27)$$

Before leaving aliphatic chemistry it should be mentioned that propylene tetramer, which is prepared by the phosphoric acid–catalyzed polymerization of propylene, used to be used very extensively to alkylate benzene. For many years the resulting dodecylbenzene was the raw material for the sodium dodecylbenzenesulfonate, which is the active ingredient in many detergents. As shown in the previous chapter, this particular isomer has been superseded by compounds possessing a more-reactive and therefore more easily biodegradable alkyl side chain.

As indicated earlier, refineries process mixtures of compounds. In the direct manufacture of aromatics, a roughly C_6–C_8 naphtha stream is used as the feed stock for the reforming process. In review, then, at least four types of reactions take place:

$$\text{cyclohexane} \rightarrow \text{benzene} + 3H_2 \quad (28)$$

$$\text{methylcyclopentane } (CH_3) \rightarrow \text{benzene} + 3H_2 \quad (29)$$

$$\text{ethylcyclohexane } (C_2H_5) \rightarrow \text{benzene} + CH_2{=}CH_2 + 3H_2 \quad (30)$$

$$CH_3(CH_2)_4CH_3 \rightarrow \text{benzene} + 4H_2 \quad (31)$$

These equations are for the preparation of benzene. Similar aliphatic raw materials containing one or two additional methyl groups give toluene or one of the xylenes.

After the cracking operation the aromatic products are extracted from the unreacted naphtha with diethylene glycol–water, polyethylene glycol, or sulfolane. Benzene, toluene, and the xylenes are separated by fractional distillation.

High-purity toluene may be obtained by azeotropic distillation with methyl ethyl ketone. This is an example of the use of a high-volatility solvent as an entrainer. Recently it has been announced that a zeolite of suitable pore size, when modified with boron and phosphorus compounds, catalyzes the reaction between toluene and methanol to give 90% *p*-xylene [3].

o-Xylene can be separated from the other two isomers by fractional distillation. However, *m*- and *p*-xylene are not that easily separated. The para isomer is isolated by crystallization, or by selective adsorption on a molecular sieve.

The uses for *m*-xylene are limited. Fortunately it can be isomerized to the other two isomers over a zeolite catalyst:

$$m\text{-}CH_3C_6H_4CH_3 \underset{}{\overset{Al_2O_3 \cdot SiO_2}{\rightleftharpoons}} o\text{-}CH_3C_6H_4CH_3 \underset{}{\overset{Al_2O_3 \cdot SiO_2}{\rightleftharpoons}} p\text{-}H_3C\text{-}C_6H_4\text{-}CH_3 \quad (32)$$

It should be remembered that *o*-xylene is the raw material for the manufacture of phthalic anhydride. *p*-Xylene is the raw material for the manufacture of terephthalic acid, one of the two ingredients in polyethylene terephthalate polymers.

The versatility of a refinery is further illustrated by the fact that aromatics can be demethylated by means of hydrogen over a silica-alumina catalyst:

$$C_6H_5CH_3 + H_2 \xrightarrow{Al_2O_3 \cdot SiO_2} C_6H_6 + CH_4 \quad (33)$$

This is most helpful, since naphtha aromatization produces more toluene and xylenes than it does benzene. Demethylation helps bring the product mix into balance.

Although not strictly a refinery operation, the manufacture of styrene is essentially one, chemically speaking. First benzene is alkylated with ethylene using an aluminum chloride catalyst in the presence of hydrogen chloride. The ethylbenzene thus produced is then dehydrogenated. Two catalysts have been employed extensively—iron oxide on magnesium oxide promoted by potassium oxide and chromia-alumina promoted in various ways. (These same catalysts also have been used to dehydrogenate propane, butane, and the butenes.)

$$C_6H_6 + CH_2{=}CH_2 \xrightarrow[HCl]{AlCl_3} C_6H_5CH_2CH_3 \quad (34)$$

$$C_6H_5CH_2CH_3 \xrightarrow[\text{or } Cr_2O_3 \cdot Al_2O_3]{Fe_2O_3 \cdot MgO \cdot K_2O} C_6H_5CH{=}CH_2 + H_2 \quad (35)$$

The styrene is separated from the unreacted ethylbenzene by fractional distillation through a column possessing of the order of 100 plates.

Recently Monsanto Company has announced a process that could be cheaper since it avoids the costly separation step [4]:

$$2C_6H_5CH_3 \xrightarrow{\text{catalyst}} C_6H_5CH{=}CHC_6H_5 \xrightarrow[\text{catalyst}]{CH_2{=}CH_2} 2C_6H_5CH{=}CH_2 \quad (36)$$

The manufacture of cumene by the alkylation of benzene with propylene has already been mentioned.

IX. PETROLEUM-INDUSTRY PRODUCTS

Although the focus of this chapter has been on how organic chemicals are obtained from petroleum, there are two other aspects of the petroleum industry that are of interest to chemists. One is the effect of molecular structure on refinery-product performance. The high-compression engine of the modern automobile requires "high octane" gasoline for its most efficient performance. "High octane" gasoline is composed largely of branched-chain paraffins and aromatics rather than straight-chain paraffins. In fact, the unit of knock intensity, known as the octane number of a fuel, is defined as the percentage by volume of isooctane (2,2,4-trimethylpentane) that must be mixed with *n*-heptane in order to match the knock intensity of the fuel undergoing testing [5]. Following are the chemical structures of the test standards:

$$CH_3C(CH_3)_2CH_2CH(CH_3)CH_3 \qquad CH_3(CH_2)_5CH_3 \quad [1]$$

From the previous discussion it can be seen that the production of branched-chain paraffins and aromatics from crudes containing mostly straight-chain compounds is expensive. Fortunately, from the point of view of the motorist, the addition of tetraethyl lead appreciably increases the octane value and lowers the knocking quality of almost any gasoline stock. Obviously its elimination will increase the cost of motor fuel to quite an extent.

Diesel fuels, on the other hand, are supposed to knock. If they do not knock hard enough, knocking promotors such as amyl nitrate are added. The species that knock the most and therefore are the worst gasoline stocks are the straight-chain paraffins.

Straight-chain paraffins of fairly high molecular weight with just enough branching to provide a low pour point are good lubricants. Some Pennsylvania crudes are very high in paraffins and at one time were the best sources of paraffinic lubricating oils, and the author believes that this has been the basis for advertising programs by some oil companies for lubricants based on these crudes. Solvent extraction now permits the use of midcontinent crudes for lubricating-oil manufacture. Naphthenes are less desirable as lubricating-oil constituents because of their poor viscosity index, and aromatics are not desired at all. Good lubricating oils are nearly free of aromatics.

The viscosity index is an arbitrary scale for comparing the rates of viscosity changes of lubricating oils with temperature. A high number (100 or more) indicates the oil is still fluid at low temperatures and retains its viscosity at high temperatures, an obviously desirable characteristic.

The other area of the petroleum industry that is of tremendous chemical interest, particularly to the synthetic organic chemist, is that of additives to gasoline and lubricating oils. Many organic chemists in the petroleum industry and in the companies supplying it are involved in meeting these needs. Even though it is beyond the scope of this book to discuss this field in detail, the beginning industrial chemist should know of its existence.

Mention has been made of the addition of tetraethyl lead to gasoline as an antiknock agent. Ethylene dibromide is added along with the tetraethyl lead to form in the cylinder lead bromide, which is sufficiently volatile to escape from the engine in the exhaust. Otherwise lead would build up as a deposit in the cylinders.

Antioxidants are added to gasoline to stabilize the diolefins that may be present as impurities and thereby reduce gum formation. These include such compounds as 2,6-di-*t*-butyl-4-methylphenol, *N*-*n*- or isobutyl-*p*-aminophenol, and *sec*-butyl derivatives of *p*-phenylenediamine:

$$H_3C-C_6H_2(C(CH_3)_3)_2-OH \qquad CH_3(CH_2)_3NH-C_6H_4-OH \qquad [2]$$

$$(CH_3)(C_2H_5)CHNH-C_6H_4-NHCH(CH_3)(C_2H_5)$$

$$[(CH_3)(C_2H_5)CH]_2N-C_6H_4-N[CH(CH_3)(C_2H_5)]_2 \qquad [3]$$

Aminoalkylphosphates are added to gasoline as rust inhibitors. Isopropanol is added to prevent icing in the carburetor. Tricresyl phosphate, among others, is added as a combustion control agent to reduce surface ignition and therefore cylinder deposits. Metal deactivators and detergents are added for much the same purpose.

Lubricating oils have received similar attention. Zinc dialkyl dithiophosphates are added as antioxidants, antiwear agents, and corrosion inhibitors. To keep sludges in suspension, complex calcium or barium sulfonates are added as detergents. Traces of silicones are added as antiforming agents. Various polymers are added to lower the pour point of the oil and improve its viscosity index. It is desirable that lubricating oils maintain a reasonably uniform viscosity over a wide temperature range and be usable at low temperatures. Various higher polymethacrylates are employed most widely.

This is only a partial list of the many compounds that are added to petroleum products for a variety of purposes. However, even from this short list the reader can see that this is a very large field in its own right.

REFERENCES

1. W. L. Nelson, *Petroleum Refinery Engineering,* Fourth Edition, McGraw-Hill, New York, 1968, p. 647.
2. Anon., *Chem. Week,* **125,** October 3, 1979, p. 33.
3. N. Platzer, *Chemtech,* **11,** February, 1981, p. 90
4. Anon., *Chem. Eng. News,* **57,** April 16, 1979, p. 36.
5. W. L. Nelson, *Petroleum Refinery Engineering,* Fourth Edition, McGraw-Hill, New York, 1968, p. 28.

3
INDUSTRIAL CHEMISTRY

I. INTRODUCTION

Now that at least some aspects of the chemical industry have been looked at broadly and the source of most industrial organic chemicals examined, it is time to look in more detail at industrial chemistry. This chapter and the one following give an altogether too brief survey, but it is hoped that they will provide the reader with an understanding of how markedly industrial chemistry differs from academic chemistry.

In the industrial world, economic considerations are preeminent. Also, technical factors other than strictly chemical ones influence the choice and conduct of processes. Therefore, this chapter attempts to characterize industrial chemistry rather than survey it.

II. ECONOMIC CONSIDERATIONS

Before some typical industrial chemical processes are examined, some economic factors will be considered very generally. The industrial chemist must be aware of these even though he or she probably will not be involved in the detailed evaluation of all of them.

When the development of a chemical process that will lead ultimately to the construction of a plant is considered, the capital expense is a major factor. It includes the costs of laboratory research and development. Next comes the pilot plant. Whenever possible only key problem steps are piloted, rather than the whole process, in order to minimize this expense.

The proposed plant location is important. Major factors include not only the cost of the land, but also the access to it. Is it handy for barge shipment or on a railroad spur? Will there be environmental problems connected with the disposal of by-products?

The cost of the proposed plant is a major item. Its size, arrangement, and detailed engineering all are important in this connection. What about materials of construction? Stainless steel is much more expensive than mild steel. Can the plant be used to make several products rather than just one, as in a hydrogenation plant?

The time it takes to build the plant also is a factor. Capital tied up in construction does not earn anything.

Working capital is very important, and an insufficient amount can wreck an operation. By working capital is meant the money that is tied up in raw materials, materials in process, finished-product inventory, and accounts due, as well as that necessary to pay current operating costs while waiting for the customers to pay their bills.

Another major financial factor is materials efficiency. All chemists who have studied organic chemistry are familiar with the concept of yield. In industry, coproducts must be sold or disposed of. Monsanto Company nitrates chlorobenzene, and for many years it had a real problem because the markets for chemicals derived from the para isomer were much larger than those derived from the ortho isomer, and the nitration produced roughly equivalent amounts of the two products.

$$C_6H_5Cl \xrightarrow[H_2SO_4]{HNO_3} \text{o-}ClC_6H_4NO_2 + O_2N\text{—}C_6H_4\text{—}Cl \quad (1)$$

One likewise has to consider the advisability of establishing a captive raw-material source compared with depending on a merchant chemical obtainable from several suppliers. Should a phenol manufacturer alkylate benzene himself or depend on the petroleum companies for cumene?

$$C_6H_6 + CH_2{=}CHCH_3 \xrightarrow{H^+} C_6H_5CH(CH_3)_2 \quad (2)$$

$$C_6H_5CH(CH_3)_2 \xrightarrow[115\text{–}130°C]{\text{air, catalyst}} C_6H_5C(CH_3)_2(OOH) \xrightarrow[60\text{–}80°C]{10\% H_2SO_4} C_6H_5OH + CH_3COCH_3 \quad (3)$$

25–30% conversion

In this book, percent conversion is percent conversion to product per pass. Percent yield is percent overall yield. Where some raw material is recovered and recycled after each pass, the percent overall yield is higher than the percent conversion per pass. When no raw material is recovered and recycled after each pass, then the percent conversion and the percent yield are the same.

Some chemists and chemical engineeers mean by percent conversion all raw material not recovered after a given pass, in other words all raw material converted to something. In this case the percent yield per pass will always be lower than the percent conversion. There are one or two examples of this.

With the sudden rise in petroleum prices, energy efficiency has assumed even greater importance. In petrochemical processes, where the cost of energy can be 60% of the product cost, it has been estimated that a 180% price increase is necessary to absorb a quadrupling of crude oil costs [1].

Operating costs are a major factor. These include labor, supervision, utilities, taxes, and so forth. Profitability is extremely important, because if a company is not profitable, it goes bankrupt. Not only do the owners (stockholders) lose their investment, but the employees lose their jobs as well. Return on investment is another major factor. There is little sense in investing money in a plant that will return 3% on capital when the same amount of money in a bank will earn 6% or more.

Government regulations are becoming increasingly important and expensive. Present emphasis is on occupational safety and avoiding environmental pollution. These economic factors are illustrated both in this chapter and in other parts of the book.

III. BASIC TECHNOLOGY

As mentioned in Chapter 1, the basic raw materials of industrial chemistry are very simple and relatively few in number—ores, inorganic salts, coal, petroleum, air, and water. As an example, one of the simplest of the key building blocks of industrial organic chemistry is carbon monoxide. It can be prepared by the controlled oxidation of coal (Equation 4), the water gas reaction (Equation 5), treating hydrocarbons with steam at high temperatures (Equation 6), and the partial oxidation of hydrocarbons (Equation 7):

$$2C + O_2 \xrightarrow{1000°C} 2CO \qquad \text{(producer gas)} \tag{4}$$

$$C + H_2O \xrightarrow{600°C} CO + H_2 \qquad \text{(water gas)} \tag{5}$$

$$CH_4 + H_2O \xrightarrow[800\text{–}1000°C]{Ni+Al_2O_3} CO + 3H_2 \tag{6}$$

$$CH_4 + 1\tfrac{1}{2}O_2 \longrightarrow CO + 2H_2O \tag{7}$$

Higher hydrocarbons (naphtha) may be substituted for methane in the methods shown in Equations 6 and 7.

In recent years there has been a big shift from coal to petroleum, natural gas, and natural gas liquids as the raw materials for industrial organic chemicals, so that the methods shown in Equations 6 and 7 are the ones presently in use. With the dramatic rise in price and the impending scarcity of petroleum and natural gas, a shift back to coal (Equations 4 and 5) perhaps may occur.

Not only do the industrial processes just described use simple raw materials—coal, naturally occurring hydrocarbons, air, and water—but they also illustrate other characteristics of industrial chemistry. They are run at high temperatures. Many are catalytic (Equation 6). They involve the handling of gases. Obviously they must be conducted in metal equipment rather than in the glass equipment of an academic laboratory. And when a given chemical, in this case carbon monoxide, is to be prepared, a choice of processes is available, whose selection is based on economic factors, here the cost and availability of the raw material.

Another example is acetylene, on which much of industrial aliphatic chemistry was based in the early days. Originally it was prepared from coal in an electric furnace. More recently, petroleum and natural gas have displaced coal as the raw material for the manufacture of organic chemicals. For acetylene:

$$\underset{}{CH_4} \xrightarrow[\text{arc}]{\text{electric}} \underset{45\%\text{ yield}}{HC{\equiv}CH} + \underset{10\%\text{ yield}}{CH_2{=}CH_2} \qquad 50\%\text{ conversion} \tag{8}$$

Here again the chemist deals with reactions conducted at high temperatures and the handling of gases on a large scale. In either case energy is a big part of the cost of manufacturing acetylene.

It should be reemphasized that hydrocarbons are the major building blocks of organic chemistry as shown in the preceding chapter. The use of methane (natural gas) was illustrated in Equation 8. The major hydrocarbon source, of course, is petroleum, but as soon as economics so justify, shale oil and tar sands can serve as well. The technology has been worked out:

$$\underset{\begin{pmatrix}\text{Petroleum}\\ \text{Shale oil}\\ \text{Tar sands}\end{pmatrix}}{\text{Hydrocarbons}} \xrightarrow[Al_2O_3\cdot SiO_2]{\text{crack}} \text{paraffins} + \text{olefins} \tag{9}$$

Fortunately coal also can be used:

$$\underset{\text{(Coal)}}{C} + H_2 \xrightarrow[450°C,\ 700\text{ atm}]{\text{Fe catalyst}} \text{hydrocarbons} \tag{10}$$

Here again the process involves catalysis at high temperatures. In this case the catalysis is also at high pressures.

In Germany during World War II, the Fischer-Tropsch process was vital. Germany has extensive reserves of coal, but essentially no petroleum.

$$CO + H_2 \xrightarrow[\substack{250\text{–}300°C\\ 20\text{ atm}}]{\text{Fe or Co catalyst}} \text{straight-chain } \underset{(C_nH_{2n+2})}{\text{paraffins}} + \underset{(\text{Mostly } C_nH_{2n+1}CH{=}CH_2)}{\text{olefins}} \tag{11}$$

It is important to note that in this process the hydrocarbons produced are straight chain, the essential characteristic of the best hydrocarbon lubricants, necessary for all machinery.

These last two processes illustrate the fact that many industrial chemical reactions are conducted at elevated pressures, some at very high pressures. In university laboratories, pressures greater than atmospheric require not only special equipment, which must be assembled at considerable trouble and expense, but also an isolated laboratory with extra safety safeguards. In industry a plant can easily be built incorporating these features. Although the capital cost is high, it can be compensated for by the use of large units that produce large amounts of product continuously.

Now that the more important financial considerations have been mentioned, and some of the more important technical factors touched on briefly, the large-scale manufacture of some of those industrial organic chemicals for which major process changes have been made over the years will be examined. These will reillustrate and reemphasize some of the points that have just been made.

In this discussion a recent article [2] in which two financial criteria were developed for estimating the value of process improvements is quoted. The first is the percentage reduction in transfer price, A, meaning how much the new process would allow the transfer (sales) price to be lowered with no reduction in profit. The second is improved profitability, B, meaning the percentage increase in profit effected by the new process based on the product's being charged at a price equivalent to that charged with current technology (the old process). Whenever possible these figures are related to the considerations just discussed, although in many of these cases the use of a new process has been dictated primarily by reduced raw-material costs. However, the discussion in the following sections indicates why the prices of most commodity chemicals have not risen from those of 1955–1960 until very recently.

IV. VINYL CHLORIDE

In 1979 U.S. production capacity of this important chemical exceeded 8.5 billion pounds per year [3]. It is used primarily to manufacture polyvinyl chloride (90%) and copolymers thereof (10%) [4]. These find application as extrusions (60%, mostly pipe), calendered sheet and film (10%), coatings (10%), and molding (10%) [5]. In addition to pipe, finished forms include siding, tile, wall coverings, draperies, upholstery, electrical insulation, phonograph records, belts, shoes, paper coatings, foil coatings, cloth coatings, plastisols for can coatings, and tarpaulins.

Three manufacturing processes have been used over the years, and a fourth has just been announced (Equations 12–15):

$$CH_2{=}CH_2 + Cl_2 \longrightarrow CH_2ClCH_2Cl \xrightarrow[\text{or pumice at } 480\text{–}510°C]{OH^- \text{ or activated carbon}} CH_2{=}CHCl + HCl \quad (12)$$

$$HC{\equiv}CH + HCl \xrightarrow[\text{carbon, } 90\text{–}140°C]{HgCl_2} CH_2{=}CHCl \quad (13)$$

These can be run in parallel, with the hydrogen chloride from the first reaction (Equation 12) used to add to the acetylene in the second reaction (Equation 13). At present essentially all vinyl chloride is made from ethylene dichloride [3].

$$CH_2{=}CH_2 \xrightarrow{\text{oxychlorination}} \underset{80\%\ \text{yield}}{CH_2{=}CHCl} \tag{14}$$

In comparison with the second process (Equation 13), it has been calculated that the third process (Equation 14) offers a 17% reduction in transfer price A or a 134% increase in profitability B because ethylene is much cheaper than acetylene [2].

$$CH_3CH_3 + HCl + O_2 \xrightarrow[KCl]{CuCl_2} \underset{80\%\ \text{yield}}{CH_2{=}CHCl} + H_2O \tag{15}$$

The fourth process (Equation 15) was announced recently by the Lummus Company. In this case there is no need to invest in an ethylene plant, so in comparison with the third process (Equation 14), it has been calculated that the fourth process offers a 16% reduction in transfer price A or a 100% increase in profitability B [2].

V. ACRYLONITRILE

In 1977 U.S. production capacity of this chemical was over 2 billion pounds [6]. Its largest use is in polyacrylonitrile fibers, which find extensive markets once held by wool. Other uses are in nitrile (Buna N) type rubbers and in acrylonitrile-butadiene-styrene (ABS) and styrene-acrylonitrile (SAN) plastics. Recently Barex (Standard Oil Company, Ohio) and Lopac (Monsanto Company) have been announced, although the future of these products appears limited. They are acrylonitrile-based resins directed toward the bottle market.

Processes for the manufacture of this chemical have changed markedly over the years, primarily as a result of new technology.

$$\underbrace{CH_2{-}CH_2}_{O} + HCN \longrightarrow HO(CH_2)_2CN \longrightarrow CH_2{=}CHCN + H_2O \tag{16}$$

$$HC{\equiv}CH + HCN \xrightarrow[70\text{–}90°C]{CuCl\cdot HCl} CH_2{=}CHCN \tag{17}$$

$$4CH_2{=}CHCH_3 + 6NO \xrightarrow[Ag]{700°C} 4CH_2{=}CHCN + 6H_2O + N_2 \tag{18}$$

The third reaction (Equation 18) was used by E. I. Du Pont de Nemours and Company for several years but is no longer competitive.

$$2CH_2{=}CHCH_3 + 2NH_3 + 3O_2 \xrightarrow[\text{425–510°C, fluid bed}]{\text{bismuth phosphomolybdate}} 2CH_2{=}CHCN + 6H_2O \quad (19)$$

In the reaction shown in Equation 19 (Standard Oil Company, Ohio) a uranium-based catalyst is believed now to be in use. The maximum yield is believed to be 74%.

In comparison with the process shown in Equation 17, it has been calculated that the process shown in Equation 19 offers a 16% reduction in transfer price *A* or a 63% increase in profitability *B* because propylene is much cheaper than acetylene [2]. Also the addition of hydrogen cyanide to acetylene is tricky, and hydrogen cyanide is a difficult chemical to handle. Parenthetically, it is made as follows:

$$2NH_3 + 3O_2 + 2CH_4 \xrightarrow[\text{1000°C}]{\text{Pt–Rh}} 2HCN + 6H_2O \quad (20)$$

It is believed that Imperial Chemical Industries, Ltd. mothballed a plant based on the second reaction (Equation 17) the day it was ready to go on stream because of the availability of the fourth-reaction (Equation 19) technology.

In this case a technological advance (Equation 19) has halved the price of a merchant chemical. Interestingly, of the two chemists most responsible for the process, one was a PhD of two years' standing and the other was still working for his PhD at night under a company program [7]. There are plenty of opportunities for young, imaginative chemists in industry.

Recently a process has been patented that if commercially feasible would eliminate the propylene manufacture step [8]:

$$CH_3CH_2CH_3 + NH_3 \xrightarrow[\text{CH}_3\text{Br promotor, O}_2\text{ from oxidized catalyst}]{\text{500°C, Sb+U oxides}} CH_2{=}CHCN \quad (21)$$

41% conversion
68% yield

Here again (Equation 21) high-temperature, vapor-phase, catalytic reactions are involved.

VI. ACETALDEHYDE

This key intermediate is used primarily for the manufacture of other chemicals such as metaldehyde, chloral, peracetic acid, acetic anhydride, pentaerythritol, and particularly acetic acid. In 1976 U.S. production capacity was about 1.5 billion pounds [9].

Originally prepared from coal via acetylene, it is now prepared from petroleum via ethylene. Some acetaldehyde also is made by the oxidation of propane and butane:

$$HC{\equiv}CH + H_2O \xrightarrow[HgSO_4,\ H_2SO_4]{100°C,\ 15\ psi} CH_3CHO \quad (22)$$

$$HC{\equiv}CH + H_2O \xrightarrow[FeSO_4,\ H_2SO_4]{94\text{–}98°C} CH_3CHO \quad (23)$$

96% yield

This reaction (Equation 23) was developed in Germany in World War II because of the shortage of mercury in that country.

$$HC{\equiv}CH + CH_3OH \xrightarrow{200°} CH_2{=}CHOCH_3 \xrightarrow[H^+]{H_2O} CH_3CHO + CH_3OH \quad (24)$$

$$\left.\begin{matrix} CH_3CH_2CH_3 \\ + \\ CH_3(CH_2)_2CH_3 \end{matrix}\right\} + O_2 \xrightarrow[7\text{–}8\ atm]{370°C} CH_3CHO + CH_2O + CH_3OH \quad (25)$$

This process (Equation 25) also yields small amounts of several other products, and separation is a big job. Only the cheapness of the raw material makes the process commercially attractive.

$$CH_2{=}CH_2 + H_2O \xrightarrow{H^+} CH_3CH_2OH \xrightarrow[270\text{–}300°C]{Cr\text{-activated } Cu} CH_3CHO + H_2 \quad (26)$$

$$CH_2{=}CH_2 + H_2O \xrightarrow{H^+} CH_3CH_2OH \xrightarrow[450°C,\ 3\ atm]{O_2,\ Ag\ gauze} CH_3CHO + H_2O \quad (27)$$

$$CH{=}CH_2 + H_2O \xrightarrow[100°C,\ 10\ atm]{O_2,\ CuCl_2,\ PdCl_2,\ HCl} CH_3CHO \quad (28)$$

95% yield

Even though the latter process (Equation 28) requires titanium equipment because of the corrosive nature of the salt mixture, it has been calculated that in comparison with the process shown in Equations 26 and 27, this process offers a 16% reduction in transfer price *A* or a 65% increase in profitability *B*, because it is one step and it eliminates the necessity of recovering the intermediate ethanol [2].

VII. ACETIC ACID

This important chemical, the U.S. production capacity of which at the end of 1978 was essentially 3.5 billion pounds [10], is used very widely as a solvent in its own right and as an intermediate for the production of vinyl acetate, acetic anhydride, ethyl acetate, cellulose acetate, and a wide variety of other acetic esters. Its manufacture is a good example of a situation in which a company's choice of available processes has depended on the raw-material situation and the technology possessed by the company in question, one basic factor being the cost of ethylene (for acetaldehyde) compared with the cost of carbon monoxide (for methanol) compared with the cost of *n*-butane.

$$CH_3(CH_2)_2CH_3 + O_2 \xrightarrow[175°C,\ 54\ atm]{Co(OCOCH_3)_2\ or\ Mn(OCOCH_3)_2} CH_3CO_2H \quad (29)$$

Here, in contrast to the oxidation of butane to produce acetaldehyde, the oxidation produces acetic acid as the major product, and the process is believed to represent half of the U.S. manufacturing capacity. When naphtha (C_4–C_8) is used in place of butane, both formic and propionic acids are appreciable byproducts for which markets must be found.

$$2CH_3CHO + O_2\ (air) \xrightarrow[40\text{–}65°C,\ 5\ atm]{Co(OCOCH_3)_2\ or\ Mn(OCOCH_3)_2} 2CH_3CO_2H \quad (30)$$

$$CH_3OH + CO \xrightarrow[175\text{–}245°C,\ 15\ atm]{iodide\text{-}promoted\ Rh} \underset{96\%\ yield}{CH_3CO_2H} \quad (31)$$

In comparison with the second process (Equation 30), it has been calculated that the third process (Equation 31) [11] offers an 8% reduction in transfer price *A* or a 39% increase in profitability *B* [2]. Presently the third route (Equation 31) is believed to be the most promising commercially [9].

By a suitable modification of reaction conditions of the third process, acetic anhydride can be produced [12]. Another modification leads to ethanol [13]:

$$CH_3OH + CO + H_2 \xrightarrow[CoI_2,\ R_3P]{200°C} CH_3CH_2OH \quad (32)$$

Recently a modification of the Willgerodt reaction has been patented, and this could be an easy way to convert ethylene to acetic acid in one step [14]:

$$CH_2{=}CH_2 + H_2O + S + NH_3 \xrightarrow{300\text{–}330°C} CH_3CO_2H \quad (33)$$

The direct oxidation of ethylene to acetic acid has also been disclosed [15]:

$$CH_2{=}CH_2 + O_2 \xrightarrow[150°C,\ 3\text{–}4\ atm]{Pd \cdot Au \cdot SO_2/C} CH_3CO_2H \quad (34)$$

70% conversion
85% yield

VIII. ETHYLENE GLYCOL

In 1976 U.S. production capacity of this important chemical was about 4.5 billion pounds [3]. It has two major uses, as an automobile radiator fluid and as one of the raw materials for polyester resins. Originally a "permanent" antifreeze only, now the efficient operation of most automobiles requires radiator temperatures above the boiling point of water under pressure year round, with ethylene glycol the important nonaqueous ingredient of the coolant.

$$CH_2(OH)_2 + CO \xrightarrow[CH_3OH]{H_2SO_4} HOCH_2CO_2CH_3 \xrightarrow[catalyst]{H_2} HO(CH_2)_2OH \quad (35)$$

$$(CH_2O + H_2O)$$

This is the old E. I. Du Pont de Nemours and Company process, which was in use until 1968 [16].

Recent Japanese patent disclosures indicate that the yields are very high [17].

$$CH_2O + CO + H_2O \xrightarrow[50\ kg/cm^2,\ 55°C]{95\%\ H_2SO_4} HOCH_2CO_2H \quad (36)$$

96% yield

If *p*-formaldehyde is used as the starting material, polyglycolides may constitute a big part of the product. These can be hydrogenated readily to ethylene glycol [18] or even oxidized to oxalic acid [19].

$$\text{Polyglycolide} \xrightarrow[CoOxide\cdot ZnO\cdot CuOxide]{H_2,\ 250°C,\ 2450\ psi} HO(CH_2)_2OH \quad (37)$$

82% conversion

$$\text{Polyglycolide} \xrightarrow[80°C]{HNO_3} HO_2CCO_2H \quad (38)$$

77% conversion
96% yield

$$CH_2{=}CH_2 + O_2(\text{Air}) \xrightarrow[260\text{–}290°C,\ 10\text{–}30\ at.]{Ag} \underset{\backslash O /}{CH_2{-}CH_2} \xrightarrow[H_2SO_4,\ 50\text{–}70°]{H_2O} HO(CH_2)_2OH \quad (39)$$

70–75% yield

The one-step conversion of ethylene has also been reported [20]:

$$CH_2{=}CH_2 + O_2 \xrightarrow[Cu^{2+}\ or\ Fe^{3+},\ 160°C,\ 20\ min]{H_2O,\ Br^-} HO(CH_2)_2OH \quad (40)$$

88% yield

In fact this type of conversion appears to be a very active field of investigation in Japan, with iodide apparently the preferred catalyst [21]:

$$CH_2{=}CH_2 + O_2 + H_2O \xrightarrow[150^\circ C,\ 30\ atm]{KI,\ HNO_3} HO(CH_2)_2OH \qquad (41)$$

(Air) 75% conversion, 96% yield

$$CH_2{=}CH_2 + \tfrac{1}{2}O_2 + 2CH_3CO_2H \xrightarrow[LiBr,\ 170^\circ C]{TeO_2,\ HBr} CH_3CO_2(CH_2)_2OCOCH_3 \xrightarrow[H^+]{H_2O}$$

97% yield

$$HO(CH_2)_2OH + 2CH_3CO_2H \qquad (42)$$

90–95% yield

The second process (Equation 39) was developed by Union Carbide Corp. The third process (Equation 42) was recently announced by Halcon International [22]. In spite of the expense of recovering and recycling the acetic acid, it has been calculated that in comparison with the second process (Equation 39), the third (Equation 42) offers a 12–16% reduction in transfer price *A* or a 47–94% increase in profitability *B*, because ethylene is 60% of the cost of the glycol [2]. However, the rise in ethylene prices may again make the first process (Equation 35) attractive. It also has spurred work on the one-step conversion of carbon monoxide to ethylene glycol (Equations 43 [23] and 44 [24]):

$$CO + H_2 \xrightarrow[pyridine,\ sulfolane,\ 240^\circ C,\ 562\ atm]{Rh\ dicarbonylacetylacetonate} HO(CH_2)_2OH + CH_3OH + HO(CH_2)_3OH \qquad (43)$$

$$CO + H_2 \xrightarrow[C_sO(CO)H,\ sulfolane]{Rh\ dicarbonylacetoacetate} HO(CH_2)_2OH + CH_3OH \qquad (44)$$

$$N{\leftarrow}[(CH_2)_2O]_3{\rightarrow}B$$

(N—(CH₂)₂O—B tricyclic borate: N bonded to three $(CH_2)_2O$ arms, each joined to B)

IX. HEXAMETHYLENEDIAMINE

Hexamethylenediamine is one of the two raw materials for nylon 6,6 and also for nylons 6,10 and 6,12 [25]. Nylon finds uses in such varied products as hosiery, garments, parachutes, tire cord, molded products, carpets, and upholstery. In 1977 U.S. manufacturing capacity for the diamine was about 1 billion pounds [25].

Hexamethylenediamine originally was prepared from adipic acid, the other raw material for nylon 6,6:

$$HO_2C(CH_2)_4CO_2H \xrightarrow{NH_3} H_2NOC(CH_2)_4CONH_2 \xrightarrow[catalyst]{H_2} H_2N(CH_2)_6NH_2 \qquad (45)$$

$$HO_2C(CH_2)_4CO_2H \xrightarrow{NH_3} NC(CH_2)_4CN \xrightarrow[catalyst]{H_2} H_2N(CH_2)_6NH_2 \quad \text{[Ref. 26]} \qquad (46)$$

The hydrogenation of adiponitrile in the presence of a cobaltous oxide on silica catalyst has been reported to be essentially quantitative (99%) [27]. The next route was from furfural:

$$\text{furfural (furan-CHO)} \xrightarrow[H_2O\ 400°C]{Zn—Cr—Mo} \text{furan} \xrightarrow[\text{Raney Ni, } 100\text{–}150\ atm]{H_2\ 120°C} \text{tetrahydrofuran} \xrightarrow{HCl} Cl(CH_2)_4Cl \xrightarrow{NaCN}$$

$$NC(CH_2)_4CN \xrightarrow[catalyst]{H_2} H_2N(CH_2)_6NH_2 \qquad (47)$$

More recently butadiene has displaced furfural as the raw material:

$$CH_2{=}CHCH{=}CH_2 + Cl_2 \xrightarrow{65\text{–}75°C} CH_2ClCHClCH{=}CH_2 + \overset{CuCl_2}{\rightleftharpoons}$$

$$CH_2ClCH{=}CHCH_2Cl \xrightarrow[80\text{–}95°C]{NaCN} NCCH_2CH{=}CHCH_2CN \xrightarrow[300°C]{H_2,\ Pd\ on\ C}$$

$$NC(CH_2)_4CN \xrightarrow[125\text{–}250°C,\ 9000\ psi]{H_2,\ Co\ catalyst} H_2N(CH_2)_6NH_2 \qquad (48)$$

$$CH_2{=}CHCH{=}CH_2 + 2HCN \xrightarrow[\text{various catalysts}]{\text{several steps}} NC(CH_2)_4CN \xrightarrow[\text{as in Equation 48}]{H_2}$$

$$H_2N(CH_2)_6NH_2 \qquad (49)$$

The process involving the addition of hydrogen cyanide to butadiene is the newer, but the E. I. Du Pont de Nemours and Company uses both butadiene processes (Equations 48 and 49). With the increase in petroleum prices, the furfural route (based on corn cobs) may quite possibly again be economically preferable. This is one of the few examples of the manufacture of a tonnage chemical from an agricultural source.

Another process is to treat hexanediol with ammonia:

$$HO(CH_2)_6OH + 2NH_3 \xrightarrow{catalyst} H_2N(CH_2)_6NH_2 + 2H_2O \qquad (50)$$

The new Monsanto Company process is technologically unique in that it is one of the very few electro-organic reactions to be used on a large scale in industry:

$$2CH_2{=}CHCN \xrightarrow[(H)]{elec} NC(CH_2)_4CN \xrightarrow[catalyst]{H_2} H_2N(CH_2)_6NH_2 \qquad (51)$$

X. TEREPHTHALIC ACID

In 1979 about 7 billion pounds of terephthalic acid or its dimethyl ester could be manufactured in the United States [9]. Either of these two compounds constitutes one of the two raw materials for the preparation of polyethylene terephthalate, and, of the two, much the larger by weight. Primary uses for this polymer are in fiber and film, for fabrics, carpeting, packaging, electrical insulation, photographic film, magnetic tapes, and so forth. Terephthalic acid itself also finds limited use in paints, coatings, printing inks, and adhesives.

A variety of methods has been used for its industrial preparation.

$$C_6H_5CO_2K + CO_2 \xrightarrow[410\text{–}430°C\ 10\text{–}15\ atm]{Cd(OCOC_6H_5)_2} HO_2C\text{—}C_6H_4\text{—}CO_2H \qquad (52)$$

This process shown in Equation 52 has found limited utility, although many patents have been issued covering its several ramifications. There has been revived interest in modifications of it recently [28].

All the other methods are based on *p*-xylene, so there has been a tremendous technical effort directed toward the preparation and purification of this hydrocarbon.

$$H_3C\text{—}C_6H_4\text{—}CH_3 \xrightarrow[150\text{–}400°C,\ 125\text{–}200\ psi]{HNO_3} HO_2C\text{—}C_6H_4\text{—}CO_2H \qquad (53)$$

$$H_3C\text{—}C_6H_4\text{—}CH_3 \xrightarrow{O_2} H_3C\text{—}C_6H_4\text{—}CO_2H \xrightarrow[H^+]{CH_3OH}$$

$$H_3C\text{—}C_6H_4\text{—}CO_2CH_3 \xrightarrow{O_2} HO_2C\text{—}C_6H_4\text{—}CO_2CH_3 \xrightarrow[H^+]{CH_3OH}$$

$$CH_3OCO\text{—}C_6H_4\text{—}CO_2CH_3 \qquad (54)$$

In spite of the many steps involved (Equation 54), this Hercules Inc. process is widely used, because its ultimate product, dimethyl terephthalate, is easy to handle and purify (in contrast to pure terephthalic acid), and many polyester plants are geared to handle the ester, rather than the pure acid.

$$H_3C\text{—}C_6H_4\text{—}CH_3 + O_2 \xrightarrow[200\text{–}450°C,\ 200\text{–}400\ psi]{Co^{++}Br^-CH_3CO_2H} HO_2C\text{—}C_6H_4\text{—}CO_2H \qquad (55)$$

The process shown in Equation 55 uses glacial acetic acid as the solvent, and this is extremely corrosive at elevated temperatures and pressures. The commercialization of the process had to await the development of techniques for forging titanium, the only commercially feasible metal to hold the reaction. All previous processes, as disclosed in the patent literature, necessitated the use of tantalum-lined equipment, a hardly economical material of construction. When the forging of titanium metal had been achieved, the process using direct oxidation to the acid offered advantages over the dimethyl terephthalate process, since the use of the free acid, obtained pure in this way, possessed cost advantages. It also possessed processing advantages in the preparation of the resin, in that the large evolution of methanol did not have to be handled and the methanol recovered.

$$H_3C-C_6H_4-CH_3 \xrightarrow[NH_3]{O_2} NC-C_6H_4-CN \xrightarrow[OH^-]{H_2O} HO_2C-C_6H_4-CO_2H \quad (56)$$

The process shown in Equation 56 has been proposed by the Lummus Company [29] and is said to offer appreciable environmental advantages in that the only by-products are water and carbon dioxide. It is discussed in more detail in Chapter 8.

XI. ANTHRAQUINONE

Traditionally this important dye intermediate has been prepared from phthalic anhydride and benzene [30]:

$$C_6H_4(CO)_2O + C_6H_6 \xrightarrow{AlCl_3} C_6H_5-CO-C_6H_4-CO_2H \xrightarrow[180°C]{H_2SO_4} C_{14}H_8O_2 \text{ (anthraquinone)} \quad (57)$$

81–90% yield

Recently Badische Anilin und Soda Fabrik has announced an alternate route with interesting possibilities [2]:

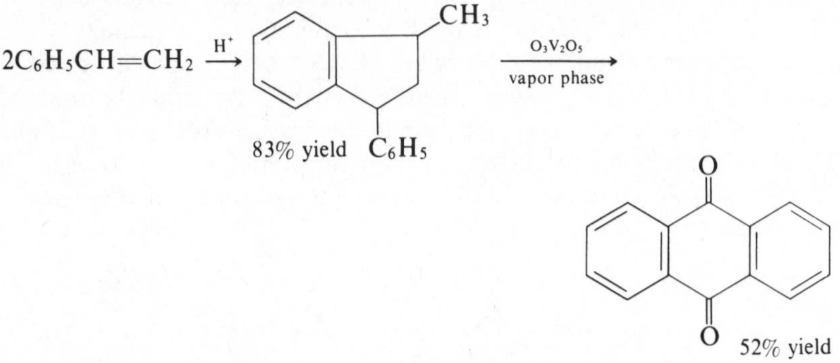

$$2C_6H_5CH{=}CH_2 \xrightarrow{H^+} \text{(1-methyl-3-phenylindane, 83\% yield)} \xrightarrow[\text{vapor phase}]{O_3V_2O_5} \text{anthraquinone (52\% yield)} \tag{58}$$

Although the yields in the first process (Equation 57) are much lower than those in the second process (Equation 58), the raw-material cost calculations are enlightening:

Compound	MW	Cost/lb [31]	Pounds Used
Phthalic anhydride	148	$0.42	148
Anhydrous aluminum chloride	133	0.46	293
Styrene	104	0.38	208
Anthraquinone	208		

Route 1:

$$148 \times 0.42 + 293 \times 0.46 = \$197.20$$

$$\text{Cost/lb: } 208 \times 0.86 = 179 \qquad \frac{197.20}{179} = \$1.10$$

Route 2:

$$208 \times 0.38 = \$82.80$$

$$\text{Cost/lb: } 208 \times 0.43 = 89.5 \qquad \frac{82.80}{89.5} = \$0.915$$

In addition, the first process (Equation 57) requires the disposal of large quantities of waste aluminum salts, whereas the second (Equation 58) does not.

However, another approach has been announced [32], and this could easily displace both processes:

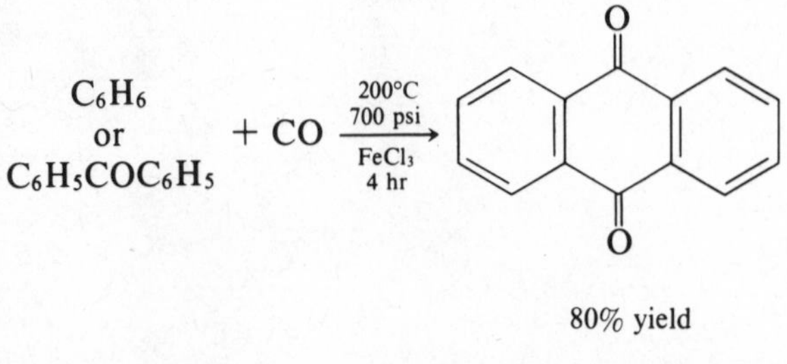

$$\begin{matrix} C_6H_6 \\ \text{or} \\ C_6H_5COC_6H_5 \end{matrix} + CO \xrightarrow[\substack{FeCl_3 \\ 4\ hr}]{\substack{200°C \\ 700\ psi}} \text{anthraquinone (80\% yield)} \tag{59}$$

Still a fourth route has been announced, this time on a commercial basis—a Bayer–Ciba-Geigy plant in West Germany plans to produce 15,000 tons per year of anthraquinone [33]:

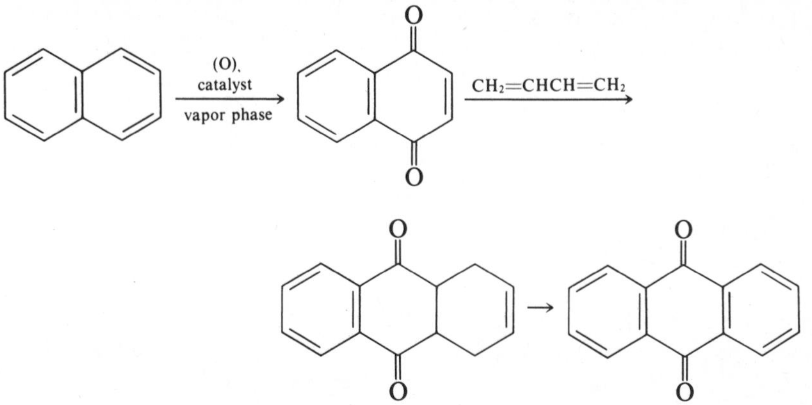

$$\text{naphthalene} \xrightarrow[\text{vapor phase}]{\text{(O), catalyst}} \text{1,4-naphthoquinone} \xrightarrow{CH_2{=}CHCH{=}CH_2}$$

$$\text{tetrahydroanthraquinone} \rightarrow \text{anthraquinone} \qquad (60)$$

XII. MALEIC ANHYDRIDE

This chemical's largest use is as a dibasic-acid component in alkyd and polyester resins. In 1978 U.S. manufacturing capacity was about 500 million pounds [34]. Traditionally it has been prepared by the vapor phase, air oxidation of benzene over a vanadium pentoxide catalyst:

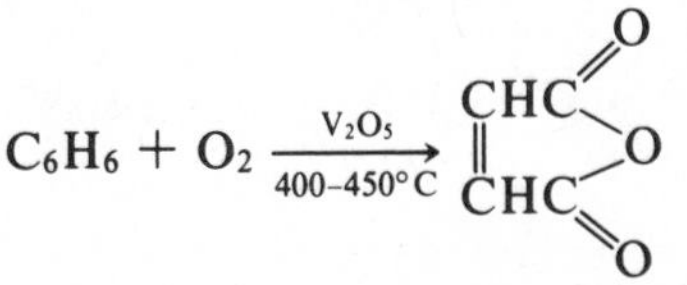

$$C_6H_6 + O_2 \xrightarrow[400\text{–}450°C]{V_2O_5} \text{maleic anhydride (CHC(=O)–O–C(=O)CH, CH=CH)} \qquad (61)$$

Recently, C_4 streams have appeared to be more-attractive raw materials than benzene. *n*-Butenes have been employed in both Germany and Japan [35]. Four companies (Amoco Chemicals Corp., Chevron Chemical Co., Monsanto Company, and Alusuisse) have announced processes based on *n*-butane [2]:

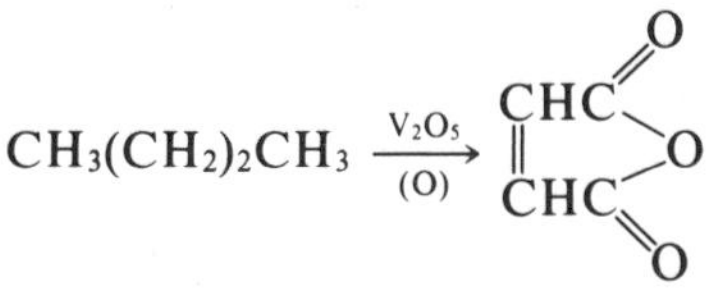

$$CH_3(CH_2)_2CH_3 \xrightarrow[(O)]{V_2O_5} \text{maleic anhydride (CHC(=O)–O–C(=O)CH, CH=CH)} \qquad (62)$$

In comparison with the process shown in Equation 61, it has been calculated that the process shown in Equation 62 offers a 12% reduction in transfer price *A* or a 40% increase in profitability *B* [2]. This is because when (ca. 1974) the calculations were made, *n*-butane was 2 cents per pound cheaper than benzene, and two carbon atoms are lost in the benzene oxidation.

XIII. PHTHALIC ANHYDRIDE

This compound is a very important intermediate for the manufacture of plasticizers and of alkyd and polyester resins. In 1977 U.S. manufacturing capacity was a little over 1 billion pounds [36]. The classical manufacturing method is the vapor-phase oxidation of naphthalene over a vanadium pentoxide catalyst:

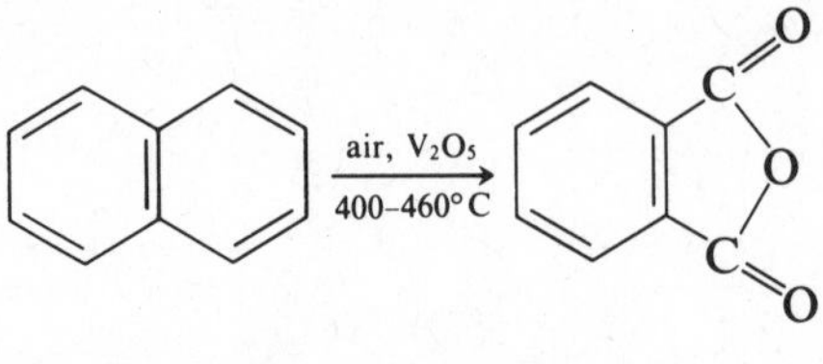

$$\text{Naphthalene} \xrightarrow[400\text{–}460^\circ C]{\text{air, } V_2O_5} \text{Phthalic anhydride} \qquad (63)$$

Originally the naphthalene came from coal tar, but now most of it comes from petroleum. More recently, *o*-xylene has taken over as the preferred feedstock for the oxidation because of lower prices and higher yields:

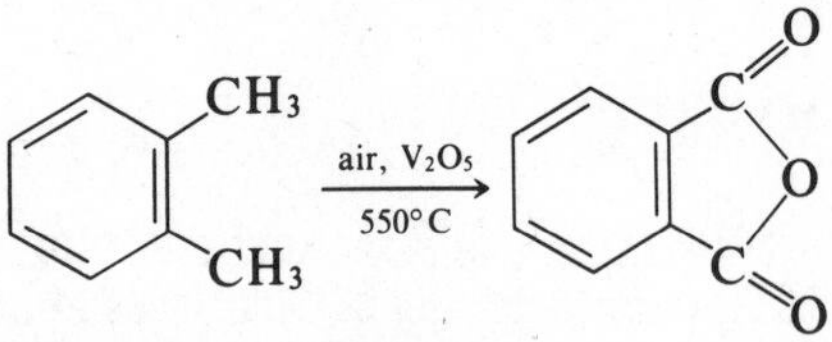

$$C_6H_4(CH_3)_2 \xrightarrow[550^\circ C]{\text{air, } V_2O_5} \text{Phthalic anhydride} \qquad (64)$$

In comparison with the first process (Equation 63), it has been calculated that the second process (Equation 64) offers a 15% reduction in transfer price *A* or a 45% increase in profitability *B* [2].

XIV. PHENOL

In 1978 U.S. production capacity of this important intermediate was about 3.5 billion pounds [35]. The production goes into phenol-formaldehyde resins, which find use as molding and laminating materials, coatings, plywood adhesives, and foundry cores. Phenol also serves as an intermediate for epoxy resins (as bisphenol A), for nylon 6, and for surface-active agents (through such further intermediates as *p*-dodecylphenol).

There are three older methods for its manufacture, two of which have been discarded. The first involves the sulfonation of benzene followed by alkali fusion of the resulting sodium sulfonate. The process was developed by the Monsanto Company and Reichold Chemicals, Inc. and was the first to be used commercially:

$$C_6H_6 + SO_3 \xrightarrow{H_2SO_4} C_6H_5SO_3H \qquad (65)$$

$$C_6H_5SO_3Na + NaOH \xrightarrow[300°C]{fuse} C_6H_5ONa + NaHSO_3 \quad (66)$$

The by-product sulfite must be disposed of. The second process involves the alkaline hydrolysis of chlorobenzene. It was commercialized by the Dow Chemical Company about 1924 and is still in use:

$$C_6H_5Cl + 2NaOH \xrightarrow[360°C,\ 5000\ psi]{C_6H_5OC_6H_5} \underset{78\%\ yield}{C_6H_5ONa + NaCl + H_2O} \quad (67)$$

The third (Raschig) process originated in Germany and was commercialized in this country by Hooker Chemical Company about 1940. It involves the oxychlorination of benzene followed by hydrolysis of the chlorobenzene thus produced:

$$C_6H_6 + HCl + \tfrac{1}{2}O_2 \xrightarrow[230°C]{CuCl_2 + FeCl_2} C_6H_5Cl + H_2O \quad (68)$$

$$C_6H_5Cl + H_2O \xrightarrow[425°C]{Ca_3(PO_4)_2\ on\ SiO_2} C_6H_5OH + HCl \quad (69)$$

This process suffers from low conversions (15% per pass) and high equipment corrosion from the temperature, chlorides, oxygen, and steam.

The process presently in widest use is based on cumene, as shown in Equations 2 and 3 of this chapter. It was discovered in Germany and first developed in this country by Allied Chemical Corp. and Hercules, Inc. Obviously the by-product acetone must be disposed of. The reader should note that sulfuric acid is used in the process, rather than the more corrosive hydrochloric acid.

The fifth process was developed by the Dow Chemical Company in 1961. It has found very limited commercial acceptance:

$$2C_6H_5CH_3 + 3O_2 \xrightarrow[110\text{–}150°C,\ 30\ psi]{Co\ salt} 2C_6H_5CO_2H + 2H_2O \quad (70)$$

$$2C_6H_5CO_2H + O_2 \xrightarrow[H_2O,\ 220\text{–}245°C]{Cu \cdot Mg\ benzoate} 2C_6H_5OH + 2CO_2 \quad (71)$$

XV. VINYL ACETATE

In 1978 the production capacity of this important vinyl monomer was about $1\frac{3}{4}$ billion pounds per year [37]. Although some of it is used in copolymers with vinyl chloride, most of it ends up as polyvinyl acetate. This polymer is used as such in adhesives, films, water-based paints, and textile and paper chemicals. Some polyvinyl acetate also is hydrolyzed to polyvinyl alcohol, which also finds use in textile and paper chemicals. Some polyvinyl alcohol is converted to polyvinyl butyral, the interleaving layer in safety glass.

Vinyl acetate is another chemical that originally was derived from coal and whose production now is based on petroleum. The original process for its manufacture involved the addition of acetic acid to acetylene:

$$HC{\equiv}CH + CH_3CO_2H \xrightarrow[75°C,\ 5\ psi]{Hg^{2+}\ \text{liquid phase}} CH_2{=}CHOCOCH_3 \quad (72)$$

80% conversion, 95% yield

Union Carbide Corp., E.I. DuPont de Nemours Company, and British Celanese Ltd. worked on the corresponding vapor-phase reaction. It was finally commercialized about 1948 by Celanese Corporation of America.

$$HC{\equiv}CH + 2CH_3CO_2H \xrightarrow[\text{vapor phase, } 200°C]{Hg(OCOCH_3)_2} CH_3CH(OCOCH_3)_2 \quad (73)$$

$$CH_3CH(OCOCH_3)_2 \xrightarrow{\Delta} CH_2{=}CHOCOCH_3 + CH_3CO_2H \quad (74)$$

During World War II, British Celanese Ltd. developed a process based on acetaldehyde and acetic anhydride. Celanese Corporation of America commercialized it in 1952.

$$CH_3CHO + (CH_3CO)_2O \longrightarrow CH_2{=}CHOCOCH_3 \quad (75)$$

Meanwhile ethylene became a more attractive raw material than acetylene. Again Celanese Corporation of America was the one to commercialize a liquid-phase process in 1965 based on Imperial Chemical Industries, Ltd. technology.

$$CH_2{=}CH_2 + CH_3CO_2H + O_2 \xrightarrow[100°C,\ 300\ psi]{PdCl_2 \cdot CuCl_2,\ \text{liquid phase}} CH_2{=}CHOCOCH_3 + CH_3CHO \quad (76)$$

This method uses acetic acid under pressure in the presence of chlorides. The corrosion problems are so bad that some of the plant must be fabricated out of titanium, which does not help the competitiveness of the process.

In comparison with the second process (Equation 73), it has been calculated that the fourth process (Equation 76) offers an 8% reduction in transfer price *A* or a 39% increase in profitability *B* [2].

In 1970 Celanese Corporation of America and National Distillers and Chemical Corp. commercialized a vapor-phase process [38]:

$$CH_2{=}CH_2 + CH_3CO_2H + O_2 \xrightarrow[175\text{–}200°C,\ 0\text{–}10\ atm]{\text{Pd catalyst, vapor phase}} CH_2{=}CHOCOCH_3 \quad (77)$$

91% yield

Although the conversion is low, catalyst life is long, and by-products apparently are not troublesome. This appears to be the preferred process for new plant construction at the present time.

If the Halcon International process (Equation 42) for ethylene glycol is commercialized, it also could be used as a source of vinyl acetate [22].

$$CH_3CO_2(CH_2)_2OCOCH_3 \xrightarrow{550°C} \underset{81\%\ \text{yield}}{CH_2{=}CHOCOCH_3} + CH_3CO_2H \quad (78)$$

XVI. METHYL METHACRYLATE

Polymethylmethacrylate dates back to 1936, when the Rohm and Haas Company first produced cast sheets of this polymer. It always has been distinguished by its superb optical properties.

In 1976 U.S. production capacity of the monomer was nearly 1 billion pounds [39]. The polymer (and therefore the monomer) finds use as cast sheet, coatings, and molding powder. Its most spectacular, although not largest, use is in airplane windshields and windows.

Methyl methacrylate is made by adding hydrogen cyanide to acetone followed by dehydration, hydrolysis, and esterification to give the final product:

$$CH_3COCH_3 + HCN \xrightarrow[25\text{–}60°C]{NaOH} CH_3\overset{\overset{CH_3}{|}}{C}OHCN \quad (79)$$

$$CH_3\overset{\overset{CH_3}{|}}{C}OHCN \xrightarrow{98\%\ H_2SO_4} CH_2{=}\overset{\overset{CH_3}{|}}{C}CONH_2\cdot\tfrac{1}{2}H_2SO_4 \quad (80)$$

$$CH_2{=}\overset{\overset{CH_3}{|}}{C}CONH_2\cdot\tfrac{1}{2}H_2SO_4 \xrightarrow{CH_3OH} CH_2{=}\overset{\overset{CH_3}{|}}{C}CO_2CH_3 \quad (81)$$

This process is believed to be the largest end use for acetone in the United States [40]. Since the biggest source of acetone is as the co-product in the manufacture of phenol from cumene, in which process it carries part of the phenol cost by being saleable, a non-aceteone-based methyl methacrylate process might appreciably change the economics of phenol production.

Even though the raw materials are cheap, it should be obvious to a chemist that the process shown (Equations 79, 80, and 81) is not an overwhelmingly attractive process. As a result, several possible alternatives have appeared in the literature. Following is one method [41]:

$$CH_3\overset{\overset{CH_3}{|}}{C}HCHO \xrightarrow[\text{then } CH_3OH]{O_2} CH_3\overset{\overset{CH_3}{|}}{C}HCO_2CH_3 \quad (82)$$

$$CH_3\overset{\overset{CH_3}{|}}{C}HCO_2CH_3 \xrightarrow[500°C]{H_2S\ S} CH_2{=}\overset{\overset{CH_3}{|}}{C}CO_2CH_3 \quad (83)$$

37% conversion, 85% yield

This method has the advantage of using as the basic raw material isobutyraldehyde, the undesired but unavoidable by-product in the oxonation of propylene.

Here is another method, this time from the Monsanto Company [42]:

$$CH_3CH_2CO_2CH_3 + CH_2O \xrightarrow[\text{silica gel, } 430°C]{KOH} CH_2{=}\overset{\overset{CH_3}{|}}{C}CO_2CH_3 + H_2O \quad (84)$$

67% conversion, 92% yield

And here is another, this one from Oxirane Corp. [40]:

$$(CH_3)_3COH \xrightarrow{(O)} CH_2{=}\overset{\overset{CH_3}{|}}{C}CHO \xrightarrow{(O)}$$

$$CH_2{=}\overset{\overset{CH_3}{|}}{C}CO_2H \xrightarrow[\text{catalyst}]{CH_3OH} CH_2{=}\overset{\overset{CH_3}{|}}{C}CO_2CH_3 \quad (85)$$

XVII. RECAPITULATION

At the start of this chapter, certain economic factors that govern the selection of industrial chemical processes were described briefly. Some of these factors were illustrated in the preparation of selected products:

1. Capital investment: vinyl chloride.
2. Materials efficiency (yield): ethylene glycol, phthalic anhydride.
3. Materials efficiency (cheaper raw material): carbon monoxide, acetic acid, hexamethylenediamine, anthraquinone, maleic anhydride, phthalic anhydride, vinyl acetate, methyl methacrylate.
4. Raw-material bind: acetaldehyde.

5. The change from coal to petroleum as the basic source of organic chemicals: carbon monoxide, acetylene, vinyl chloride, acrylonitrile, acetaldehyde, phthalic anhydride, vinyl acetate.
6. Energy efficiency: acetylene.
7. Profitability: vinyl chloride, acrylonitrile, acetaldehyde, acetic acid, ethylene glycol, terephthalic acid, anthraquinone, maleic anhydride, phthalic anhydride, vinyl acetate.
8. Co-product disposal: the nitrochlorobenzenes, acetaldehyde, acetic acid, phenol.
9. Environmental considerations: terephthalic acid, anthraquinone.
10. Toxicity and safety: hydrogen cyanide, acrylonitrile, hexamethylenediamine, methyl methacrylate.
11. Raw-material cost calculations: anthraquinone.

Certain characteristics of industrial chemistry that differentiate it sharply from academic chemistry were illustrated

1. The use of high temperatures: carbon monoxide, acetylene, coal hydrogenation, Fischer-Tropsch, vinyl chloride, acrylonitrile, hydrogen cyanide, acetaldehyde, acetic acid, ethylene glycol, hexamethylenediamine, terephthalic acid, anthraquinone, maleic anhydride, phthalic anhydride, phenol, vinyl acetate, methyl methacrylate.
2. The use of high pressures: coal hydrogenation, Fischer-Tropsch, acetaldehyde, acetic acid, ethylene glycol, hexamethylenediamine, terephthalic acid, phenol.
3. Catalytic processes: Carbon monoxide, coal hydrogenation, Fischer-Tropsch, vinyl chloride, acrylonitrile, hydrogen cyanide, acetaldehyde, acetic acid, ethylene glycol, hexamethylenediamine, terephthalic acid, anthraquinone, maleic anhydride, phthalic anhydride, phenol, vinyl acetate, methyl methacrylate.
4. The handling of gases and vapor-phase reactions: carbon monoxide, acetylene, vinyl chloride, acrylonitrile, hydrogen cyanide, acetaldehyde, acetic acid, ethylene glycol, hexamethylenediamine, terephthalic acid, anthraquinone, maleic anhydride, phthalic anhydride, phenol, vinyl acetate, methyl methacrylate.
5. Materials-of-construction problems: acetaldehyde, terephthalic acid, phenol, vinyl acetate.

The purpose of this chapter has been to show the reader some of the differences between industrial and academic chemistry. Because of the author's own background, the examples have been chosen from organic chemistry. Although this treatment is expanded in the next chapters, the book hardly constitutes a survey of industrial chemistry. The reader who wishes a more detailed picture may consult several available references [43] [44] [45] [46].

REFERENCES AND NOTES

1. M. J. Montet, *Chemtech*, **4**, May, 1974, p. 268.
2. M. Sherwin, *Chemtech*, **4**, April, 1974, p. 225.
3. Anon., *Chem. Eng. News*, **57**, June 18, 1979, p. 9.
4. N. Platzer, *Chemtech*, **11**, February, 1981, p. 90.
5. B. F. Greek and W. F. Fallwell, *Chem. Eng. News*, **55**, October 6, 1980, p. 13.
6. Chemical Profiles, *Chem. Market. Report.*, **212**, July 1, 1977.
7. E. C. Hughes, private communication.
8. R. H. Kahney and T. D. McMinn, Jr., U.S. Pat. 4,000,178; *C.A.*, **86**, 89213t (1977).
9. Anon., *Chem. Eng. News*, March 5, 1979, p. 10.
10. Anon., *Chem. Week*, **124**, March 28, 1979, p. 30.
11. F. E. Paulik, A. Hershman, W. R. Knox, and J. F. Roth, U.S. Pat. 3,769,329.
12. Anon., *Chem. Week*, **127**, November 19, 1980, p. 41.
13. Anon., *Chemtech*, **7**, March, 1977, p. 141.
14. S. Suzuki, U.S. Pat. 3,806,544; *C.A.*, **81**, 3375 (1974).
15. J. A. Scheben, J. A. Hinnenkamp, and I. L. Mador, Ger. Offen. 2,529,365; *C.A.*, **84**, 121170y (1976).
16. R. Hughes, *Chemtech*, **6**, September, 1976, p. 581.
17. T. Isshiki, H. Yoshino, T. Kato, and T. Kondo, Jap. Kokai Tokkyo Koho ('77) 71,417; *C.A.*, **87**, 200801v (1977); ('78) 44,454; *C.A.*, **90**, 186385r (1979).
18. R. G. Wall, U.S. Pat. 4,113,662; *C.A.*, **90**, 38546d (1979).
19. T. Isshiki, W. Yoshino, Y. Kijima, and Y. Miyauchi, Jap. Kokai Tokkyo Koho ('78) 44,516; *C.A.*, **89**, 75261g (1978).
20. I. Hirose and H. Okitsu, Ger. Offen. 2,607,039; *C.A.*, **85**, 176829w (1976).
21. T. Okano, N. Wada, and Y. Kobayashi, Jap. Kokai Tokkyo Koho ('77) 91,810, 89,605; *C.A.*, **88**, 37244a, 37246c (1978). See also: H. Okitsu, K. Yamamoto, H. Sakai, and I. Hirose, Jap. Kokai Tokkyo Koho ('77) 151,107; *C.A.*, **88**, 120604p (1978). T. Okano, N. Wada, and Y. Kobayashi, Jap. Kokai Tokkyo Koho ('78) 23,909; *C.A.*, **89**, 23760z (1978). I. Hirose, M. Suzuki, H. Sakai, H. Okitsu, and K. Kato, Jap. Kokai Tokkyo Koho ('78) 25,509; *C.A.*, **89**, 23761a (1978). H. Okitsu, I. Hirose, M. Suzuki, H. Sakai, and K. Kato, Jap. Kokai Tokkyo Koho ('78) 25,508; *C.A.*, **89**, 23762b (1978).
22. R. Hughes, *Chemtech*, **4**, September, 1974, p. 516.
23. L. Kaplan, Ger. Offen. 2,559,057; *C.A.*, **85**, 93806k (1976).
24. R. Hughes, *Chemtech*, **6**, December, 1976, p. 731.
25. Anon., *C.E.H. Marketing Report*, August, 1977.
26. The hydrogenation has been reported to proceed in 99% yield when conducted in the presence of ammonia and an iron on alumina catalyst. T. Toshimitsu, S. Ono, K. Nishihira, K. Nishimura, and K. Matsui, Jap. Kokai ('74) 06,907; *C.A.*, **84**, 164138m (1976).
27. G. Frank, K. Kaempfer, and M. Schwarzmann, Ger. Offen. 2,654,028; *C.A.*, **89**, 75243c (1978).
28. Anon., *Chem. Week*, **119**, October 13, 1976, p. 67.
29. A. P. Gelbein, M. C. Sze, and R. T. Whitehead, *Chemtech*, **3**, August, 1973, p. 479.
30. The yields quoted and the ratios used here are for *p*-methylanthraquinone in *Organic Syntheses, Coll. Vol. I.*, pp. 353 and 517, and therefore may be slightly optimistic.
31. These figures are taken from *Chem. Market. Report.*, **220**, October 12, 1981.
32. W. S. Durrell and J. H. Bateman, *Chemtech*, **7**, January, 1977, p. 5.

33. Anon., *Chem. Eng. News,* **56,** March 20, 1978, p. 13.
34. Chemical Profiles, *Chem. Market. Report.,* **214,** July 1, 1978.
35. A. L. Waddams, *Chemicals from Petroleum,* Wiley, New York, 1973, p. 161.
36. Chemical Profiles, *Chem. Market. Report.,* **211,** February 7, 1977.
37. Chemical Profiles, *Chem. Market. Report.,* **214,** October 9, 1978.
38. *Chem. Eng. News,* **50,** November 27, 1972, p. 14.
39. Chemical Profiles, *Chem. Market. Report.,* **210,** October 1, 1976.
40. Anon., *Chem. Eng. News,* **56,** August 28, 1978, p. 8.
41. A. Brownstein, *Chemtech,* **3,** March, 1973, p. 131.
42. A. J. C. Pearson, Ger. Offen. 2,339,243; *C.A.,* **80,** 132838r (1974).
43. A. L. Waddams, *Chemicals from Petroleum,* Wiley, New York, 1973.
44. B. G. Reuben and M. L. Burstall, *The Chemical Economy,* Longman, London, 1973.
45. A series of articles by G. T. Austin in *Chem. Engineer.,* **81,** 1974: (2), p. 127; (4), p. 125; (6), p. 87; (8), p. 86; (9), p. 143; (11), p. 101; (15), p. 107; (16), p. 96.
46. *Chemistry in the Economy,* American Chemical Society, Washington, D.C., 1973.

4

PETROCHEMISTRY

I. INTRODUCTION

The chemistry described in this chapter is essentially industrial aliphatic chemistry. As an unorthodox approach, the author took the 1977–1978 Union Carbide Corp. catalogue [1] and speculated about the methods for manufacturing some 124 of the monomeric compounds listed therein. In many cases this was not inordinately difficult because the route is suggested if a company also sells the intermediates that have to be prepared by the company along the way to make the product. A Union Carbide Corp. publication that shows a few of the "trick" syntheses also was helpful [2]. Needless to say, this publication divulged no process details.

In the case of the large-tonnage products, cheaper processes obviously are being sought continually, and usually several synthetic routes are possible in each case. Wherever hints are mentioned in the literature, the attempts so described have been indicated. However, as the basis for the chapter, reliance has been placed on the known classical industrial methods for preparing the compounds in question.

In the chapter immediately preceding, certain aspects of industrial organic chemistry were emphasized. They are reemphasized here. In addition, certain other facets will become evident. Almost the entire Union Carbide Corp. monomeric organic chemical product line, as described herein, is very closely integrated chemically. The line is based on a relatively few simple, and for the most part captive (Union Carbide Corp. owned or manufactured), raw materials. These are transformed into the various chemical products by relatively few reactions. In other words, Union Carbide Corp. Manufacturing is expert at conducting certain reactions by which several products can be made in a single plant.

In the equations that follow, products marked by an asterisk are listed in the Union Carbide Corp. catalogue.

What are the raw materials? There are not many (Table 1). The list of reactions is equally short (Table 2).

TABLE 1. Union Carbide Corporation's Raw Materials

CO O_2 H_2 NH_3 Cl_2 H_2O HCN HC≡CH
$CH_2{=}CH_2$ Simple olefins Acids Bases
Catalysts Energy

TABLE 2. Union Carbide Corporation's Synthetic Reactions

Oxo and related carbon monoxide chemistry
Olefin hydration
Catalytic hydrogenation
Oxidation and dehydrogenation
Addition of chlorine and hypochlorous acid to olefins
Aldol and related condensations
Conversion of acids to ketones
Ethoxylation and propoxylation
Amination of alcohols and alkyl halides

Now what are some of the products? This goes right back to the kind of chemistry that was involved in the preceding chapter.

Olefin hydration is a starter:

$$CH_2{=}CH_2 + H_2O \longrightarrow CH_3CH_2OH^* \quad (1)$$

$$CH_3CH{=}CH_2 + H_2O \longrightarrow CH_3CHOHCH_3^* \quad (2)$$

These reactions have been effected in a number of ways, in the liquid phase through the acid sulfate, in the liquid phase catalyzed by an ion-exchange resin, and in the vapor phase over a supported phosphoric acid catalyst. Temperatures have run as high as 260°C and pressures as high as 3000 psi. In this case it is not possible to guess which process Union Carbide Corp. is using. The company probably uses several, depending on the economically most favorable technology at the time the plant in question was constructed.

II. ACETALDEHYDE AND ALDOL CHEMISTRY

In the preceding chapter individual compounds were discussed. Now their integration will be examined. The acetylene–acetaldehyde complex, must have been very important to Union Carbide Corp. at one time (Fig. 1).

The ketene used in Fig. 1 can be prepared from either acetic acid or acetone:

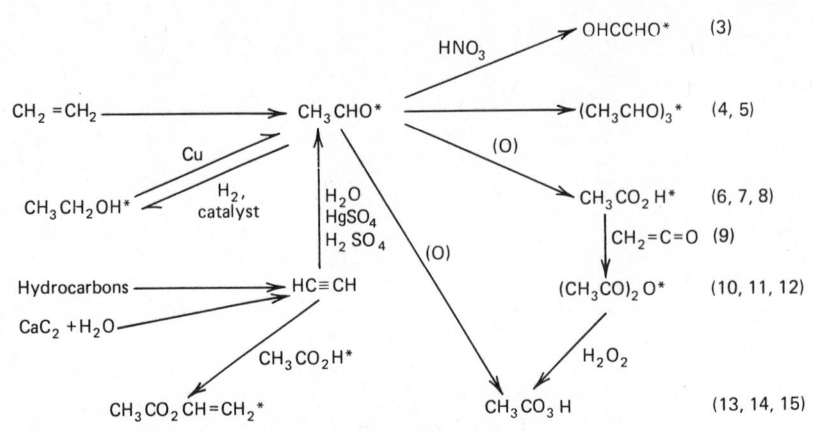

FIG. 1. Acetaldehyde chemistry. This was and still is one of the most important bases of Union Carbide Corporation's tonnage organic chemicals business.

$$CH_3CO_2H^* \xrightarrow[700°C,\ 200\ mm]{(C_2H_5O)_3PO} CH_2{=}C{=}O + H_2O \quad (16)$$

$$CH_3COCH_3^* \xrightarrow{600°} CH_2{=}C{=}O + CH_4 \quad (17)$$

Acetates are valuable chemicals, and Union Carbide Corp. makes a lot of them. Either of two methods can be used:

$$ROH + CH_3CO_2H^* \xrightarrow{H^+} ROCOCH_3 + H_2O \quad (18)$$

$$ROH + (CH_3CO)_2O^* \longrightarrow ROCOCH_3 + CH_3CO_2H^* \quad (19)$$

Probably Union Carbide Corp. uses a combination of both methods. In the first case (Equation 18) the water produced is removed continuously. Table 3 shows the alcohols that Union Carbide Corp. acetylates.

Next comes aldol chemistry, the classical method for preparing higher alcohols, aldehydes, and acids from lower homologues. Figure 2 shows how a number of compounds can be prepared from acetaldehyde.

Yields by the aldol condensation are known to be very high, as illustrated in a process patent application [3]:

$$CH_3CH_2CHO^* \xrightarrow[90°C]{1\%\ NaOH} CH_3CH_2CH{=}\underset{\displaystyle CH_3}{\underset{|}{C}}CHO \quad 98\%\ \text{yield} \quad (29)$$

Although the oxonation of propylene probably has taken over as the method of choice for producing the C_4 compounds, the fact that Union Carbide Corp. sells almost everything else shown in Fig. 2 means that most of these reactions must still be in commercial use.

TABLE 3. Alcohols Acetylated by Union Carbide Corporation

CH_3CH_2OH* $CH_3(CH_2)_2OH$* $(CH_3)_2CHOH$*

$CH_3(CH_2)_3OH$* $(CH_3)_2CHCH_2OH$*

$CH_3(CH_2)_3CH(CH_2CH_3)CH_2OH$* C_6H_5OH* $CH_3O(CH_2)_2OH$*

$CH_3CH_2O(CH_2)_2OH$* $CH_3(CH_2)_3O(CH_2)_2OH$*

$CH_3CH_2O(CH_2CH_2O)_2H$*

$CH_3(CH_2)_3O(CH_2CH_2O)_2H$* $HO(CH_2)_2OH$*

$CH_2OHCHOHCH_2OH$*

$HO(CH_2CHO)_3H$* $(CH_3)_2COHCH_2CHOHCH_3$*

Using similar techniques of aldol chemistry, the compounds shown in Table 4 can be prepared by a mixed aldol condensation between acetaldehyde and propionaldehyde (Table 4). All of them can equally well be based on the oxonation of 1-butene.

Two other mixed aldol condensations are between isobutyraldehyde and formaldehyde and between acetaldehyde and formaldehyde:

$$(CH_3)_2CHCHO^* + CH_2O \longrightarrow HOCH_2C(CH_3)_2CHO \qquad (30)$$

$$HOCH_2C(CH_3)_2CHO \xrightarrow{(H)} HOCH_2C(CH_3)_2CH_2OH \qquad (31)$$

$$HOCH_2C(CH_3)_2CHO \xrightarrow{OH^-} HOCH_2C(CH_3)_2CO_2CH_2C(CH_3)_2CH_2OH^* \qquad (32)$$

TABLE 4. Acetaldehyde-Propionaldehyde Mixed Aldol Products

$CH_3(CH_2)_4OH$* $CH_3(CH_2)_3CHO$* $CH_3(CH_2)_3CO_2H$*

$CH_3CH_2CH(CH_3)CH_2OH$*

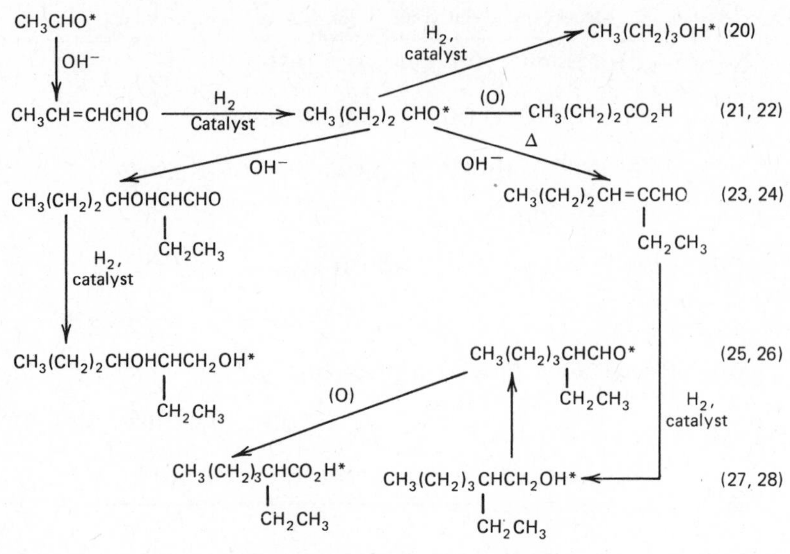

FIG. 2. Aldolization of acetaldehyde.

$$CH_3CHO^* + 3CH_2O \longrightarrow HOCH_2\overset{CH_2OH}{\underset{CH_2OH}{C}}CHO \xrightarrow[\text{catalyst}]{H_2} HOCH_2\overset{CH_2OH}{\underset{CH_2OH}{C}}CH_2OH \qquad (33)$$

The aldolization of propionaldehyde gives two more compounds that Union Carbide Corp. markets:

$$CH_3(CH_2)_2\underset{CH_3}{CH}CH_2OH^* \qquad CH_3(CH_2)_2\underset{CH_3}{CH}CHO^* \qquad [1,2]$$

The chemist should have no trouble delineating for himself the detailed preparations of these last two compounds.

An alternative route to several of the alcohols is the Guerbet reaction [4]:

$$CH_3(CH_2)_3OH^* + NaOH \xrightarrow{\Delta} CH_3(CH_2)_3ONa + H_2O \qquad (34)$$

$$CH_3(CH_2)_3ONa + CH_3(CH_2)_3OH^* \longrightarrow CH_3(CH_2)_3\underset{CH_2CH_3}{CH}CH_2OH^* + NaOH \qquad (35)$$

Its main disadvantage is that it tends to keep on going. (Obviously the alcohols and alcoholates are in equilibrium.)

$$\underset{\displaystyle CH_2CH_3}{CH_3(CH_2)_3CHCH_2ONa} + CH_3(CH_2)_3OH^* \longrightarrow CH_3(CH_2)_3\underset{\displaystyle CH_3CH_2}{CH}CH_2\underset{\displaystyle CH_2CH_3}{CH}CH_2OH \tag{36}$$

However, the basic reaction is the heating of an alcohol with sodium hydroxide. It is inconceivable to the author that a scientifically alert company like Union Carbide Corp. has not investigated this reaction in detail, and perhaps the company has found ways to minimize by-product formation, so the reaction might be commercially feasible in certain instances.

Certainly there are claims that yields are very good [5].

$$CH_3(CH_2)_5OH^* \xrightarrow[\text{CuCO}_3\cdot\text{Cu(OH)}_2\cdot\text{FeSO}_4,\ \text{separate evolved H}_2\text{O}]{\text{KOH, 10–40\% residue from previous reaction}} \underset{\text{95\% yield}}{CH_3(CH_2)_3\underset{\displaystyle (CH_2)_5CH_3}{CH}CH_2OH} \tag{37}$$

III. THE OXO REACTION

Next comes the oxo reaction and related carbon monoxide–olefin chemistry. Probably the biggest application is in the case of propylene:

$$CH_3CH{=}CH_2 + CO + H_2 \xrightarrow{[Co(CO)_4]_2} CH_3(CH_2)_2CHO^* + (CH_3)_2CHCHO^* \tag{38}$$

$$CH_3(CH_2)_2CHO^* \xrightarrow{(O)} CH_3(CH_2)_2CO_2H \tag{39}$$

$$CH_3(CH_2)_2CHO^* \xrightarrow[\text{catalyst}]{H_2} CH_3(CH_2)_3OH^* \tag{40}$$

$$(CH_3)_2CHCHO^* \xrightarrow{(O)} (CH_3)_2CHCO_2H \tag{41}$$

$$(CH_3)_2CHCHO^* \xrightarrow[\text{catalyst}]{H_2} (CH_3)_2CHCH_2OH^* \tag{42}$$

Although yields are not highly publicized, the overall yield is known to be very high. A clue can be found in a German patent application [6]:

$$CH_3(CH_2)_7CH{=}CH_2 + CO + H_2 \xrightarrow[(n\text{-}C_8H_{17})_2POH]{[Co(CO)_4]_2} \underset{\text{97\% yield}}{\text{undecanal} + \text{undecanol}} \tag{43}$$

TABLE 5. **1-Butene Oxonation Products**

$CH_3(CH_2)_4OH^*$	$CH_3(CH_2)_3CHO^*$	$CH_3(CH_2)_3CO_2H^*$
A	B	C
$CH_3CH_2CH(CH_3)CH_2OH^*$	$CH_3CH_2CH(CH_3)CHO$	$CH_3CH_2CH(CH_3)CO_2H^*$
D	E	F

The classical oxonation technique using the cobalt carbonyl catalyst requires temperatures of the order of 145–180°C and pressures of 1500–6000 psi. *n*-Butyraldehyde is the desired product, and by this process its ratio to the iso isomer is about 3:1 to 4:1.

Recently Union Carbide Corp. announced a new process in which the catalyst is believed to be $H_2Rh(CO)_2$ [7]. Conditions are much milder than in the classical case—a temperature of about 100°C and pressures of 100–350 psi. In addition the isomer ratio is better, being 8:1 to 16:1 in favor of *n*-butyraldehyde. Thus not only is the yield of desired product higher, but the capital and operating costs also are lower than in the older process.

The oxonation of 1-butene leads to similar products (Table 5). Compound D can also be obtained from 2-butene.

The oxonation of ethylene is very important also:

$$CH_2{=}CH_2 + CO + 2H_2 \xrightarrow{Fe(CO)_4} CH_3(CH_2)_2OH^* \tag{44}$$

$$CH_2{=}CH_2 + CO + H_2 \xrightarrow{[Co(CO)_4]_2} CH_3CH_2CHO^* \xrightarrow[\text{catalyst}]{H_2} CH_3(CH_2)_2OH^* \tag{45}$$

$$CH_2{=}CH_2 + CO + H_2 \xrightarrow{[Co(CO)_4]_2} CH_3CH_2CHO^* \xrightarrow{(O)} CH_3CH_2CO_2H^* \tag{46}$$

Here again Union Carbide Corp. has applied the rhodium technique [8].

The oxonation of ethylene can go directly to propionic acid in very high yield [9]:

$$CH_2{=}CH_2 + CO + H_2O \xrightarrow[\substack{CH_3CO_2H,\\ Br:Rh\,=\,182:1,\\ 175°C,\ 400\ psi}]{\substack{RhCl_3\cdot 3H_2O,\\ 48\%\ HBr}} CH_3CH_2CO_2H^* \quad 99\%\ \text{yield} \tag{47}$$

As a matter of interest there are other routes to propionic acid:

$$CH_2{=}CH_2 + CO + H_2O \xrightarrow[Ni(CO)_4]{BF_3\ \text{or}} CH_3CH_2CO_2H^* \tag{48}$$

$$CH_3CH_2OH + CO \xrightarrow[125\text{–}130°C]{BF_3,\ 13{,}500\ psi} CH_3CH_2CO_2H^* \quad (49)$$

$$CH_3CH_2ONa + CO \xrightarrow{\text{then } H^+} CH_3CH_2CO_2H^* \quad (50)$$

IV. ACETONE AND SIMILAR KETONE CHEMISTRY

Another tonnage petrochemical is acetone. There are lots of ways to make it:

$$CH_3CHOHCH_3^* \xrightarrow{Cu} CH_3COCH_3^* + H_2 \quad (51)$$

$$2CH_3CO_2H^* \xrightarrow{MnO} CH_3COCH_3^* \quad (52)$$

$$2CH_3CH_2OH^* \xrightarrow[80\%]{SnO_2,\ 400°C} CH_3COCH_3^* \quad (53)$$

$$2CH_3CH_2OH^* \xrightarrow[400°C,\ H_2O]{Fe\cdot Fe_2O_3} CH_3COCH_3^* \quad (54)$$

$$C_6H_5C(CH_3)_2(OOH) \xrightarrow[\text{co-product}]{H_2SO_4,\ C_6H_5OH^*} CH_3COCH_3^* \quad (55)$$

$$\text{Corn starch or molasses} \xrightarrow[\text{co-product}]{\text{ferment } CH_3(CH_2)_3OH^*} CH_3COCH_3^* \quad (56)$$

$$CH_3CHO^* + H_2O \xrightarrow[90\%\ \text{conversion},\ 78\%\ \text{yield}]{Cr\cdot Mn\cdot Zn,\ 300\text{–}350°C} CH_3COCH_3^* \quad (57)$$

The fermentation route was the original one and is now obsolete. With the adoption of the cumene process for manufacturing phenol, a great deal of acetone is produced in this way. The acetaldehyde-based route (Equation 57) was announced recently in the Soviet Union [10].

Still more recently a new route from isobutyraldehyde was announced [11]:

$$(CH_3)_2CHCHO^* + O_2 + H_2O \xrightarrow[CuO\cdot MnO\cdot ZnO \text{ on } Al_2O_3]{140\text{–}160°C} CH_3COCH_3^* \quad (58)$$

99% yield

Since this reaction uses up the unwanted isomer from the oxonation of propylene, it could have an appreciable effect on the economics of that process.

The self-condensation of acetone leads to a number of industrially interesting products:

$$2CH_3COCH_3^* \longrightarrow (CH_3)_2COHCH_2COCH_3^* \quad (59)$$

$$(CH_3)_2COHCH_2COCH_3^* \xrightarrow[\text{catalyst}]{H_2} (CH_3)_2COHCH_2CHOHCH_3^* \quad (60)$$

$$(CH_3)_2COHCH_2COCH_3^* \xrightarrow{-H_2O} (CH_3)_2C{=}CHCOCH_3^* \quad (61)$$

$$(CH_3)_2C{=}CHCOCH_3^* \xrightarrow[\text{catalyst}]{H_2} (CH_3)_2CHCH_2COCH_3^* \quad (62)$$

$$(CH_3)_2CHCH_2COCH_3^* \xrightarrow[\text{catalyst}]{H_2} (CH_3)_2CHCH_2CHOHCH_3^* \quad (63)$$

Barium hydroxide is the classical catalyst for the initial condensation. However, if the desired product is methyl isobutyl ketone, the route shown in Equations 61 and 62 is tedious. Following are three direct methods:

$$CH_3COCH_3^* + H_2 \xrightarrow[120\text{–}160°C]{Zr_3P_4 \cdot Pd,\ 300\text{–}450\ psi} CH_3COCH_2CH(CH_3)_2^* \quad \text{[Ref. 12]} \quad (64)$$

$$CH_3CHOHCH_3^* \xrightarrow[\text{Co, Cu, or Raney Ni}]{ZnO \text{ or } Cr_2O_3 \text{ on } ZnCrO_4} CH_3CO(CH_2)CH(CH_3)_2^* \quad \text{[Ref. 13]} \quad (65)$$

$$CH_3COCH_3^* + CH_3CH{=}CH_2 \xrightarrow[CH_3CHO^* \text{ (initiator)}]{390\text{–}540°C,\ 700\text{–}150\ psi} CH_3CO(CH_2)CH(CH_3)_2^* \quad \text{[Ref. 14]} \quad (66)$$

The first method (Equation 64) has been the subject of some study involving various catalysts. With palladium on a KU-2 catalyst at 120°C, a 50% conversion and a 95% yield have been reported [15]. A mixed catalyst based on palladium, zirconium, and phosphorus yielded 91% methyl isobutyl ketone [16].

The method shown in Equation 66 is in a patent assigned to Union Carbide Corp.

Useful products are also derived from the self-condensation of three or four molecules of acetone (Fig. 3).

Two other acetone derivatives are important:

$$CH_3COCH_3^* \xrightarrow[\text{alkali}]{\text{strong}} \text{Isophorone} \quad (74)$$

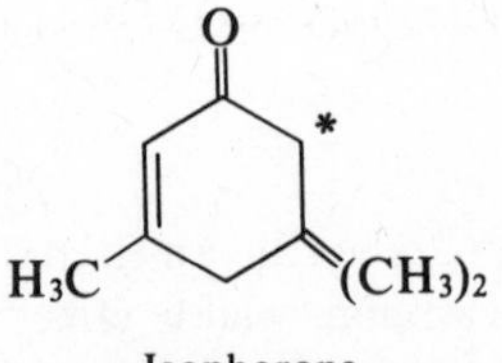

O
*
H₃C (CH₃)₂
Isophorone

$$CH_3COCH_3^* \xrightarrow[CH_3CH_2ONa]{CH_3CO_2CH_2CH_3^*} CH_3COCH_2COCH_3^* \quad (75)$$

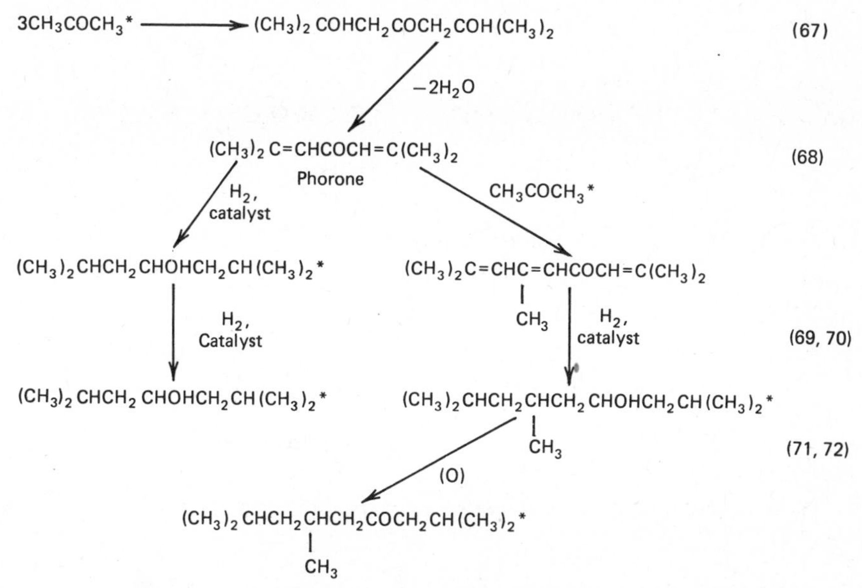

FIG. 3. Self-condensation of three or four molecules of acetone.

However, Union Carbide Corp. makes acetylacetone by isomerizing isopropenyl acetate [2]:

$$CH_2{=}\underset{\displaystyle CH_3}{\underset{|}{C}}OCOCH_3^* \longrightarrow CH_3COCH_2COCH_3^* \tag{76}$$

Isopropenyl acetate is obtained by treating acetone with ketene or with acetic anhydride in the presence of sulfuric acid:

$$CH_3COCH_3^* + CH_2{=}C{=}O \xrightarrow{H_2SO_4} CH_2{=}\underset{\displaystyle CH_3}{\underset{|}{C}}OCOCH_3^* + H_2O \tag{77}$$

Certain other ketones of commerce cannot be prepared from acetone, although in two cases acetone could be one of the intermediates. Methyl ethyl ketone may be prepared by the oxidation/dehydrogenation of *sec*-butyl alcohol:

$$CH_3CHOHCH_2CH_3^* \xrightarrow{Cu} CH_3COCH_2CH_3^* \tag{78}$$

Union Carbide Corp. gets methyl ethyl ketone as a by-product in the oxidation of *n*-butane to acetic acid [2].

There are two routes to methyl *n*-propyl ketone:

$$CH_3CHOH(CH_2)_2CH_3^* \xrightarrow{Cu} CH_3CO(CH_2)_2CH_3^* + H_2 \quad (79)$$

$$CH_3CO_2H^* + CH_3(CH_2)_2CO_2H \xrightarrow{MnO} CH_3CO(CH_2)_2CH_3^* \quad (80)$$

The second method is the classical one for preparing ketones.

With a slightly different catalyst, the second route appears to be an extremely effective way to make diisopropyl ketone [17]:

$$(CH_3)_2CHCO_2H \xrightarrow[H_2O]{Ga_2O_3} (CH_3)_2CHCOCH(CH_3)_2 \quad (81)$$

90% conversion, 97% yield

An alternate route to this ketone also appears interesting [18]:

$$(CH_3)_2CHCO_2CH_2CH(CH_3) \xrightarrow[ZrO_2 \cdot Li_2O]{440°C} (CH_3)_2CHCOCH(CH_3)_2 \quad (82)$$

93% conversion, yield

There are several routes to diethyl ketone:

$$2CH_3CH_2CO_2H^* \xrightarrow{MnO} CH_3CH_2COCH_2CH_3^* \quad (83)$$

$$2CH_3(CH_2)_2OH^* \xrightarrow[H_2O,\ 400°C]{Fe \cdot Fe_2O_3} CH_3CH_2COCH_2CH_3^* \quad (84)$$

$$2CH_2{=}CH_2 + CO + H_2 \xrightarrow{[Co(CO)_4]_2} CH_3CH_2COCH_2CH_3^* \quad (85)$$

$$2CH_2{=}CH_2 + 2CO + H_2 \xrightarrow[CO_2\ \text{by-product}]{RuCl_2 \cdot 2H_2O,\ C_6H_5CH_3,\ 1500\ \text{psi},\ 100°C,\ 6\ \text{hr}} CH_3CH_2COCH_2CH_3^* \quad (86)$$

81% conversion, 98% yield

This first (Equation 83) is, of course, the classical method for preparing symmetrical ketones from carboxylic acids. In this case the yield has been reported to be 90% [19]. The method shown in Equation 86 appears to be particularly promising [20].

A rather unusual synthesis of methyl *t*-butyl ketone appeared in the literature recently [21]:

$$(CH_3)_2C{=}CHCH_3 + CH_2O \xrightarrow[5\text{–}6\ \text{hr, boil 2 hr}]{\text{conc HCl},\ 50\text{–}55°C} (CH_3)_3CCOCH_3 \quad (87)$$

75% yield

Two higher ketones, methyl *n*-amyl and methyl isoamyl, may be prepared by the aldol method with acetone as one of the raw materials:

$$CH_3(CH_2)_2CHO^* + CH_3COCH_3^* \xrightarrow{OH^-} CH_3COCH{=}CH(CH_2)_2CH_3 + H_2O \quad (88)$$

$$CH_3COCH{=}CH(CH_2)_2CH_3 \xrightarrow[\text{catalyst}]{H_2} CH_3CO(CH_2)_4CH_3^* \quad (89)$$

$$(CH_3)_2CHCHO^* + CH_3COCH_3^* \xrightarrow{OH^-} CH_3COCH{=}CHCH(CH_3)_2 + H_2O \quad (90)$$

$$CH_3COCH{=}CHCH(CH_3)_2 \xrightarrow[\text{catalyst}]{H_2} CH_3CO(CH_2)_2CH(CH_3)_2^* \quad (91)$$

Cyclohexanone is prepared by the oxidation of the corresponding alcohol and is the raw material for *e*-caprolactone:

$$C_6H_5\text{—}OH^* \xrightarrow[\text{catalyst}]{H_2} C_6H_{11}\text{—}OH \xrightarrow{(O)} C_6H_{10}{=}O^* \quad (92)$$

$$C_6H_{12} \xrightarrow[\text{catalyst}]{\text{air}} C_6H_{11}\text{—}OH + C_6H_{10}{=}O \xrightarrow{(O)} C_6H_{10}{=}O \quad (93)$$

$$C_6H_{10}{=}O^* + CH_3CO_3H \longrightarrow (CH_2)_5\text{—}O^*\text{—}CO \text{ (ring)} \quad (94)$$

V. VINYL MONOMERS

Vinyl ether chemistry is of interest, although apparently no longer of much interest to Union Carbide Corp.:

$$HC{\equiv}CH + CH_3CH_2OH^* \xrightarrow{KOH} CH_2{=}CHOCH_2CH_3 \quad (95)$$

An interesting alternative appeared in Japan [22]:

$$CH_2{=}CHOCOCH_3^* + CH_3CH_2OH^* \xrightarrow[Na_2WO_4]{PdCl_2} \underset{96\%\ \text{yield}}{CH_2{=}CHOCH_2CH_3} \quad (96)$$

Vinyl ethers add to α,B-unsaturated carbonyl compounds in Diels–Alder fashion to give dihydropyrans. Union Carbide Corp. evidently conducted this reaction at one time with ethyl vinyl ether and acrolein, since the company then marketed essentially all the intermediates [23]. These are marked with a circled asterisk.

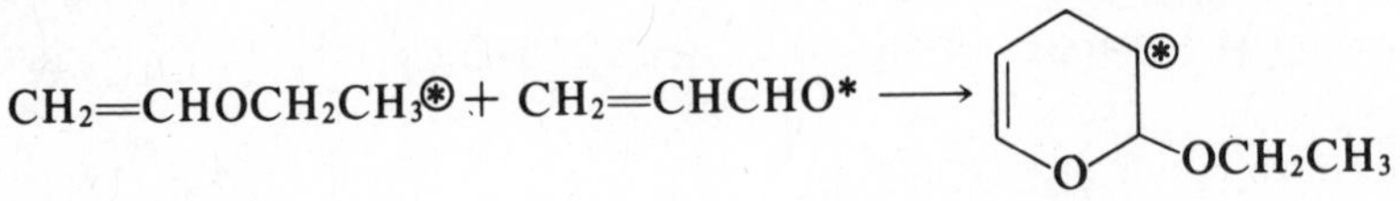

$$CH_2{=}CHOCH_2CH_3^{\circledast} + CH_2{=}CHCHO^{*} \longrightarrow \text{(dihydropyran)}^{\circledast}\text{—}OCH_2CH_3 \quad (97)$$

$$\text{(dihydropyran)}^{\circledast}\text{—}OCH_2CH_3 \xrightarrow[OH^-]{H^+} OHC(CH_2)_3CHO^{*} \quad (98)$$

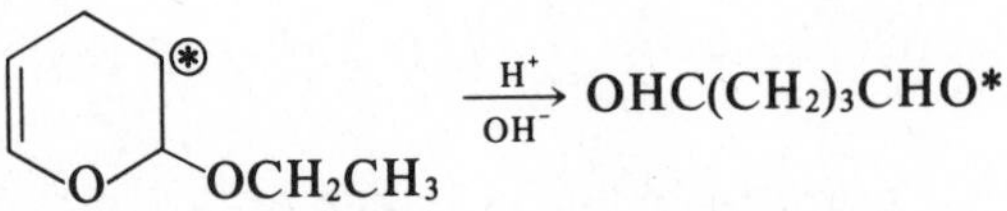

$$OHC(CH_2)_3CHO^{*} \xrightarrow[\text{yield} > 90\%]{H_2 + \text{catalyst}} HO(CH_2)_5OH^{*} \quad (99)$$

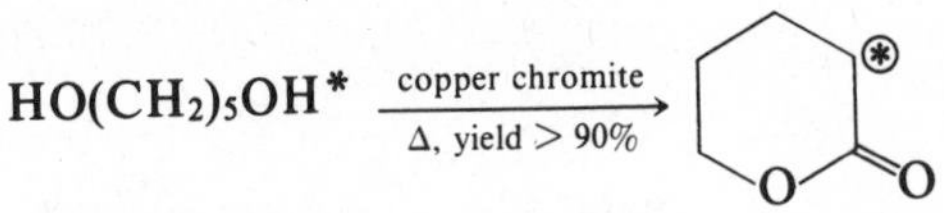

$$HO(CH_2)_5OH^{*} \xrightarrow[\Delta,\ \text{yield} > 90\%]{\text{copper chromite}} \text{(lactone)}^{\circledast} \quad (100)$$

$$OHC(CH_2)_3CHO^{*} \xrightarrow{OH^-} \text{(lactone)}^{\circledast} \quad (101)$$

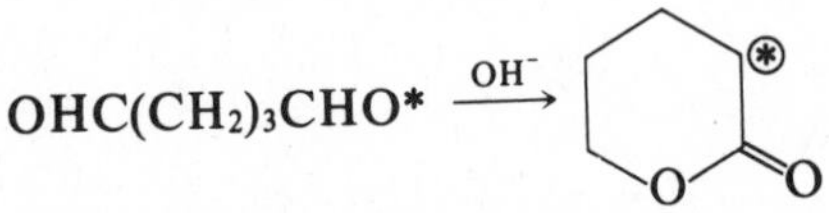

$$OHC(CH_2)_3CHO^{*} \xrightarrow{(O)} HO_2C(CH_2)_3CO_2H \quad (102)$$

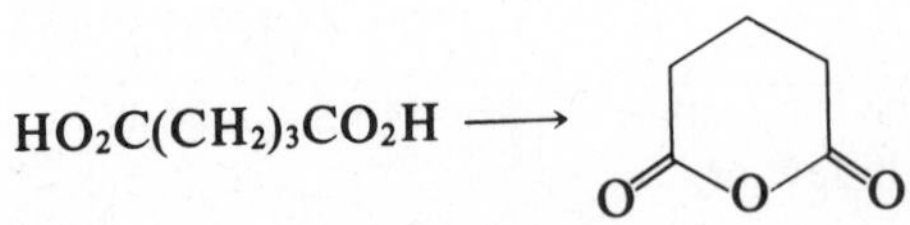

$$HO_2C(CH_2)_3CO_2H \longrightarrow \text{(anhydride)} \quad (103)$$

Union Carbide Corp. manufactures butadiene and conducts two other Diels–Alder reactions therewith:

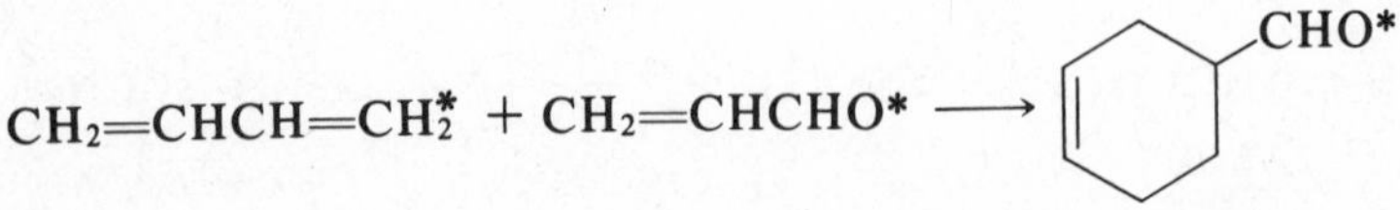

$$CH_2{=}CHCH{=}CH_2^{*} + CH_2{=}CHCHO^{*} \longrightarrow \text{(cyclohexene)—}CHO^{*} \quad (104)$$

$$CH_2{=}CHCH{=}CH_2^* + \text{cyclopentadiene} \longrightarrow \text{bicyclo (CH}_2\text{ bridge)}^*\text{–}CH{=}CH_2 \quad (105)$$

$$\xrightarrow{\text{isomerization}} \text{bicyclo (CH}_2\text{ bridge)}^*{=}CHCH_3 \quad (106)$$

Acrolein can be made in two ways:

$$CH_2{=}CHCH_3 \xrightarrow[\text{catalyst}]{O_2} CH_2{=}CHCHO \quad (107)$$

$$HC{\equiv}CH + CO + H_2 \xrightarrow{[Co(CO)_4]_2} CH_2{=}CHCHO \quad (108)$$

Even if the new Union Carbide Corp. rhodium catalyst is effective in the second reaction (Equation 108), the first method would appear to be preferable. This is another example of the change from coal to petroleum as the basic source of organic chemicals.

The corresponding acid, acrylic acid, can of course be prepared by the oxidation of acrolein, but several other methods also are available:

$$CH_2{=}CHCHO^* \xrightarrow{(O)} CH_2{=}CHCO_2H^* \quad (109)$$

$$HC{\equiv}CH + CO + H_2O \xrightarrow{Ni(CO)_4} CH_2{=}CHCO_2H^* \quad (110)$$

$$2CH_2{=}CH_2 + 2CO + O_2 \xrightarrow[\text{Fe or group II metal}]{\Delta,\ ReCl_3 + Li,\ Cu,\ Mn} 2CH_2{=}CHCO_2H^* \quad (111)$$

[Ref. 24]

$$\underset{\diagdown O \diagup}{CH_2{-}CH_2} + CO \xrightarrow{Ni(CO)_4} CH_2{=}CHCO_2H^* \quad (112)$$

$$CH_2{=}C{=}O + CH_2O \longrightarrow \begin{matrix} CH_2{-}CH_2 \\ | \qquad | \\ O{-}{-}CO \end{matrix} \xrightarrow{H_3PO_4} CH_2{=}CHCO_2H^* \quad (113)$$

The last reaction (Equation 113) can be conducted in one step in passable conversion and reasonable yield [25]:

$$CH_3CO_2H^* + CH_2O \xrightarrow[430°C,\ 1\text{-}2\ atm]{\text{KOH on silica gel}} CH_2{=}CHCO_2H^* \quad (114)$$

58% conversion, 78% yield

Acrylic esters are prepared in much the same way as the free acid.

$$HC{\equiv}CH + CO + ROH \xrightarrow[HCl]{Ni(CO)_4} CH_2{=}CHCO_2R^* \qquad (115)$$

$$\underset{\diagdown O \diagup}{CH_2{-}CH_2^*} + HCN \longrightarrow HO(CH_2)_2CN \qquad (116)$$

$$HO(CH_2)_2CN \xrightarrow[H^+]{ROH} HO(CH_2)_2CO_2R \qquad (117)$$

$$HO(CH_2)_2CO_2R \xrightarrow[acetate]{crack} CH_2{=}CHCO_2R^* \qquad (118)$$

$$CH_2{=}C{=}O + CH_2O \longrightarrow \begin{array}{l} CH_2{-}CH_2 \\ | \qquad\ \ | \\ O{-}{-}{-}CO \end{array} \qquad (119)$$

$$\begin{array}{l} CH_2{-}CH_2 \\ | \qquad\ \ | \\ O{-}{-}{-}CO \end{array} \xrightarrow[ROH]{H^+} CH_2{=}CHCO_2R^* \qquad (120)$$

$$CH_2{=}CHCN + H_2SO_4 \xrightarrow{H_2O} CH_2{=}CHCONH_2 \cdot \tfrac{1}{2}H_2SO_4 \qquad (121)$$

$$CH_2{=}CHCONH_2 \cdot \tfrac{1}{2}H_2SO_4 \xrightarrow{ROH} CH_2{=}CHCO_2R^* \qquad (122)$$

Just to give the reader an idea of the variety of acrylates that find commercial utility, in Table 6 are listed the alcohols from which the acrylates that Union Carbide Corp. markets are made.

TABLE 6. Alcohols of Union Carbide Corporation Acrylates

$CH_3CH_2OH^*$ $CH_3(CH_2)_3OH^*$ $(CH_3)_2CHCH_2OH^*$

$$\begin{array}{c} CH_3(CH_2)_3CHCH_2OH^* \\ | \\ CH_2CH_3 \end{array} \qquad \begin{array}{c} CH_3 \\ | \\ HOCH_2CCH_2OH \\ | \\ CH_3 \end{array}$$

$$\begin{array}{c} CH_2OH \\ | \\ HOCH_2CCH_2OH \\ | \\ CH_2OH \end{array}$$

(Triacrylate)

VI. ETHYLENE OXIDE AND RELATED CHEMISTRY

The classic route to the simplest oxide, ethylene oxide, was through ethylene chlorohydrin (Equations 125 and 126). Union Carbide Corp. pioneered the direct, catalytic oxidation of ethylene (Equation 123). According to a recently quoted Imperial Chemical Industries Ltd. patent, the oxidation of ethylene to the oxide proceeds in 94% yield when the silver-on-alumina catalyst is promoted with potassium and rubidium at 240°C and 16 atm [26].

$$2CH_2{=}CH_2 + O_2 \xrightarrow[225°C]{Ag} 2\,\underset{\diagdown O \diagup}{CH_2{-}CH_2^*} \tag{123}$$

$$\underset{\diagdown O \diagup}{CH_2{-}CH_2^*} \xrightarrow[H^+]{H_2O} HO(CH_2)_2OH^* \tag{124}$$

$$CH_2{=}CH_2 + Cl_2 + H_2O \longrightarrow Cl(CH_2)_2OH^* + HCl \tag{125}$$

$$Cl(CH_2)_2OH^* \underset{HCl}{\overset{Ca(OH)_2}{\rightleftharpoons}} \underset{\diagdown O \diagup}{CH_2{-}CH_2^*} \tag{126}$$

$$Cl(CH_2)_2OH^* \xrightarrow[OH^-]{H_2O} HO(CH_2)_2OH^* \tag{127}$$

$$HO(CH_2)_2OH^* \xrightarrow[H_2SO_4]{\Delta} \text{1,4-dioxane}^* \tag{128}$$

There are alternative routes to dioxane. At 190°C such catalysts as silicomolybdic acid [27] and aluminum sulfate [28] have been used to convert ethylene glycol to dioxane.

Ethylene oxide is a very reactive chemical and condenses with such compounds as aldehydes, ketones, hydrogen sulfide, and mercaptans:

$$\underset{\diagdown O \diagup}{CH_2{-}CH_2^*} + CH_2O \longrightarrow \text{1,3-dioxolane } (CH_2{-}CH_2,\ O\ \ O,\ CH_2) \tag{129}$$

$$\underset{\diagdown O \diagup}{CH_2{-}CH_2^*} + H_2S \longrightarrow HS(CH_2)_2OH \tag{130}$$

$$\underset{\diagdown O \diagup}{CH_2—CH_2^*} + HS(CH_2)_2OH \longrightarrow HO(CH_2)_2S(CH_2)_2OH^* \quad (131)$$

$$\underset{\diagdown O \diagup}{CH_2—CH_2^*} + CH_3COCH_2CH_3^* \longrightarrow \begin{array}{c} CH_2—CH_2 \\ | \quad\quad | \\ O \quad\quad O \\ \diagdown \diagup \\ C \\ \diagup \diagdown \\ H_3C \quad\quad CH_2CH_3 \end{array} \quad (132)$$

Probably the largest use of ethylene oxide outside ethylene glycol manufacture is its condensation with alcohols to give ethers:

$$\underset{\diagdown O \diagup}{CH_2—CH_2^*} + ROH \xrightarrow{OH^-} RO(CH_2)_2OH \quad (133)$$

Union Carbide Corp. markets a wide variety of these products including polymers of varying molecular weight. In Table 7 is a list of the alcohols that Union Carbide Corp. ethoxylates. Obviously the ethoxylation of an alcohol continues, so the higher derivatives appear as by-products in the manufacture of the lower homologues.

$$CH_3OH^* \xrightarrow{\underset{\diagdown O \diagup}{CH_2—CH_2^*}} CH_3O(CH_2)_2OH^* \xrightarrow{\underset{\diagdown O \diagup}{CH_2—CH_2^*}} CH_3O(CH_2CH_2O)_2H^* \xrightarrow{etc.} \quad (134)$$

However, according to a recent Japanese patent application, this need not be a grievous disadvantage if there is any market for the higher ethoxylated derivative [29].

$$C_2H_5OH^* + \underset{\diagdown O \diagup}{CH_2—CH_2^*} \xrightarrow[100°C]{Zr \cdot Si \cdot Zn\ oxide} \underset{92\%\ yield}{C_2H_5O(CH_2)_2OH^*} + \underset{100\%\ conversion,\ 6\%\ yield}{C_2H_5O(CH_2)_2O(CH_2)_2OH^*} \quad (135)$$

Etherification of an ethoxylation product can also be effected by means of an alkyl chloride [2]:

$$CH_3CH_2O(CH_2CH_2O)_2H^* + CH_3CH_2Cl \longrightarrow CH_3CH_2O(CH_2CH_2O)_2CH_2CH_3^* \quad (136)$$

$$CH_3(CH_2)_3O(CH_2CH_2O)_2H^* + CH_3(CH_2)_3Cl \longrightarrow CH_3(CH_2)_3O(CH_2CH_2O)_2(CH_2)_3CH_3^* \quad (137)$$

TABLE 7. Alcohols Ethoxylated by Union Carbide Corporation

CH_3OH* CH_3CH_2OH* $CH_3(CH_2)_3OH$*
$CH_3(CH_2)_5OH$* $CH_3O(CH_2)_2OH$* $CH_3CH_2O(CH_2)_2OH$*
$CH_3(CH_2)_3O(CH_2)_2OH$* $CH_3(CH_2)_5O(CH_2)_2OH$* $CH_3O(CH_2CH_2O)_2H$*
$CH_3CH_2O(CH_2CH_2O)_2H$* $CH_3(CH_2)_3O(CH_2CH_2O)_2H$*
$HO(CH_2)_2OH$* $HO(CH_2CH_2O)_2H$* $HO(CH_2CH_2O)_3H$*
Polyether alcohols* CH_3O—Polyether alcohols*
$CH_3(CH_2)_5O(CH_2CH_2O)_2H$

Union Carbide Corp. offers methyl chloride and *n*-butyl chloride. Both are prepared by chlorination of the corresponding hydrocarbons. In the case of *n*-butane, the primary chloride is obtained by a "molecular sieve" process [2].

Propylene oxide is also important. Because of the sensitivity to oxidation of the methyl group in propylene, essentially all the propylene oxide used to be made from the chlorohydrin (Equation 138). The use of ethylbenzene hydroperoxide (Equation 140) and of *t*-butyl hydroperoxide are both commercial now, and the peracetic acid route (Equation 139) would appear to be a good possibility, particularly for Union Carbide Corp. In the case of the ethylbenzene hydroperoxide route (Equation 140), the co-product methyl phenyl carbinol is dehydrated to styrene:

$$CH_3CH{=}CH_2 + Cl_2 + H_2O \xrightarrow{35°C} CH_3CHOHCH_2Cl \underset{HCl}{\overset{Ca(OH)_2,\ 25°C}{\rightleftharpoons}} CH_3\underset{\diagdown O \diagup}{CH{-}CH_2} \qquad (138)$$

$$CH_3CHO^* + O_2 \longrightarrow CH_3CO_3H \xrightarrow[CH_3CO_2H^*\ \text{co-product}]{CH_3CH=CH_2} CH_3\underset{\diagdown O \diagup}{CH{-}CH_2} \qquad (139)$$

$$C_6H_5CH_2CH_3^* \xrightarrow{O_2} C_6H_5\underset{OOH}{\underset{|}{C}}HCH_3 \xrightarrow[C_6H_5CHOHCH_3\ \text{co-product}]{CH_3CH=CH_2} CH_3\underset{\diagdown O \diagup}{CH{-}CH_2} \qquad (140)$$

$$CH_3\underset{\diagdown O \diagup}{CH{-}CH_2} \xrightarrow{H_2O} CH_3CHOHCH_2OH^* \qquad (141)$$

Again, propylene oxide is used to make propylene glycol (Equation 141) and to propoxylate alcohols, seven of which, besides polymers, are shown in Table 8.

$$CH_3\underset{\diagdown O \diagup}{CH{-}CH_2^*} + ROH \longrightarrow RO\underset{CH_3}{\underset{|}{C}}HCH_2OH \qquad (142)$$

TABLE 8. Alcohols Propoxylated

CH_3OH* $CH_3(CH_2)_2OH$* $CH_3(CH_2)_3OH$*
CH_3OCHCH_2OH* (with CH_3 branch) $CH_3(CH_2)_3O(CH_2)_2OH$*
$CH_3CHOHCH_2OH$* $HOCHCH_2OCHCH_2OH$* (with two CH_3 branches) Polyether alcohols*

VII. ALIPHATIC AMINES

By and large the simpler aliphatic amines are prepared by heating alcohols with ammonia in the presence of a catalyst:

$$ROH + NH_3 \xrightarrow{\text{catalyst}} RNH_2 + R_2NH + R_3N \quad (143)$$

All three products are obtained unless steric hindrance minimizes tertiary amine formation. The primary amine can be caused to be the preponderant product by using a large excess of ammonia, and the tertiary amine can be heavily favored by using a large excess of the alcohol. The three products must be separated by fractional distillation, which is not too easy in the case of the lower homologues. Union Carbide Corp. markets two sets of these amines. Following are the relevant equations:

$$CH_3CH_2OH^* + NH_3 \longrightarrow CH_3CH_2NH_2^* + (CH_3CH_2)_2NH^* + (CH_3CH_2)_3N^* \quad (144)$$

$$CH_3CHOHCH_3^* + NH_3 \longrightarrow (CH_3)_2CHNH_2^* + [(CH_3)_2CH]_2NH^* \quad (145)$$

Union Carbide Corp. also conducts a more complex alcohol amination [2]:

$$HO(CH_2)_2O(CH_2)_2OH + 2(CH_3)_2NH \longrightarrow (CH_3)_2N(CH_2)_2O(CH_2)_2N(CH_3)_2^* + 2H_2O \quad (146)$$

Amines react with phosgene to give isocyanates:

$$CH_3NH_2 + COCl_2 \longrightarrow CH_3NCO^* + 2HCl \quad (147)$$

A large number of commercially useful amines can be prepared by treating ammonia or an aliphatic amine with ethylene oxide:

$$\underbrace{CH_2{-}CH_2^*}_{O} + R_2NH \longrightarrow HO(CH_2)_2NR_2 \quad (148)$$

Union Carbide Corp. markets a number of these chemicals. Their preparation is shown (Equations 149–155.)

$$\underset{\diagdown O \diagup}{CH_2\text{—}CH_2^*} + NH_3 \longrightarrow HO(CH_2)_2NH_2^* + [HO(CH_2)_2]_2NH^* + [HO(CH_2)_2]_3N^* \quad (149)$$

$$\underset{\diagdown O \diagup}{CH_2\text{—}CH_2^*} + CH_3NH_2 \longrightarrow HO(CH_2)_2NHCH_3^* + [HO(CH_2)_2]_2NCH_3^* \quad (150)$$

Actually Union Carbide Corp. uses the following reaction [2]:

$$CH_3OH + [HO(CH_2)_2]_2NH^* \longrightarrow [HO(CH_2)_2]_2NCH_3^* + H_2O \quad (151)$$

$$\underset{\diagdown O \diagup}{CH_2\text{—}CH_2^*} + (CH_3)_2NH \longrightarrow HO(CH_2)_2N(CH_3)_2^* \quad (152)$$

$$\underset{\diagdown O \diagup}{CH_2\text{—}CH_2^*} + (CH_3CH_2)_2NH^* \longrightarrow HO(CH_2)_2N(CH_2CH_3)_2^* \quad (153)$$

$$\underset{\diagdown O \diagup}{CH_2\text{—}CH_2^*} + H_2N(CH_2)_2NH_2^* \longrightarrow HO(CH_2)_2NH(CH_2)_2NH_2^* \quad (154)$$

$$\underset{\diagdown O \diagup}{CH_2\text{—}CH_2^*} + HN\langle\text{piperazine}\rangle NH^* \longrightarrow HO(CH_2)_2N\langle\text{piperazine}\rangle NH^* \quad (155)$$

The reaction of ethylene oxide with ethylenediamine has been reported to proceed in 94% yield [30].

Diethanolamines can be cyclized, presumably by the same techniques used in preparing dioxane. The products are morpholines:

$$HN[(CH_2)_2OH]_2^* \longrightarrow HN\langle\text{morpholine}\rangle O + H_2O \quad (156)$$

$$CH_3N[(CH_2)_2OH]_2^* \longrightarrow CH_3N\langle\text{morpholine}\rangle O^* + H_2O \quad (157)$$

Both Union Carbide Corp. and Dow Chemical Co. used to employ the first method (Equation 156) to manufacture morpholine itself, but both manufacturing units were shut down in the mid 1970s [31]. Texaco Inc., the only present producer, treats diethylene glycol with ammonia at 700–3000 psi [31]. Recently, Air Products and Chemicals Inc. has been reported to have developed a low-pressure process in which one of the hydroxyl groups in diethylene glycol is converted to an amino group before cyclization [32].

Diamine chemistry is similar to that which was discussed earlier in this chapter. Most ethylenediamine derivatives probably are made from ethylene dichloride rather than from ethylene chlorohydrin or ethylene glycol. The ethylene dichloride can come from two sources:

$$CH_2{=}CH_2 + Cl_2 \longrightarrow Cl(CH_2)_2Cl \tag{158}$$

$$CH_2{=}CH_2 + Cl_2 + H_2O \xrightarrow{\text{by-product}} Cl(CH_2)_2Cl \tag{159}$$

Ethylene dichloride reacts with ammonia to give ethylenediamine:

$$Cl(CH_2)_2Cl + NH_3 \longrightarrow H_2N(CH_2)_2NH_2^* + 2HCl \tag{160}$$

Since the dichloride reacts more readily with the diamine than with ammonia, higher diamines are by-products. Fortunately, they are commercially useful chemicals. A type reaction is shown in Equations 161 and 162:

$$H_2N(CH_2)_2NH_2^* + Cl(CH_2)_2Cl \longrightarrow H_2N(CH_2)_2NH(CH_2)_2Cl \xrightarrow{NH_3} \tag{161}$$

$$H_2N(CH_2)_2NH(CH_2)_2NH^* \xrightarrow[NH_3]{Cl(CH_2)_2Cl} H_2N(CH_2)_2NH(CH_2)_2NH(CH_2)_2NH_2^* \tag{162}$$

This can continue to give

$$H_2N(CH_2)_2NH(CH_2)_2NH(CH_2)_2NH(CH_2)_2NH_2^* \qquad [3]$$

$$H_2N(CH_2)_2NH(CH_2)_2NH(CH_2)_2NH(CH_2)_2NH(CH_2)_2NH_2^* \qquad [4]$$

Obviously, certain products can be favored by appropriate adjustment of stoichiometry and reaction conditions. However, the yield of ethylenediamine itself has been reported to be as high as 80% [33].

Recently it has been reported that Union Carbide Corp. has a new catalyst that enables the production of ethylenediamine and its homologues to be made directly in one step from ethylene oxide [34]:

$$\underbrace{CH_2{-}CH_2^*}_{O} + NH_3 \xrightarrow{\text{catalyst}} H_2N(CH_2)_2NH_2^*, \text{etc.} \tag{163}$$

The author hates to rule out the idea that the corresponding alcohols, either as such or through their chloro derivatives, may not also be intermediates. Following are some examples:

$$HO(CH_2)_2NH_2^* + NH_3 \longrightarrow H_2N(CH_2)_2NH_2^* + H_2O \quad (164)$$

$$H_2N(CH_2)_2NH_2^* + \underset{\diagdown O \diagup}{CH_2{-}CH_2^*} \longrightarrow H_2N(CH_2)_2NH(CH_2)_2OH^* \quad (165)$$

$$H_2N(CH_2)_2NH(CH_2)_2OH^* + NH_3 \longrightarrow H_2N(CH_2)_2NH(CH_2)_2NH_2^* + H_2O \quad (166)$$

$$HO(CH_2)_2NH_2^* + \underset{\diagdown O \diagup}{CH_2{-}CH_2^*} \longrightarrow HO(CH_2)_2NH(CH_2)_2OH^* \quad (167)$$

$$HO(CH_2)_2NH(CH_2)_2OH^* + 2NH_3 \longrightarrow H_2N(CH_2)_2NH(CH_2)_2NH_2^* + 2H_2O \quad (168)$$

In fact the preparation of ethylenediamine from ethanolamine and ammonia was recently patented with a yield of 43% and a conversion of 67% [35]. A Leonard Process Company method is reported to be better (93% conversion) [36].

Union Carbide Corp. sells three trimethylenediamine derivatives. The first one **[5]** is made from the corresponding glycol and dimethylamine [2]:

$$CH_3\underset{\displaystyle N(CH_3)_2}{\underset{|}{C}}H(CH_2)_2N(CH_3)_2^* \qquad [5]$$

The other two are prepared by cyanoethylation of an amine followed by hydrogenation [2]:

$$CH_3NH_2 + 2CH_2{=}CHCN \longrightarrow NC(CH_2)_2\underset{\displaystyle CH_3}{\underset{|}{N}}(CH_2)_2CN \quad (169)$$

$$\xrightarrow[\text{catalyst}]{H_2} H_2N(CH_2)_3\underset{\displaystyle CH_3}{\underset{|}{N}}(CH_2)_3NH_2^* \quad (170)$$

$$(CH_3)_2NH + CH_2{=}CHCN \longrightarrow (CH_3)_2N(CH_2)_2CN \quad (171)$$

$$\xrightarrow[\text{catalyst}]{H_2} (CH_3)_2N(CH_2)_3NH_2^* \quad (172)$$

Finally, these diamines can be cyclized to form piperazine derivatives. Union Carbide Corp.'s contributions are shown in Equations 173–177.

$$2Cl(CH_2)_2Cl + 2NH_3 \longrightarrow HN\bigcirc NH^* + 4HCl \qquad (173)$$

$$HO(CH_2)_2NH(CH_2)_2NH_2^* + NH_3 \longrightarrow HN\bigcirc NH^* \qquad (174)$$

$$H_2N(CH_2)_2NH(CH_2)_2NH_2^* \cdot 2HCl \xrightarrow{\Delta} HN\bigcirc NH^* \qquad (175)$$

Piperazine can be alkylated:

$$HN\bigcirc NH^* + CH_3OH \longrightarrow H_3CN\bigcirc NH^* + H_2O \qquad (176)$$

$$HN\bigcirc NH^* + \underset{\diagdown O \diagup}{CH_2{-}CH_2^*} \longrightarrow HN\bigcirc N(CH_2)_2OH^* \qquad (177)$$

VIII. RECAPITULATION

The emphasis in this chapter has been entirely on chemistry. No attempt has been made to mention the many industrial uses for the various chemicals described. This omission has been deliberate.

What the author hopes has been accomplished is to show the very close chemical integration that one company has achieved in a major part of its product line. Very few basic raw materials are involved. Likewise, very few chemical reactions are employed, although these few reactions have been used to manufacture a large number of chemicals, hence the lists in the tables, which at first glance may appear to be overly long.

As in the last chapter, certain characteristics of industrial organic chemistry have appeared again. These include catalysis, high reaction temperatures, high pressures, the handling of gases, and often a choice of routes whose selection is based on both the technology and the economics at the time of plant construction.

REFERENCES

1. *Chemicals and Plastics Physical Properties,* 1977–1978 edition, Union Carbide Corporation, New York.
2. *Union Carbide Chemicals and Plastics, How They Are Manufactured,* Union Carbide Corporation, New York, 1975.
3. W. Schoenleben, H. Hoffmann, W. Lengsfeld, and H. Mueller, Ger. Offen. 2,727,330; *C.A.* **90,** 86745g (1979).
4. M. Guerbet, *Compt. rend.,* **128,** 511 (1899).
5. K. Reinhold, Ger. Offen. 2,703,746; *C.A.,* **89,** 179535r (1978). See also: Henkel und Cie, G.m.b.H., Ger. Offen. 2,634,676; *C.A.,* **86,** 155192p (1977). K. Ota and T. Kito, Jap. Kokai Tokkyo Koho ('77) 77,002; *C.A.,* **87,** 200775q (1977).
6. M. Matsumoto and M. Tamura, Ger. Offen. 2,931,883; *C.A.,* **93,** 25886b (1980).
7. R. Fowler, H. Connor, and R. A. Baehl, *Chemtech,* **6,** December, 1976, p. 772.
8. Anon., *Chem. Eng. News,* **53,** October 13, 1975, p. 6. Anon., *Chem. Week,* **117,** October 15, 1975, p. 38.
9. J. H. Craddock, J. F. Roth, A. Hershman, and F. E. Paulik, U.S. Pat. 3,989,747; *C.A.,* **86,** 54988x (1977). See also: Anon., *Chemtech,* **7,** March, 1977, p. 140. F. E. Paulik, A. Hershman, J. F. Roth, and J. H. Craddock, U.S. Pat. 3,989,748; *C.A.,* **86,** 54989y (1977).
10. A. Y. Karmil'chik, V. Stonkus, M. V. Shimanskaya and S. Hillers, *C.A.,* **84,** 73590m (1976).
11. H. Scharf, Ger. Offen. 2,802,672; *C.A.,* **91,** 157277x (1979).
12. Y. Onone, Y. Mizutani, S. Akiyama, Y. Izumi, and Y. Watanabe, *Chemtech,* **7,** January, 1977, p. 36.
13. T. Naito, T. Imai, and H. Ebisawa, Jap. Pat. ('73) 35,245; *C.A.,* **80,** 120274r (1974).
14. N. S. Aprahamian, U.S. Pat. 3,772,367; *C.A.,* **80,** 36711y (1974).
15. S. I. Guseva, A. A. Grigor'eva, and T. G. Prozorovskaya, *C.A.,* **89,** 90039 (1978).
16. A. Izumi, Y. Watanabe, and M. Sato, Jap. Kokai Tokkyo Koho ('79) 52,023; *C.A.,* **91,** 157276w (1979).
17. M. Fukui, S. Furukawa, I. Koga, and T. Inoi, Jap. Kokai ('74) 30,309; *C.A.,* **81,** 77474n (1974).
18. M. Fukui, S. Hayashi, T. Okamoto, I. Koga, and T. Inoi, Jap. Kokai ('74) 61,110; *C.A.,* **81,** 119973g (1974).
19. V. A. Shmelev, V. K. Parshikov, N. I. Kosolopova, V. V. Zhitneva, and A. B. Letunova, *C.A.,* **88,** 6282g (1978).
20. R. Hughes, *Chemtech,* **6,** December, 1976, p. 731.
21. W. Merz and D. Nachtsheim, Ger. Offen. 2,461,503; *C.A.,* **85,** 176844x (1976).
22. K. Takagi and T. Motohashi, Jap. Kokai ('74) 43909, 43910; *C.A.,* **81,** 119950x, 119951y (1974).
23. *Chemicals and Plastics Physical Properties,* 1973–1974 edition, Union Carbide Corporation, New York.
24. Rohm, Belgian Pat., *Chemtech,* **4,** April, 1974, p. 195.
25. R. Hughes, *Chemtech,* **4,** December, 1974, p. 708.
26. Anon., *Chemtech,* **9,** August, 1979, p. 465.
27. Y. Fujita and T. Morimoto, Jap. Pat. ('74) 16,433; *C.A.,* **82,** 16850f (1975).
28. Y. Fujita and T. Morimoto, Jap. Pat. ('74) 16,434; *C.A.,* **82,** 43431b (1975).
29. N. Kanetaka and K. Marumo, Jap. Kokai Tokkyo Koho ('79) 106,409; *C.A.,* **92,** 22039w (1980).

30. R. Schubart, Ger. Offen. 2,716,946; *C.A.*, **90,** 38523u (1979).
31. Anon., *Chem. Week*, **126,** January 30, 1980, p. 30.
32. Anon., *Chem. Eng. News*, **58,** January 28, 1980, p. 32.
33. T. Sunaga, Jap. Kokai Tokkyo Koho ('79) 32,405; *C.A.*, **91,** 38902n (1979).
34. Anon., *Chem. Week*, **123,** August 9, 1978, p. 13.
35. C. E. Habermann, U.S. Pat. 4,153,581; *C.A.*, **91,** 38905r (1979). C. E. Habermann, U.S. Pat. 4,152,353; *C.A.*, **91,** 19885t (1979).
36. *Chemtech*, **8,** June, 1978, p. 327.

5
POLYMER CHEMISTRY

I. INTRODUCTION

It has been stated that 50% of the industrial chemists in this country work in some area of polymer chemistry. Upon examination of all the ramifications of what is called polymer chemistry, this is not surprising. On the other hand, the author believes that even today polymer chemistry itself receives limited attention in many college and university chemical curricula.

This has placed him in a bit of a quandary as to how to write this chapter. The best approach appears to be to cover industrial polymer chemistry broadly and introduce the beginning chemist to as many basic concepts and as much polymer vocabulary as possible. In this way he or she will at least know what people are talking about on going to work, and it is hoped that he or she will have some understanding of the chemistry and technology that will be faced.

The approach will be to look at the most important polymers (industrially speaking). Not only will theoretical considerations be brought into the picture, but the related chemistry will also be discussed. Mention will be made of some of the formulations that are used commercially, and a little space will be devoted to the characterization (analysis) of polymers, since many of the methods are different from those used with monomeric compounds and will be new to the beginning chemist.

However, polymers are basically simply macromolecules and should be considered as such. Almost all of them comprise very simple (molecularly speaking), small building blocks, which are put together in a multitude of repeating

Most of the data for this chapter were gleaned from B. G. Reuben and M. L. Burstall, *The Chemical Economy,* Longman, London, 1973; R. N. Shreve, *Chemical Process Industries,* McGraw-Hill, New York, 1967; G. B. Butler and K. D. Berlin, *Fundamentals of Organic Chemistry,* Ronald Press, New York, 1972, Chapter 34; W. R. Sorenson and T. W. Campbell, *Preparative Methods of Polymer Chemistry,* Interscience, New York, 1961; and *Chemistry in the Economy,* Americam Chemical Society, Washington, D.C., 1973.

units. Likewise, nearly every commercial polymer product also includes a wide variety of additives, which perform necessary functions in relation to the particular application in question.

One important polymer class is the vinyl polymers. These all are based on the fact that the ethylenic double bond in an organic molecule can be opened, usually with the help of some kind of intramolecular activation, but even without it, to undergo a chain addition reaction, which leads to the formation of a molecule of high molecular weight:

$$CH_2{=}CHR \xrightarrow{\text{catalyst}} (-CH_2-CHR-)_n \tag{1}$$

Unless side reactions occur, these long chains are not joined to each other. (The polymers are not crosslinked.) The mechanical properties of these polymers depend on temperature. They are thermoplastic. They can be melted and resolidified without an appreciable change in their physical properties. When softened by heat, they can be extruded as films or fibers or molded into all kinds of shapes.

This first section concerns those polymers in which R = H, CH_3, C_6H_5, Cl, OR′, CN, $CONH_2$, and CO_2R''. It is also possible to prepare polymers from monomers of the structure $CH_2{=}CRR'''$, two types of which are quite important commercially, one in which R, R‴ = Cl, Cl (Saran), and the other in which R, R‴ = CH_3, CO_2R'' (the polymethacrylates). Both of these will be mentioned.

II. POLYETHYLENE

The simplest polymer of the vinyl type, polyethylene, is also the largest volume. In 1979 its production capacity in the U.S. was a little over six million tons [1]. Because of its cheapness, ease of fabrication, excellent chemical resistance, excellent moisture-vapor resistance, and good electrical insulating properties, it finds use in packaging films and sheets, extrusion coating of containers, wire and cable insulation, pipe, bottles, and many other applications.

There are two types of polyethylene, low-density polyethylene prepared at very high pressures, and high-density polyethylene prepared at low pressures. In the high-pressure product the chains are branched, whereas in the low-pressure product they are essentially linear.

$$CH_2{=}CH_2 \xrightarrow[\sim 20{,}000\text{–}30{,}000\ \text{psi}]{O_2\ 190°C} (-CH_2-CH_2-)_n \tag{2}$$

The mechanism of polymerization is free radical. Any chemist can readily appreciate that a great deal of mechanical engineering research and development was involved in designing and building the equipment for this high-pressure manufacturing process. Also, because of the pressures involved, the reaction is run behind reinforced concrete. This process was pioneered by Imperial Chemical Industries, Ltd. in Great Britain and by E. I. DuPont de Nemours and Company and Union Carbide Corp. in this country.

$$CH_2{=}CH_2 \xrightarrow[\substack{Cr_2O_3 \text{ on } SiO_2\cdot Al_2O_3,\text{ cyclohexane,}\\ 450\text{–}600\text{ psi, }100\text{–}150^\circ C}]{Al(C_2H_5)_3\cdot TiCl_4,\text{ at pressure, or}} (-CH_2-CH_2-)_n \qquad (3)$$

The $Al(C_2H_5)_3$ process was discovered by Karl Ziegler in Germany and by Giulio Natta in Italy. The obvious disadvantage is that the catalyst is spontaneously inflammable in air. That is why the Cr_2O_3 (Phillips Petroleum Co.) process has largely taken over in this country.

III. POLYPROPYLENE

Polypropylene is another very large volume polymer, with U.S. production capacity in 1979 at a little over 2 million tons [1]. As would be expected, its physical properties are somewhat similar to those of polyethylene. However, it has a higher (about 40°C) melting point and greater strength and hardness. It can be formed into filaments and fibers, which constitute its largest uses, although an appreciable quantity is used in such packaging applications as film, bottles, and closures, in molding applications such as automotive and appliance parts, and many others.

Most polypropylene is made by the Ziegler process, similar to that for polyethylene:

$$\underset{\displaystyle\ \ \ \ \ \ \ \ |\ \ \atop \displaystyle\ \ \ \ \ \ \ \ CH_3}{CH_2{=}CH} \xrightarrow[\substack{70^\circ C,\ 50\text{ psi}\\ n\text{-heptane}}]{Al(C_2H_5)_3\cdot TiCl_4} \left(-CH_2-\underset{\displaystyle |\atop \displaystyle CH_3}{CH}-\right)_n \qquad (4)$$

The chemist can readily see that the carbon atom carrying the methyl group in polypropylene is asymmetric. In the isotactic form all the methyl groups are on the same side of the chain, and in the syndiotactic form they are on alternate sides. This stereoregularity means that the chains pack together more closely than they would if the methyl groups were located at random (atactic form). As a result, polypropylene is more crystalline than polyethylene, and this is why it can be formed into filaments and fibers.

In polypropylene the tertiary carbon atom is sensitive to oxidation, so the polymer has to be formulated with antioxidants and light stabilizers. Like polyethylene, polypropylene has little affinity for dyes, so special techniques have had to be developed to color these polymers.

IV. POLYSTYRENE

Polystyrene actually embraces a group of polymers. In 1979 U.S. production capacity of polystyrene itself was about 2.5 million tons [1]. Polystyrene itself is low in cost and finds extensive use in all kinds of molded products, particularly since it can be obtained in a crystal clear, glassy form. It also can be foamed (by a suitable blowing agent), and because of the structural and insulating properties of the foam, it finds use in construction and packaging applications.

In the United States most polystyrene is prepared by continuous bulk polymerization in which the monomeric styrene also serves as the vehicle. Free-radical catalysts such as benzoyl peroxide are used.

$$\underset{\displaystyle C_6H_5}{CH_2{=}\underset{|}{C}H} \xrightarrow{(C_6H_5CO_2)_2} \left(-CH_2-\underset{\underset{\displaystyle C_6H_5}{|}}{CH}- \right)_n \qquad (5)$$

Actually styrene can be polymerized in many ways. It can be polymerized in a solvent such as toluene. Another commercial method is polymerization in aqueous suspension in which drops of the monomer are converted to beads of polystyrene. Each drop constitutes a tiny bulk polymerization. Styrene can also be polymerized as an almost colloidal emulsion in water. The polystyrene forms as a latex, which must be broken up to separate it from the water and other chemicals present.

All these methods proceed by free-radical mechanisms. Styrene can also be polymerized by a cationic mechanism, in which the active intermediate is a carbonium ion produced by a proton or by a Lewis acid such as stannic chloride:

$$CH_2{=}\underset{\underset{\displaystyle C_6H_5}{|}}{CH} \xrightarrow{H^+} CH_3-\underset{\underset{\displaystyle C_6H_5}{|}}{CH^+} \qquad (6)$$

Polystyrene can also be prepared by an anionic mechanism, in which the active intermediate is a carbanion:

$$CH_2{=}\underset{\underset{\displaystyle C_6H_5}{|}}{CH} \xrightarrow{R^-} RCH_2-\underset{\underset{\displaystyle C_6H_5}{|}}{CH^-} \qquad (7)$$

Pure polystyrene is brittle. In order to improve its toughness, impact polystyrene was developed. This is prepared by copolymerizing styrene with a small amount of butadiene or of a styrene-butadiene rubber, thus incorporating tiny pieces of rubber in the polystyrene matrix. Its properties are indicated by the fact that it is being used to replace wood in furniture.

Another polystyrene product possessing great toughness, hardness, and chemical resistance is acrylonitrile-butadiene-styrene (ABS). In the United States annual production capacity is about 800 million pounds [1]. It finds extensive use in such forms as sheet and pipe, and even in such specialty applications as radio housings, color telephones, and golf club heads.

Originally, ABS was prepared by copolymerizing the three monomers:

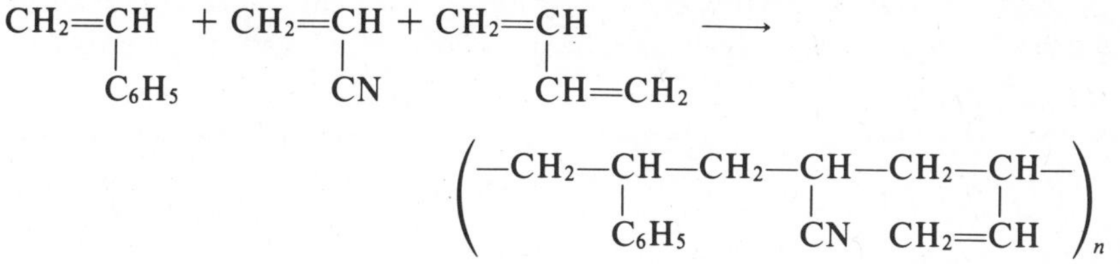

$$\underset{\displaystyle C_6H_5}{\underset{|}{CH_2{=}CH}} + \underset{\displaystyle CN}{\underset{|}{CH_2{=}CH}} + \underset{\displaystyle CH{=}CH_2}{\underset{|}{CH_2{=}CH}} \longrightarrow \left(-CH_2-\underset{\displaystyle C_6H_5}{\underset{|}{CH}}-CH_2-\underset{\displaystyle CN}{\underset{|}{CH}}-CH_2-\underset{\displaystyle CH_2{=}CH}{\underset{|}{CH}}- \right)_n \quad (8)$$

Presently it is manufactured by graft polymerization of acrylonitrile and styrene on a performed polybutadiene elastomer.

Until the reaction shown in Equation 8, the discussion has been about homopolymers, which are prepared by polymerizing one monomer. Copolymers are prepared from two or more monomers, as shown in Equation 8 for ABS. If the polarities of the monomers involved are reasonably similar, the different monomeric units are distributed at random throughout the polymer chain. However, these monomers may polymerize at rates so different that the faster-polymerizing monomer has to be added gradually during the course of the polymerization in order to keep the ratios of the two monomers reasonably constant throughout the final polymer.

If the polarities of the two monomers are so different that each reacts preferentially with the other rather than with itself, an alternating copolymer is obtained. The paper and textile size styrene-maleic anhydride (SMA) is a good example.

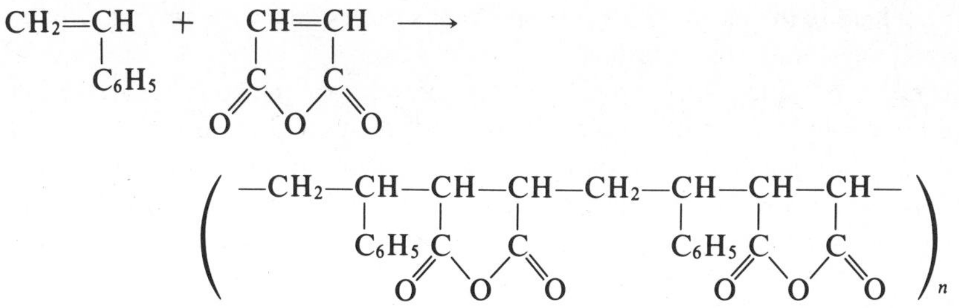

$$CH_2{=}CH(C_6H_5) + \text{maleic anhydride } (CH{=}CH \text{ bridged by } C(=O)-O-C(=O)) \longrightarrow \left(-CH_2-CH(C_6H_5)-CH-CH-CH_2-CH(C_6H_5)-CH-CH- \right)_n \quad (9)$$

(each $-CH-CH-$ pair in the product bears a $C(=O)-O-C(=O)$ anhydride ring)

Grafting means that side chains of a different polymer are attached to a preformed polymer backbone. ABS is a good example.

V. POLYVINYL CHLORIDE

Polyvinyl chloride is another tremendous volume polymer. In 1979 its U.S. production capacity exceeded 3 million tons [2]. It is extremely versatile, and by suitable plasticization its properties can be varied from rigidity to great pliability. It also is cheap. This property-price combination has led to all kinds of applications, such as pipe and conduit, sheet and film (upholstery), flooring, wallpaper, fabric and paper coating, shoes, and wire and cable coating, to name a few.

Polyvinyl chloride is prepared by emulsion or suspension polymerization using a free-radical catalyst such as benzoyl peroxide or a redox system:

$$CH_2{=}\underset{\displaystyle Cl}{\underset{|}{CH}} \xrightarrow{(C_6H_5CO_2)_2} \left(-CH_2-\underset{\displaystyle Cl}{\underset{|}{CH}}- \right)_n \tag{10}$$

A redox system is a mixture of an oxidizing agent and a reducing agent that produces a free-radical intermediate in the course of their interaction. A typical example is hydrogen peroxide and ferrous ammonium sulfate:

$$Fe^{2+} + H_2O_2 \longrightarrow Fe^{3+} + OH\cdot + OH^- \tag{11}$$

As in the case of polypropylene, there are stability problems in connection with polyvinyl chloride. In this case, trouble starts with the loss of a molecule of hydrogen chloride. More hydrogen chloride can be lost to set up a conjugated system. The double bonds also activate the substituents on adjacent carbon atoms. Therefore, polyvinyl chloride is formulated with heat stabilizers, light stabilizers, and antioxidants. These include such chemicals as epoxidized oils; barium, cadmium, or zinc soaps; and dibutyltin dilaurate.

For many applications a plasticizer is added. A plasticizer is a polymer additive (usually a high-boiling oil) that softens the polymer. For many applications unplasticized polyvinyl chloride is far too rigid, so a plasticizer is added. Plasticizers are held in the polymer by van der Waal's forces and serve to separate the polymer chains and lubricate them.

For polyvinyl chloride the commonest plasticizers are the phthalate esters, of which the 2-ethylhexyl is the most widely used. Adipates, epoxidized oils, and low-molecular-weight polymers are among others employed. It is very important that the plasticizer be held in by van der Waals forces, so that it does not migrate (exude from the polymer). It would be most disconcerting to the occupant of an automobile seat if the plasticizer in the polyvinyl chloride upholstery migrated. Other desirable properties of a plasticizer include low price (the plasticizer usually costs more than the polymer), heat and light stability, wide compatibility range, and high efficiency (so that you use as little as possible).

A coating technique discovered by industrial polymer chemists is the use of a plastisol. A plastisol is a suspension of beads of a polymer such as polyvinyl chloride in a plasticizer. The article to be coated (such as a glove) is dipped in the plastisol and then passed through an oven. The polymer dissolves in the plasticizer at the elevated temperature, so upon cooling the article is covered by a smooth coating of the plasticized polymer.

So far mention has been made of external plasticizers only. Plasticization can be effected internally. For example, incorporation of vinyl acetate groups in the polyvinyl chloride chain gives a flexible product. Obviously an internal plasticizer has the advantage that it will not migrate, nor will it be lost through evaporation or dissolved out by organic solvents.

VI. POLYVINYL ACETATE

Polyvinyl acetate is an important vinyl polymer (1978 U.S. production capacity equalled nearly 700 million pounds [3]), which finds use in latex paints and adhesives, and as the raw material for the manufacture of other vinyl polymers. It can be prepared by polymerizing vinyl acetate with a free-radical catalyst in solution, in emulsion, or in suspension.

$$CH_2{=}\underset{\displaystyle OCOCH_3}{\underset{|}{CH}} \xrightarrow{(CH_3CO_2)_2} \left(-CH_2-\underset{\displaystyle OCOCH_3}{\underset{|}{CH}}- \right)_n \qquad (12)$$

VII. POLYACRYLONITRILE

The use of acrylonitrile in styrene copolymers has been mentioned. Polyacrylonitrile itself finds extensive use as a textile fiber. Fabrics made therefrom are similar to wool in physical properties and find widespread use in carpets, blankets, sweaters, and outerwear. The U.S. production capacity exceeded 850 million pounds in 1978 [4]. Orlon and Acrilan are familiar trade names. Polyacrylonitrile usually is spun from dimethylformamide solution, since it is insoluble in most other solvents and begins to decompose before it is soft enough to spin from a melt (see Section XIX).

Acrylonitrile can be polymerized by free-radical or anionic catalysts in solution, suspension, or emulsion:

$$CH_2{=}\underset{\displaystyle CN}{\underset{|}{CH}} \xrightarrow[NaHSO_3]{K_2S_2O_8} \left(-CH_2-\underset{\displaystyle CN}{\underset{|}{CH}}- \right)_n \qquad (13)$$

For fiber use, a small amount of a basic comonomer, such as 2-vinylpyridine, usually is included so that the final fabric can be dyed with acid dyes.

Reactive monomers such as acrylonitrile, styrene, and vinyl acetate usually contain a polymerization inhibitor, such as hydroquinone, as they are received from the manufacturer. This is to prevent spontaneous polymerization in transit or on storage. In the laboratory it often is most convenient to distill the monomer from the inhibitor and use it freshly distilled. Commercially, to remove the inhibitor the monomer usually is passed through a column packed with an active adsorbent such as silica gel.

Since monomers such as vinyl chloride and acrylonitrile possess some toxicity, there recently has been interest in removing the last traces of these monomers from the polymer, particularly when the polymer will be used as film to wrap foods or as a bottle to contain an edible liquid. That traces should be present is not surprising, since small monomer molecules tend to become mechanically trapped in the network of growing polymer chains. Standard methods such as steam distillation, drying *in vacuo,* or even solvent extraction are used to remove these traces.

VIII. POLYACRYLATES

Polyacrylates are mentioned here primarily for completeness, since they are large-volume polymers with U.S. production capacity at nearly 1 million pounds in 1974 [5]. The commonest ones—methyl, ethyl, *n*-butyl, and 2-ethylhexyl—find use in a variety of copolymers, particularly in latex paints and in textile applications. In general, their properties are inferior to those of the polymethacrylates, which will be mentioned next. Heat, light, or free radicals may be used as polymerization catalysts, and the polymerization can be carried out en masse, in solution, in emulsion, or in suspension.

$$CH_2{=}\underset{\displaystyle CO_2CH_3}{\underset{|}{CH}} \xrightarrow{(NH_4)_2S_2O_8} \left(-CH_2-\underset{\displaystyle CO_2CH_3}{\underset{|}{CH}}- \right)_n \tag{14}$$

IX. POLYMETHACRYLATES

Polymethyl methacrylate and its homologues constitute another high-volume family of polymers with 1976 U.S. production of polymethyl methacrylate at nearly 900 million pounds [6]. Homopolymers of the lowest member of the series, polymethyl methacrylate, have excellent optical clarity and brilliance, good outdoor durability, and convenient handling properties. As a result, they find use as cast sheet (Lucite, Plexiglas) for advertising signs and displays, lighting fixtures, and airplane windshields, as coatings, and as molding powders. Higher polymethacrylates (particularly 2-ethylhexyl) are used as viscosity-index improvers in lubricating oils.

Methyl methacrylate can be polymerized by either a free-radical or an anionic mechanism. Usually benzoyl peroxide is used to initiate free-radical polymerization, either in bulk or in suspension.

$$CH_2{=}\underset{CO_2CH_3}{\overset{CH_3}{\overset{|}{\underset{|}{C}}}} \xrightarrow{(C_6H_5CO_2)_2} \left[-CH_2-\underset{CO_2CH_3}{\overset{CH_3}{\overset{|}{\underset{|}{C}}}}- \right]_n \tag{15}$$

X. POLYVINYL ALCOHOL

Polymers are simply macromolecules. They undergo the same kinds of chemical reactions as their monomeric counterparts. An excellent example is the hydrolysis (really transesterification) of polyvinyl acetate to give polyvinyl alcohol, which is subsequently converted to polyvinyl butyral:

$$\left(-CH_2-\underset{OCOCH_3}{\underset{|}{CH}}-CH_2-\underset{OCOCH_3}{\underset{|}{CH}}- \right)_n + CH_3OH \xrightarrow[\text{or } H^+]{OH^-}$$

$$(-CH_2CHOHCH_2CHOH-)_n + CH_3CO_2H\uparrow \xrightarrow[H_2SO_4]{CH_3(CH_2)_2CHO}$$

$$\left[\begin{array}{c} -CH_2CHCH_2CH- \\ / \quad \backslash \\ O-CH-O \\ | \\ (CH_2)_2CH_3 \end{array} \right]_n \tag{16}$$

Polyvinyl alcohol, a water-soluble polymer, finds use in such formulations as adhesives and textile sizes. Polyvinyl butyral, because of its elasticity and strong adherence to glass, is the essentially universal interleaving layer in safety glass for automobile windshields and windows. In this application a plasticizer such as triethylene glycol 2-ethylbutyrate **[1]** is needed.

$$CH_3CH_2\underset{C_2H_5}{\underset{|}{C}}HCO_2-[(CH_2)_2O]_3-CO\underset{C_2H_5}{\underset{|}{C}}HCH_2CH_3 \tag{[1]}$$

XI. DIENE POLYMERS AND RUBBER

The whole story of the development of the rubber industry is fascinating; it is described in detail in *Chemistry in the Economy* [7]. However, this chapter is limited to the main outlines of the polymer chemistry involved. This means omission of a description of many of the specialty rubbers even though they are tonnage articles of commerce.

The rubber industry is a huge one. In 1969 its volume in the United States embraced about 2.7 million tons of product with a monetary value of about $17 billion. Some 500,000 people were employed. The largest-volume product is automobile tires, but everyone is familiar with many other rubber applications.

XII. NATURAL RUBBER

The rubber industry started with natural rubber, which is produced as a latex by a tree native to tropical America, but now raised primarily in the tropical orient—Malaysia, Indonesia, and so forth. Although synthetic rubbers have taken over about 50% of the market, natural rubber still is used extensively in tires, particularly for trucks, because of its outstanding resistance to heat buildup. In 1975 this use amounted to nearly 500,000 tons [8]. Also, in many applications it is one of several rubbers that are compounded together to achieve maximum performance properties for the use in question.

Chemically, natural rubber is primarily *cis*-polyisoprene:

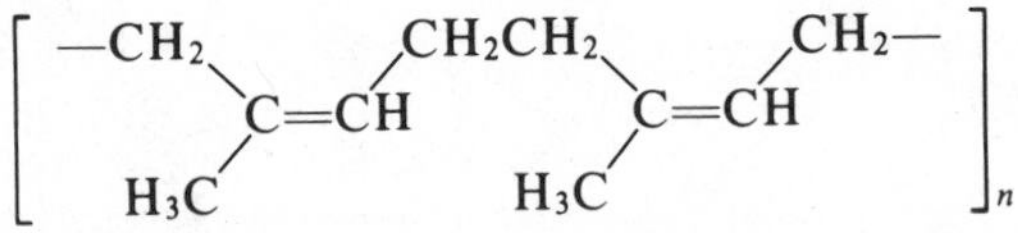

$$\left[-CH_2-C(CH_3)=CH-CH_2CH_2-C(CH_3)=CH-CH_2- \right]_n \quad [2]$$

Its unique physical properties and industrial usefulness depend on the fact that it can be crosslinked, with sulfur essentially universally used as the crosslinking agent. This crosslinking is sufficiently sparse and yet firm enough to enable the polydiene network to remain elastomeric, but the crosslinking compels it to return to its original configuration after stretching or compression.

The crosslinking (vulcanization) of polydienes is the basis of the rubber industry. For instance, butyl rubber, which is basically polyisobutylene, must contain a small amount of a diene (isoprene is used mostly at present) to be vulcanized and therefore to possess rubbery properties.

Polydiene processing and use require a tremendous variety of chemicals–vulcanization accelerators and controllers, age resistors (antioxidants and antiozone agents), and a variety of other stabilizers. Polydienes that are used in automobile tires need fillers such as silica or carbon black (3 billion pounds of the latter in 1969) and adhesives to enable the polydiene to stick to the tire cord, which may be nylon, polyester, steel, and so forth. Recently it has been found that the interstices in the polydiene matrix can be filled with high-boiling oils, up to one third of the weight of the tire tread. This oil extension makes the rubber more processible, so it can be extruded (a bonus in tire-tread manufacture) rather than having to be molded. Even the physical properties of the polydiene are improved. And, of course, the cost is improved also.

XIII. POLYISOPRENE

Recently it has been found that a Ziegler-type catalyst employed in a low-boiling hydrocarbon solvent polymerizes isoprene to the cis configuration. The product is indistinguishable from natural rubber.

$$CH_2{=}\overset{\displaystyle CH_3}{\overset{|}{C}}{-}CH{=}CH_2 \xrightarrow{\text{catalyst}} \left[\begin{array}{c} -CH_2\diagdown\quad\;\; \diagup CH_2CH_2\diagdown\quad\;\; \diagup CH_2- \\ C{=}CH \qquad\qquad C{=}CH \\ H_3C\diagup \qquad\qquad H_3C\diagup \end{array} \right]_n \qquad (17)$$

Isoprene can also be polymerized by a catalyst such as butyllithium in pentane or by lithium metal en masse, in the latter case also to produce *cis*-polyisoprene. In 1970 about 120,000 tons of synthetic *cis*-polyisoprene were produced [7].

All diene monomers must be inhibited to prevent premature polymerization prior to use. All the commercial isoprene polymerizations proceed by a 1,4, not a 1,2 mechanism. In polydienes, stereochemistry is extremely important, as evidenced by the great difference in valuable properties between natural rubber (*cis*-polyisoprene) and gutta percha (*trans*-polyisoprene). Lastly, all these diene polymerizations are highly exothermic, so special precautions must be taken to control the heat of reaction in commercial processes.

XIV. POLYBUTADIENE

The simplest diene, butadiene, also is polymerized commercially. In fact, in 1979 the U.S. production capacity amounted to more than 400 thousand tons [9]. Its primary use is in automobile and truck tires. Here again a Ziegler-type catalyst in a light hydrocarbon medium also gives the desired cis configuration:

$$CH_2{=}CHCH{=}CH_2 \xrightarrow{\text{catalyst}} \left[\begin{array}{c} \diagup CH{=}CH\diagdown \qquad \diagup CH{=}CH\diagdown \\ -CH_2 \qquad\qquad CH_2CH_2 \qquad\qquad CH_2- \end{array} \right]_n \qquad (18)$$

Butadiene can also be polymerized by metallic sodium in a medium such as toluene.

XV. SBR

Commercially, the most important diene polymer actually is a copolymer of butadiene and styrene (SBR). In 1979 U.S. production capacity was about 1.5

million tons [10]. (SBR) finds use in tires for automobiles, farm machinery, and tractors; in mechanical goods; in hoses and belting, including conveyor belting; as sponge and foam; in shoe soles and heels; in overshoes, rubbers, and raincoats; in adhesives and sealants; and in wire insulation, just to name a few applications.

SBR contains three parts of butadiene to one of styrene, which means about 6 mol of butadiene to 1 mol of styrene. The diene again is largely cis and is polymerized by a 1,4 mechanism. Commercially the polymerization is carried out continuously in aqueous emulsion (to absorb the heat) using potassium persulfate or a redox system as a catalyst. A C_{12} mercaptan is added to regulate chain growth. The polymerization is taken to about 60% conversion, since after that the rate slows down and the product quality is inferior.

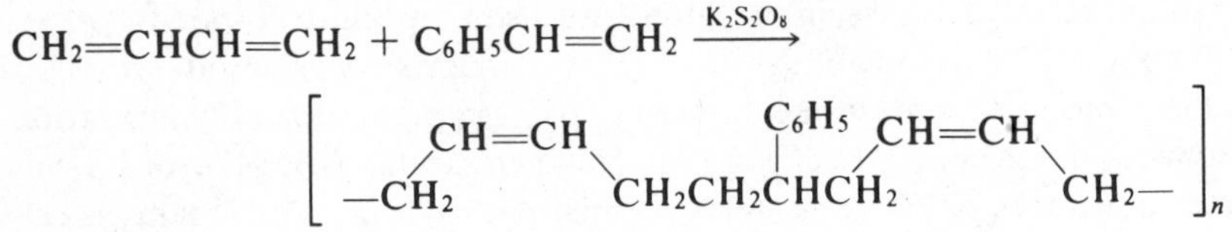

$$CH_2{=}CHCH{=}CH_2 + C_6H_5CH{=}CH_2 \xrightarrow{K_2S_2O_8} \left[-CH_2-CH{=}CH-CH_2CH_2\underset{\displaystyle |}{\overset{C_6H_5}{C}}HCH_2-CH{=}CH-CH_2- \right]_n \tag{19}$$

The monomer distribution is random.

Recently the Ziegler-catalyst system has been adapted to SBR production. It produces almost entirely the cis diene configuration. In 1970 about 280,000 tons of stereo SBR was produced commercially.

XVI. POLYCHLOROPRENE (NEOPRENE)

Polychloroprene is one of the very earliest of the synthetic rubbers. In 1977 U.S. production capacity was about 4.4 million pounds [11]. Even though it is more expensive than the general-purpose synthetics, polychloroprene has maintained its position as an important specialty rubber because it resists most oils and greases as well as many chemicals, because it has a good temperature-utility spread, because it resists oxygen, ozone, and ultraviolet light, and because it does not propagate a flame. Its principal uses are in industrial rubber goods, automotive applications, and wire and cable.

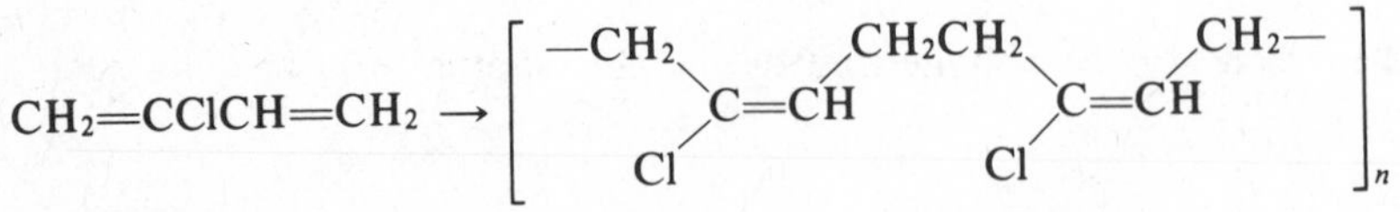

$$CH_2{=}CClCH{=}CH_2 \rightarrow \left[-CH_2-\underset{\displaystyle Cl}{C}{=}CH-CH_2CH_2-\underset{\displaystyle Cl}{C}{=}CH-CH_2- \right]_n \tag{20}$$

XVII. POLYMER BACKBONES CONTAINING ATOMS BESIDES CARBON

So far polymers whose backbones consist entirely of carbon atoms have been considered. These polymers are prepared entirely by addition reactions.

However, many polymers of tremendous commercial importance have atoms other than carbon, particularly oxygen and nitrogen, incorporated in their backbones. Most of them are prepared by condensation, rather than by addition, reactions.

Actually, any organic chemical reaction that can be made to proceed in essentially quantitative yield can be used to prepare high polymers. The two reactants merely have to be difunctional rather than monofunctional. An essentially quantitative yield is necessary or else the chains stop growing and a low-molecular-weight product results.

There are certain other corollaries to the manufacture of condensation polymers. The two monomers that are to be condensed must be used in as close to a stoichiometric ratio as possible. This means that both monomers must be extremely pure, far purer than most organic chemicals of commerce. Secondly, in many condensation reactions two products are produced. For instance, in the reaction between an acid and an alcohol, water is produced as well as the ester. In preparing a polyester this water must be removed. In an industrial plant, removing the last traces of such a co-product from a very viscous although molten polymer can be difficult.

XVIII. PROTEINS

Proteins are the building blocks of all mammalian bodies. They are polyamides based on the condensation of some 26 different α-amino acids.

$$\left[-\mathrm{NH}\underset{\mathrm{A}}{\underset{|}{\mathrm{C}}}\mathrm{HCONH}\underset{\mathrm{B}}{\underset{|}{\mathrm{C}}}\mathrm{HCONH}\underset{\mathrm{D}}{\underset{|}{\mathrm{C}}}\mathrm{HCONH}\underset{\mathrm{E}}{\underset{|}{\mathrm{C}}}\mathrm{HCO}-\right]_n \qquad [3]$$

Protein structures are extremely complex and of endless variety, since there is essentially no limit to the order in which the various amino acids can occur in the chain. The order and the nature of the amino acids determine the function of the specific protein in the mammalian body.

The chemist who does not become involved in protein chemistry as such is unlikely to have to deal with this class of polymers. However, it should be remembered that wool and silk are and have been for centuries very important articles of commerce.

XIX. SYNTHETIC POLYAMIDES

Everyone is familiar with nylon. This is the generic name for a class of synthetic polyamides in which relatively small (up to about 11 carbon atoms) and mostly aliphatic hydrocarbon chains are spaced regularly between amide linkages.

Polyamides of this type are tough, strong, easily molded, light in weight, chemically and abrasion resistant, and can be drawn into fibers that have good stretch properties. These polyamides find extensive use in clothing, carpets, tire cord, gears and bearings, electronic connectors, appliance housing, fishing lines, and rope. In 1978 U.S. production capacity of all types exceeded 3 billion pounds [12,13].

Commercially, two nylons are most important—nylon 6,6 and nylon 6. Nylon 6,6 is a condensation polymer prepared from adipic acid and hexamethylenediamine:

$$HO_2C(CH_2)_4CO_2H + H_2N(CH_2)_6NH_2 \xrightarrow{280°C} \left[-\overset{O}{\overset{\|}{C}}(CH_2)_4\overset{O}{\overset{\|}{C}}NH(CH_2)_6NH- \right]_n \tag{21}$$

Commercially, the preparation is essentially that shown in Equation 21. In order to ensure equal molarity between the two monomers, the hexamethylenediamine salt of adipic acid usually is prepared and purified prior to the final polymerization. In this typical condensation polymerization, water has to be removed continuously.

The Schotten–Bauman reaction is a good example of the fact that any organic chemical reaction that proceeds in essentially quantitative yield can be used to prepare high polymers. The interfacial technique for preparing polyamides that are unstable when molten utilizes this reaction. A water solution of the diamine and an alkali is stirred violently with a water-immiscible solution (often a chlorinated hydrocarbon) of the diacid chloride in question. The reaction proceeds so rapidly at the aqueous-organic interface that there is minimal hydrolysis of the acid chloride. Very high molecular weight polyamides can be obtained by this technique.

The other commercially most important polyamide is nylon 6. It is based on caprolactam:

$$\text{(caprolactam ring: } CH_2{-}CH_2{-}C{=}O,\ NH,\ CH_2{-}CH_2,\ CH_2) \xrightarrow[250°C]{H_2O,\ CH_3CO_2H} \left[-NH(CH_2)_5\overset{O}{\overset{\|}{C}}- \right]_n \tag{22}$$

The commercial process is the water-initiated addition of caprolactam molecules to the growing polymer chain. Since this is addition polymerization, no small molecule such as water is emitted during the process. Acetic acid is added to control the chain length. In the laboratory, metallic sodium can also be used as the initiator.

The largest use of the nylons is as fibers. For this purpose they are melt spun, that is, the molten polymer is forced through a spinnerette and is air cooled as it emerges. To achieve maximum physical properties, the resulting fiber is stretched (cold drawn) to three to six times its original length. This cold drawing orients the chains, sets up crystalline latices, and markedly improves the strength of the fiber.

XX. POLYESTERS

Of the vast number of polyesters that are chemically possible, the one that has achieved greatest commercial significance is polyethylene terephthalate, known commercially as Dacron. In 1978 U.S. production capacity was nearly 6.5 billion pounds [1,13]. It finds use in fibers (primarily apparel), film, tire cord, and carpets, to name a few applications.

Commercially it is prepared either by the direct esterification of terephthalic acid with ethylene glycol or by the trans esterification of dimethyl terephthalate with the glycol. The latter method is the older process.

$$CH_3OCO{-}C_6H_4{-}CO_2CH_3 + HO(CH_2)_2OH \xrightarrow[\textit{in vacuo}]{250\text{–}300°C} \left[{-}\overset{O}{\overset{\|}{C}}{-}C_6H_4{-}\overset{O}{\overset{\|}{C}}O(CH_2)_2O{-} \right]_n + 2CH_3OH \quad (23)$$

Here again a small, volatile co-product, methyl alcohol, must be removed from an increasingly viscous reaction mixture.

As was true with nylon 6,6, polyethylene terephthalate fibers are melt spun.

XXI. POLYURETHANES

Chemically related to the polyamides and to the polyesters are the polyurethanes. They are prepared by addition polymerization. They are extremely versatile and have good physical, chemical, and electrical properties. Commercially their largest use is as both flexible and rigid foams for such things as mattresses, furniture, rug underlay, building insulation, and various automotive and appliance applications. For flexible foams, 2,4-toluene diisocyanate is caused to react with a polyether (usually polypropylene glycol) of about 1000–2000 molecular weight:

$$\text{CH}_3\text{C}_6\text{H}_3(\text{NCO})_2 + \text{HO}(-\underset{\text{CH}_3}{\text{CHCH}_2\text{O}})_x\text{H} \xrightarrow{115°\text{C}}$$

$$\left[\text{CH}_3\text{C}_6\text{H}_3 \begin{matrix} \text{NHCO}_2(\underset{\text{CH}_3}{\text{CHCH}_2\text{O}})_x- \\ \text{NHCO}_2(\underset{\text{CH}_3}{\text{CHCH}_2\text{O}})_x- \end{matrix} \right]_n \quad (24)$$

Rigid foams are made from polyols based on treating glycerol or pentaerythritol with propylene oxide. Their rigidity stems from the built-in cross linking.

Foaming is achieved by treating a prepolymer prepared from the diisocyanate and the glycol and containing excess isocyanate groups with water or by the use of organic blowing agents:

$$\text{RNCO} + \text{H}_2\text{O} \longrightarrow \text{RNH}_2 + \text{CO}_2\uparrow \quad (25)$$

XXII. CELLULOSICS

At one time cellulose and its derivatives constituted by far the most important group of commercial polymers. Cellulosics still are of great importance in that rayon and cellulose acetate still rank in volume with nylon and polyester in the fiber field. Paper and cotton must not be forgotten.

From the point of view of the polymer chemist, the development of the cellulose-based industries is a superb example of macromolecular chemistry. Cellulose is a macromolecule, and the products derived from it are all based on treating it as a reactive organic chemical.

Cellulose is a β-D-glucopyranoside. In the following structural formula, the stereochemistry has been deliberately ignored in order to emphasize the chemical functionality of the molecule.

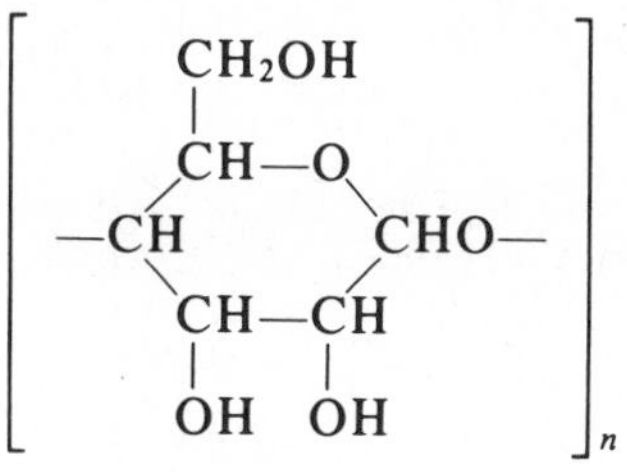

$$\left[\begin{array}{c} \mathrm{CH_2OH} \\ | \\ \mathrm{CH{-}O} \\ \diagup \qquad \diagdown \\ \mathrm{{-}CH} \qquad\qquad \mathrm{CHO{-}} \\ \diagdown \qquad \diagup \\ \mathrm{CH{-}CH} \\ | \qquad | \\ \mathrm{OH} \quad \mathrm{OH} \end{array} \right]_n \tag{4}$$

Cellulose possesses three reactive hydroxyl groups.

As the polymer itself, cellulose finds tremendous use as paper and paperboard and as the textile fiber cotton. The many chemical reactions performed upon the molecule in adapting it to the various uses in these two applications are way beyond the scope of this chapter.

Very early on it was found that for certain applications the physical properties of cellulose could be maximized by utilizing it in regenerated form. The commercial process (Viscose) involves solubilizing the cellulose as the xanthate and then regenerating it:

$$ROH + CS_2 \xrightarrow{NaOH} ROCSSNa \xrightarrow[H^+]{H_2O} ROH + CS_2 \tag{26}$$

As a result of these reactions, some degradation of the cellulose chains occurs so that their average molecular weight is reduced by at least half.

In the commercial process the cellulose xanthate solution is forced through spinnerettes into the acid bath to produce the rayon fiber. Rayon fiber finds use in apparel and as cords for tires, hose, and belting. On the other hand, if the hydrolysis is conducted at the outlet of a slit, a smooth, transparent sheet of cellophane is produced.

Again, very early on, it was found possible to nitrate cellulose to produce nitrate esters:

$$ROH \xrightarrow[H_2SO_4]{HNO_3} RONO_2 \tag{27}$$

The trinitrate, guncotton, was the original ingredient of smokeless powder. When only two of the hydroxyl groups are nitrated, the resulting polymer can be plasticized with camphor to give celluloid, the earliest broad-gauge thermoplastic. Because of its light sensitivity and obvious fire hazard, celluloid has been almost completely replaced by other synthetic polymers.

On the other hand, cellulose acetate still is a large article of commerce. Besides its extensive use as a fiber in apparel, its mechanical strength, impact resistance, hardness, toughness, and transparency still lead to its usefulness in many other applications. For instance, for a long time most photographic film was made of cellulose acetate. It is made as follows:

$$\mathrm{ROH + (CH_3CO)_2O \xrightarrow[CH_3CO_2H]{H_2SO_4} ROCOCH_3 + CH_3CO_2H} \tag{28}$$

In commercial practice the cellulose is acetylated completely and then hydrolyzed back with a little water to about the $2\frac{1}{2}$-acetate.

Finally, one more reaction of cellulose should be mentioned, namely, ether formation. A good example is the preparation of carboxymethyl cellulose, which finds use as a paper and textile size and as an antiredeposition agent in detergent formulations:

$$\mathrm{ROH + ClCH_2CO_2H \xrightarrow{NaOH} ROCH_2CO_2H} \tag{29}$$

XXIII. THERMOSETTING POLYMERS

The polymers in this group are characterized by a physically similar final form in which they are highly crosslinked and therefore hard and infusible. All of them can be prepared in a liquid or soft solid "prepolymer" state, which is molded, usually by heat and under pressure, into the finished shape. These polymers also often can be cast directly in the final form.

XXIV. PHENOLICS

The oldest and largest-volume group of thermosetting polymers is the phenolics. They are the condensation products of a phenol and an aldehyde. These polymers are strong, heat stable, impact and chemically resistant, and machinable. The most widely used member of the group, phenol-formaldehyde, finds use in electrical components, glues, binders, laminates, varnishes, and as impregnating resins. Resorcinol-formaldehyde finds particular use as an adhesive for plywood and tire cord.

One phenol-formaldehyde prepolymer is called a resol. It is prepared by causing about 1 mol of phenol to react with about 1.5 mol of formaldehyde at about 70°C in the presence of an alkaline catalyst. Its molecular weight is low, in the 500 range, and the chains are terminated by $\mathrm{—CH_2OH}$ groups.

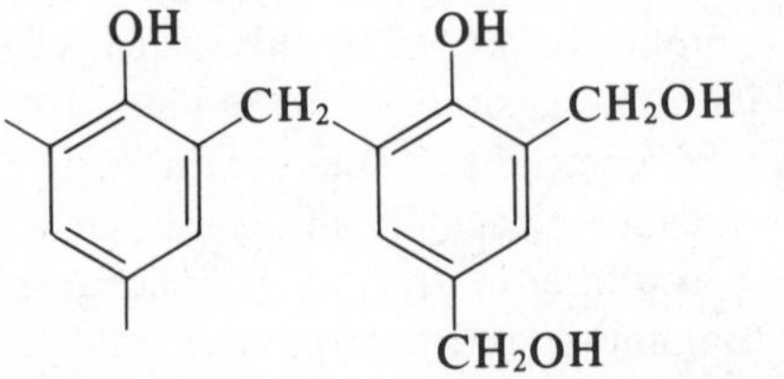

OH OH
$\mathrm{CH_2}$ $\mathrm{CH_2OH}$
$\mathrm{CH_2OH}$

[5]

The resol usually is mixed with a filler such as wood flour, more formaldehyde and catalyst are added, and the mixture is molded at about 100°C to achieve the final cure. The rings are held together by some —CH_2OCH_2— groups as well as by —CH_2— linkages.

Another phenol-formaldehyde prepolymer is called a novolak. Here the formaldehyde is heated with a slight excess of phenol at about 100°C in the presence of an acid catalyst such as oxalic acid. The molecular weight of the novolak is about 1200–1500, and it contains no free —CH_2OH groups.

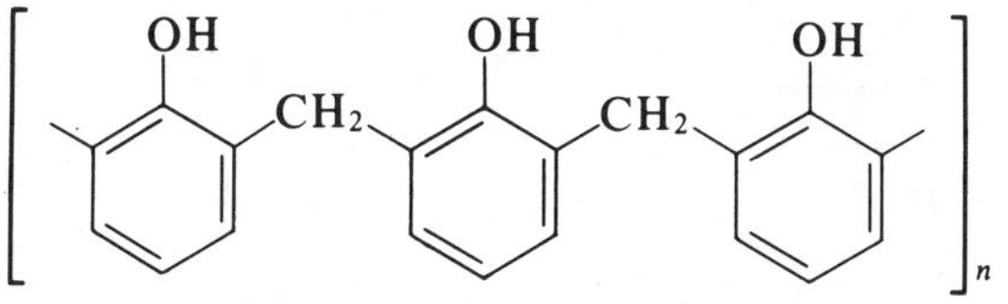

[6]

It is formulated and cured in the same way as the resol in the presence of an alkaline catalyst. In many of these cures, hexamethylenetetramine is used both as the formaldehyde source and as the alkaline catalyst. Under these circumstances there are some —CH_2NHCH_2— linkages in the final resin.

A cast phenolic may be prepared by heating phenol and formaldehyde in the presence of sodium hydroxide at about 70–80°C until the resin is barely pourable. It is then poured into a mold and cured at about 80°C.

The final phenol-formaldehyde resin has some such structure as

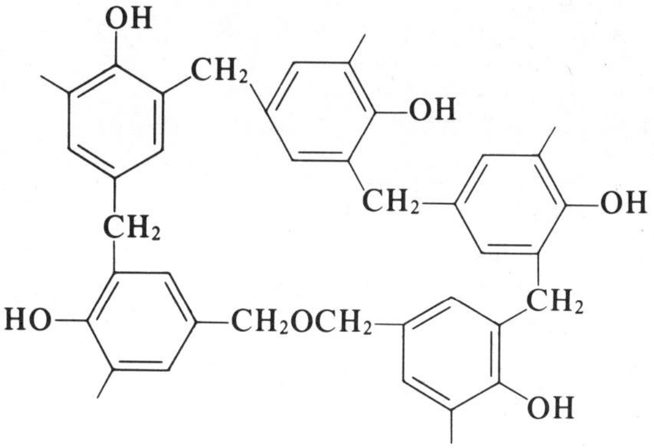

[7]

From the preceding discussion it must be obvious to any chemist that there are all kinds of recipes for preparing phenolic resins. Not only can other phenols such as the cresols and resorcinol be used alone or in combination, but other aldehydes such as acetaldehyde and furfural can be employed also. The reaction conditions can be varied widely, depending on the process being used and the application in mind.

XXV. UREAS

Urea-formaldehyde resins have chemical and solvent resistance, extreme surface hardness, and resistance to discoloration. They are used widely as plywood and furniture adhesives, as paper-coating agents, as textile-treating agents, and in molded dinnerware.

As in the case of the phenolics, a soluble form is first prepared, such as by treating urea with formaldehyde in the presence of an alkaline catalyst like sodium hydroxide at about 100°C. This form contains terminal —CH_2OH groups:

$$HOCH_2NHCONHCH_2OH \qquad [8]$$

As before, this soluble form is mixed with a filler such as cellulose and molded under pressure at about 150°C. In this case an acid catalyst is used for the final cure. The final polymer looks something like this:

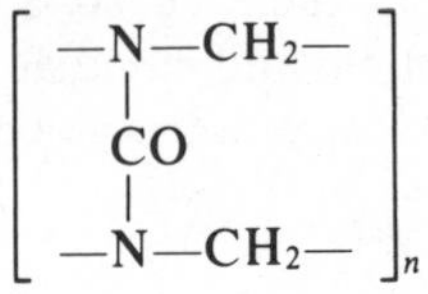

$$\left[\begin{array}{c} -N-CH_2- \\ | \\ CO \\ | \\ -N-CH_2- \end{array}\right]_n \qquad [9]$$

XXVI. MELAMINES

Melamine-formaldehyde resins are similar to the ureas in properties. They find especial use in tableware and decorative structures.

As in the case of the ureas, a soluble form containing —CH_2OH groups is first produced by treating melamine with formaldehyde in the presence of an alkaline catalyst at about 100°C. Again the final cure is effected in the presence of an acid catalyst at about 150°C. The final melamine-formaldehyde resin has much the following structure:

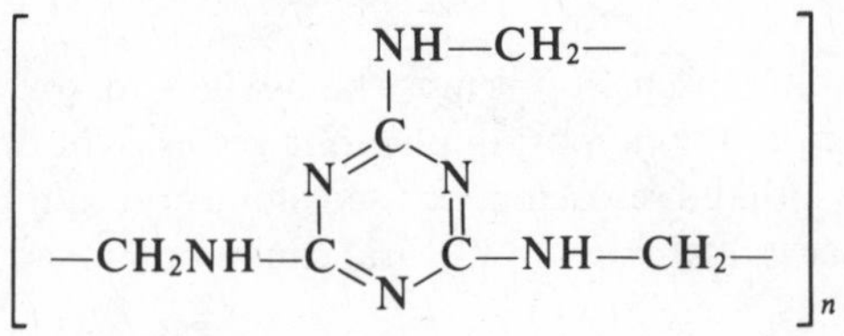

$$\left[\begin{array}{ccc} & NH-CH_2- & \\ & | & \\ & C & \\ N\!\!\!\!& & \!\!\!\!N \\ | & & \| \\ -CH_2NH-C & \!\!\!N\!\!\! & C-NH-CH_2- \end{array}\right]_n \qquad [10]$$

XXVII. EPOXIES

These resins have excellent chemical resistance, good adhesion, and excellent electrical properties. They are strong and tough and exhibit low shrinkage during cure. They find use as adhesives and coatings and in laminates and flooring.

The basis for most epoxies is the reaction product of bisphenol A and epichlorhydrin. For simplicity only the hydroxyl of the bisphenol A is shown in the following equation:

$$\mathrm{ROH} + \underset{\backslash\ \mathrm{O}\ /}{\mathrm{CH_2{-}CHCH_2Cl}} \xrightarrow{\mathrm{OH^-}} \mathrm{ROCH_2CHOHCH_2Cl} \xrightarrow{\mathrm{NaOH}} \mathrm{ROCH_2}\underset{\backslash\ \mathrm{O}\ /}{\mathrm{CH{-}CH_2}} \qquad (30)$$

More bisphenol A reacts with the epoxide group, and the reactions continue with a buildup of alternate groups along the chain. Since an excess of epichlorhydrin is used, the final fusible prepolymer is terminated by epoxide groups. It looks something like this:

$$\underset{\backslash\ \mathrm{O}\ /}{\mathrm{CH_2{-}CHCH_2}}{-}\left[{-}\mathrm{O}{-}\mathrm{C_6H_4}{-}\mathrm{C(CH_3)_2}{-}\mathrm{C_6H_4}{-}\mathrm{OCH_2CHOHCH_2}{-}\right]_n$$

$$-\mathrm{O}{-}\mathrm{C_6H_4}{-}\mathrm{C(CH_3)_2}{-}\mathrm{C_6H_4}{-}\mathrm{OCH_2}\underset{\backslash\ \mathrm{O}\ /}{\mathrm{CH{-}CH_2}} \qquad [11]$$

Curing is effected by means of a dibasic acid or anhydride or by an amine. Dibasic acids react both with the epoxide rings and with the hydroxyl groups along the chains to crosslink the chains together. Primary amines also react with the epoxide groups to effect crosslinking, and all three types of amines catalyze the reaction of the epoxides with the hydroxyl groups on other chains.

XXVIII. SILICONES

The final class of polymers to be considered comprises the silicones, which possess an inorganic backbone of alternating silicon and oxygen atoms. They can be both flexible and crosslinked. They have excellent electrical properties, are excellent repellents, and are inert and heat resistant. Silicones find use as rubbers, in laminates, and in water-repellent applications.

Flexible silicones are prepared essentially by hydrolyzing such compounds as $(CH_3)_2SiCl_2$:

$$(CH_3)_2SiCl_2 + H_2O \rightarrow HO\text{—}\left[\begin{array}{c} CH_3 \\ | \\ \text{—Si—O—} \\ | \\ CH_3 \end{array}\right]_n\text{—H} \qquad (31)$$

Crosslinked silicones are obtained by adding compounds such as CH_3SiCl_3 or $SiCl_4$ to the hydrolysis mixture.

$$(CH_3)_2SiCl_2 + CH_3SiCl_3 + H_2O \longrightarrow \begin{array}{ccccc} CH_3 & & CH_3 & & CH_3 \\ | & & | & & | \\ \text{—Si} & \text{—O—} & \text{Si} & \text{—O—} & \text{Si—} \\ | & & | & & | \\ CH_3 & & O & & CH_3 \\ CH_3 & & | & & CH_3 \\ | & & | & & | \\ \text{—Si} & \text{—O—} & \text{Si} & \text{—O—} & \text{Si—} \\ | & & | & & | \\ CH_3 & & CH_3 & & CH_3 \end{array} \qquad (32)$$

XXIX. POLYMER ANALYSIS [14]

In a pure sample of a monomeric chemical, all the molecules are identical, no matter how the sample was prepared. All the molecules have the same molecular weight, and every pure sample of the chemical in question has the same freezing point; the same boiling point; the same infrared, the same ultraviolet, and the same nuclear magnetic resonance spectra; the same crystalline form; the same elemental analysis; and the same solubility in various solvents.

Polymers are different. For instance, although all vinyl polymer samples have the same elemental analysis as the monomers from which they were prepared, the length and the structure of the macromolecular chains may vary markedly from sample to sample in accordance with the method of preparation. This means that the average molecular weight, the softening point, the spectra, the crystalline form, the solubility in various solvents, and the various physical properties may vary widely from sample to sample. Since the uses of polymers depend on their physical as much as on their chemical properties, it is important to be able to determine certain of these physical properties, so that a polymer with the desired characteristics can be manufactured reproducibly.

Obviously it is impossible to discuss in detail all the methods for characterizing polymers in one short chapter. All that can be done is to acquaint the beginning chemist with some of those key tests whose use and significance would probably not be familiar from the chemist's monomeric training. Fortunately, infrared and nuclear magnetic resonance spectroscopy differ little in their application between monomers and polymers, although reflectance spectroscopy as well as absorption spectroscopy is used occasionally in measuring the infrared spectra of polymers. Likewise, such polymer physical properties as tensile strength, elongation at break, creep, impact strength, and flexural strength are measured by mechanical instruments whose function should be obvious to any chemist or chemical engineer starting work in polymer chemistry.

In polymer characterization one of the most important values to determine is the molecular weight. When macromolecular chains are formed in a polymerization reaction, the molecular weights follow the statistical pattern shown in Fig. 1, where $\bar{M}_n$ is the number-average molecular weight. It is defined as

$$\bar{M}_n = \frac{\Sigma N_1 M_1}{\Sigma N} \tag{33}$$

It is a simple numerical average, where N_1 is the number of molecules of molecular weight M_1 and N is the total number of molecules present. The weight-average, or second-power number-average, molecular weight is defined as

$$\bar{M}_w = \frac{\Sigma w_1 M_1}{\Sigma w_1} = \frac{\Sigma N_1 M_1^2}{\Sigma N_1 M_1} \tag{34}$$

Here w_1 is the weight of particles of molecular weight M_1. Whether a number-average or a weight-average molecular weight is obtained depends on the method of measurement.

For low-molecular-weight polymers (molecular weight below ~25,000), end-group analysis can be used. It gives a number-average molecular weight. An end group that can be readily analyzed quantitatively must be present on each chain, and there should be no chain branching or loss of end groups during polymerization.

Number-average molecular weights of higher-molecular-weight polymers can be determined by osmometry. This can be done relatively easily by measuring the osmotic pressure of several dilute polymer solutions of different concen-

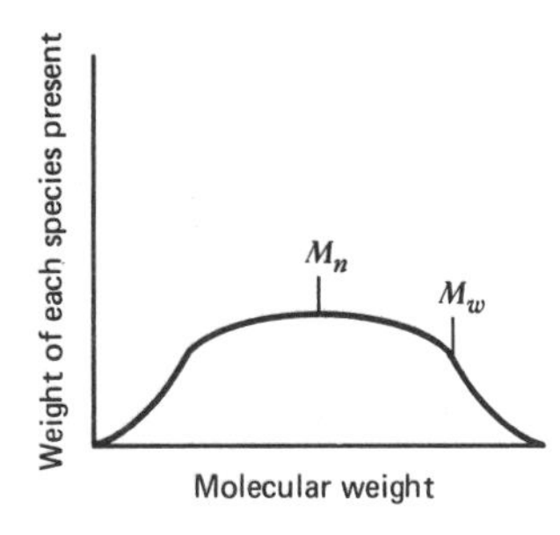

FIG. 1. Polymer molecular weights. Reprinted with permission from E. M. McCaffery, *Laboratory Preparation for Macromolecular Chemistry,* McGraw-Hill, New York, 1970, p. 174. Copyright 1970 by McGraw-Hill, Inc.

trations in an automatic recording osmometer. These pressures are plotted and extrapolated to zero concentration, from which value the number-average molecular weight can be calculated, since it is related directly to the osmotic pressure.

Gel-permeation chromatography can be used not only to fractionate polymers according to molecular weight, but also to measure number-average and weight-average molecular weights. The technique is based on the fact that the pore size of a crosslinked resin can be controlled. Thus by absorbing a polymer sample from solution on resins with different known pore sizes and then eluting, the molecular-weight distribution of the polymer can be determined and its number-average and weight-average molecular weights calculated.

Polymer fractionation and weight-average-molecular-weight measurement also can be effected by means of an ultracentrifuge, in which the fractions of different molecular weights are separated by centrifugal force. Polymers also can be fractionated by fractional precipitation, although, as any chemist knows, this is a very tedious procedure. A typical technique is to dissolve the entire polymer sample in a good solvent and then precipitate fractions by the stepwise addition of portions of a nonsolvent.

Light scattering, which depends on the size and shape of molecules, also can be used to obtain an absolute value of the weight-average molecular weight. The technique is based on the principle that when a beam of monochromatic light strikes a large particle in solution, not only is the light reemitted at different angles, but its intensity is also reduced markedly. The weight-average molecular weight can be computed by comparing the intensities of the incident and the scattered light.

The methods for determining polymer molecular weights described thus far require specialized expensive instrumentation. Although this apparatus is available, it often is more convenient in the laboratory to employ viscosimetry. This technique involves comparing the rate of flow through a capillary of dilute solutions of the polymer in question with that of the pure solvent. From these comparisons (relative viscosities) a molecular weight can be calculated. This molecular weight lies between the number-average and the weight-average molecular weight shown in Fig. 1. It usually is closer to the weight-average molecular weight. As background, it is necessary for one to determine the solubility of the particular polymer in a variety of solvents. This information, of course, is necessary for casting films and other applications involving shaping from solution.

Polymers can be amorphous, partly amorphous and partly crystalline, or almost entirely crystalline. Usually they are either wholly amorphous or partly crystalline. In general, polar groups and a regular backbone encourage crystallinity by facilitating packing and attraction between chains, whereas chain branching, crosslinking, and nonpolar groups tend to make a polymer amorphous.

The best way to determine crystallinity is by x-ray diffraction, a highly specialized technique involving expensive equipment. In the case of crystalline polymers, it is important to know the crystalline melting point, at which all crystallinity disappears. In the laboratory the easiest way to determine this is to use a hot stage polarizing microscope and see at what temperature the crystalline birefringence disappears. This will be over a small temperature range, since polymers are mixtures of macromolecular chains of varying molecular weight.

In the case of an amorphous polymer, it is very important to know the glass transition temperature. Below this temperature the polymer is a hard and brittle glass. Above it the polymer usually is soft and rubbery and can be fabricated. Crystalline polymers also have a glass transition temperature.

Both the glass transition temperature and the crystalline melting point can be determined by means of a differential thermal analyzer. In this apparatus the sample to be analyzed is heated next to an inert reference material. Heating is conducted at a constant rate, and the difference in temperature between the samples is recorded as heating progresses. Discontinuities in the resulting thermogram occur whenever a physical or chemical change takes place in the polymer, and show both the glass transition temperature and the crystalline melting point.

At the present time there is considerable interest in designing polymers for use at elevated temperatures. Therefore, one needs to know the behavior and, one hopes, the mechanism of what happens when a polymer is heated. This is accomplished by thermogravimetric analysis, in which loss in weight of a polymer as it is heated is plotted against the temperature. For mechanistic studies an inert atmosphere often is used; for applications studies the hostile air atmosphere (containing oxygen) usually is used. Obviously, the mechanism of decomposition can vary all the way from the simple unzipping of a vinyl polymer such as polymethyl methacrylate to an exceedingly complex group of reactions.

This last section of the chapter has deliberately been made simple. The mathematics and the detailed procedures can be found in reference 15 as well as in standard textbooks on polymer chemistry.

REFERENCES AND NOTES

1. Anon., *Mod. Plast.*, **56,** January, 1979, p. 66.
2. Anon., *Mod. Plast.*, **56,** January, 1979, p. 69.
3. Anon., *SPI Statistical Report,* October, 1978.
4. Anon., *Chem. Eng. News,* **56,** December 4, 1978, p. 12.
5. Anon., *Chem. Market. Report.*, **205,** April 8, 1974.
6. Anon., *Chem. Market. Report.*, **209,** April 12, 1976.
7. *Chemistry in the Economy,* American Chemical Society, Washington, D.C., 1973, p. 116.
8. Anon., *Chem. Week,* **121,** November 16, 1977, p. 45.
9. Anon., *Chem. Eng. News,* March 5, 1979, p. 11.
10. Anon., *Chem. Eng. News,* March 5, 1979, p. 10.
11. Anon., *Chem. Market. Report.*, **211,** January 15, 1977.

12. Anon., *Chem. Eng. News,* **56,** December 4, 1978, p. 11.

13. Anon., *Monthly Statistical Report,* SPI Committee on Resin Statistics, January, 1978.

14. Anon., *Chem. Eng. News,* **56,** December 4, 1978, p. 10.

15. Details can be found in W. R. Sorenson and T. W. Campbell, *Preparative Methods of Polymer Chemistry,* Interscience, New York, 1961, and E. M. McCaffery, *Laboratory Preparation for Macromolecular Chemistry,* McGraw-Hill, New York, 1971.

6

INDUSTRIAL CHEMICAL RESEARCH

I. INTRODUCTION

The next few chapters deal with industrial chemical research and industrial chemical development. It is hoped that this will give the reader some idea of how research discoveries in industry and even in academia have led to important commercial developments.

Research is a very much overworked word. Some companies have used it to include all their technical efforts, particularly to impress potential investors. In this book, the very simple definition *technical effort directed toward the discovery of new scientific facts* is used. No attempt is made to try to classify research as basic, fundamental, applied, pure, or impure.

Research is subjective, inspirational, and often irrational [1]. Much of the thinking occurs in the subconscious mind. All chemists have woken up in the middle of the night with a good idea or with the solution to some problem. Chemical research is creative. The emotional involvement is similar to that of the writer and of the artist.

Although the descriptions of research investigations in scientific journals read as though the authors had planned their entire study in advance and gone directly and logically to their goal ($A \rightarrow B$), the path usually was anything but that, and the ultimate goal achieved often was quite different from that originally planned ($A \rightarrow C$) (see Fig. 1). That does not mean that C was not as great an achievement as B would have been.

Research produces scientific knowledge, not products and processes. It is the use of research results through development and innovation that creates the new products and the new processes and therefore the profits. Very high caliber research is conducted in the chemical industry.

Since the results of a research investigation are seldom known in advance, research is very high risk. A research program is difficult to control and even more difficult to measure in any quantitative fashion. It is dynamic and creates

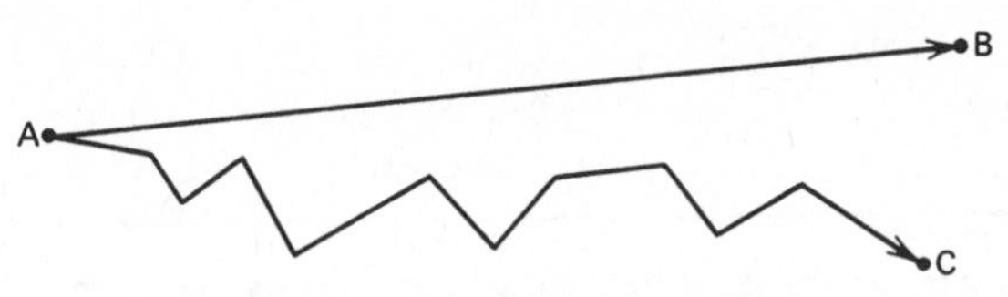

FIG. 1. Paths of research progress.

change, a phenomenon not always welcome in a company. In research, negative results can be very important, and this can be quite confusing to business executives.

Research depends on accurate observation and often can and should be very opportunistic. Years ago in Monsanto Company's then central research laboratory, a friend of the author's prepared a compound designed to compete with another company's water repellent. The report from the applications laboratory showed it was not a water repellent, but a very good detergent. At that time it had no commercial promise because it was sudsless. Then the big use for detergents was in dishwashing formulations, in which it was necessary to have an inch of suds on top of the dishwater so that the dishwasher could not see the food scraps floating around. Two or three years later a representative of Westinghouse Electric Corp. came to Monsanto Company with an automatic washing machine. They could not sell it because after a few minutes of use with a standard detergent there were several inches of suds on the laundry floor. The experimental nonsudsing detergent was pulled off the shelf and was developed into the commercially important product All.

It is people and not money that is the real guarantee of success in research. Projects are best evaluated on the past performance of the individual involved, not by some systematic and supposedly logical means. Vision, enthusiasm, dedication, and persistence are extremely important attributes of a good research worker.

II. DEFENSIVE RESEARCH

Research can be divided into two types—offensive and defensive. Defensive research is directed toward maintaining the present business. It is much the commoner of the two in industry. In very large companies it usually is found in divisional laboratories. Business executives support it because they have to and because they can easily see its value if successful.

In Chapter 3 the effect of new processes on the cost and profitability of several tonnage chemicals was shown. Following are two specific examples in which it might prove profitable to study the mechanism of a reaction in detail or even try to find a new reaction.

In 1978 U.S. production capacity for phenol was 3.46 billion pounds. The selling price was $0.19 per pound [2]. If a company possessed 10% of the market, a not unreasonable assumption since five of them did [2], that would mean 346 million pounds of sales. A saving of $0.01 per pound would be worth $346,000 per year. In 1976 U.S. production capacity for terephthalic acid and its dimethyl ester was about 5.85 billion pounds. The selling price was of the order of $0.21 per pound [3]. If a company possessed 20% of the market, a not unreasonable assumption since three of them did [3], that would mean 1.17 billion pounds of sales. A saving of $0.01 per pound would be worth $1.17 million per year.

The rewards for success in such a research investigation would be high and would justify an appreciable effort. In contrast, success on the part of a competitor could be extremely costly.

III. OFFENSIVE RESEARCH

Offensive research is directed toward company expansion by means of new products and product lines, and perhaps even through new businesses. In very large companies offensive research usually is found in central research laboratories. To be successful commercially it must have the support of top management, corporate planning, and central marketing.

Over the years there has been an interesting evolution in the basis for selecting offensive research projects, which has very materially increased the chances for commercial success of such investigations.

A. Serendipitous Approach

Originally reliance was placed on serendipity. A brilliant scientist was turned loose in the laboratory pretty much with the instruction to "discover something." The best known example of this approach is the work of Wallace H. Carothers on polyamides, which led to the development of nylon.

B. Raw-Material Approach

The next approach to offensive research was to start from a cheap and readily available raw material. Basically the approach was to make a tank-car amount of each promising derivative, advertise its availability, and hope the potential customers would find the uses. As shown in Chapter 4, Union Carbide Corp. exploited this approach very successfully, starting with acetylene and then with ethylene, so that the company markets more than 124 pure organic chemicals based largely on these two raw materials and the technology developed in this field [4].

As mentioned in Chapter 3, Monsanto Company nitrates chlorobenzene. This has led to a wide variety of raw-material-based products as shown schematically in Figs. 2–4. Uses are listed under some of the products [5,6].

Of the four approaches to selecting offensive research projects described in this chapter, the author followed the last three (not the first) successively in his

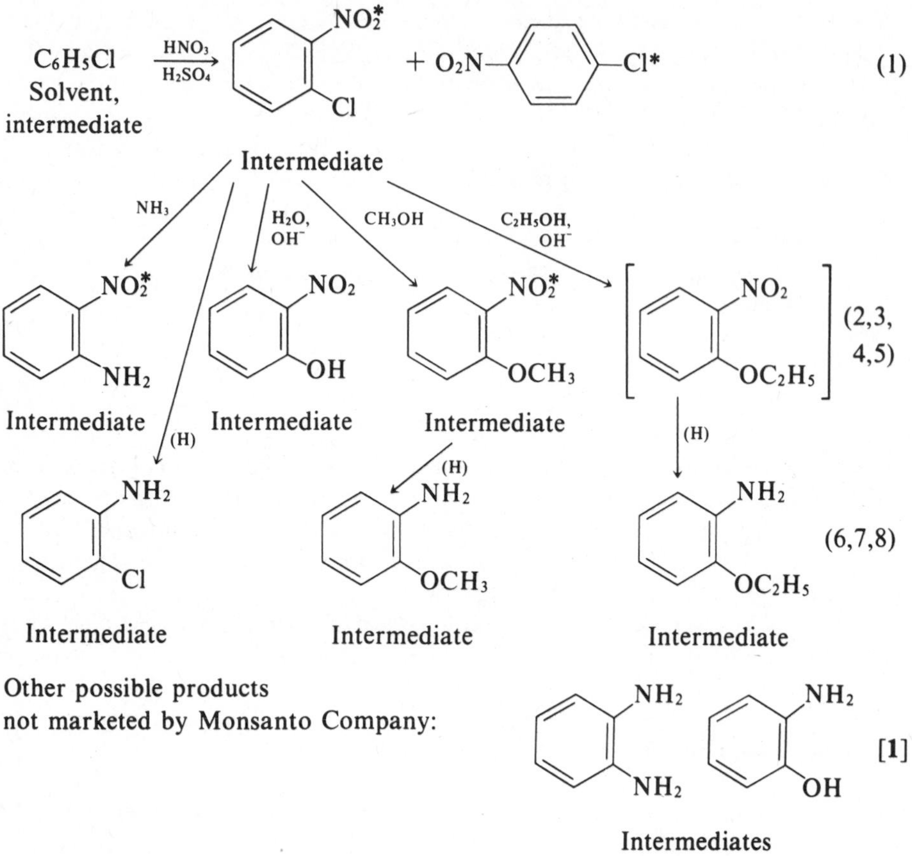

C_6H_5Cl $\xrightarrow[H_2SO_4]{HNO_3}$ (o-nitrochlorobenzene: NO_2^*, Cl) + O_2N–C_6H_4–Cl^* (1)

Solvent, intermediate

Intermediate

NH_3 → (NO_2^*, NH_2) Intermediate

(H) → (NH_2, Cl) Intermediate

H_2O, OH^- → (NO_2, OH) Intermediate

CH_3OH → (NO_2^*, OCH_3) Intermediate; (H) → (NH_2, OCH_3) Intermediate

C_2H_5OH, OH^- → [(NO_2, OC_2H_5)] (2,3, 4,5); (H) → (NH_2, OC_2H_5) (6,7,8) Intermediate

Other possible products not marketed by Monsanto Company: (NH_2, NH_2) (NH_2, OH) [1]

Intermediates

FIG. 2. *o*-Nitrochlorobenzene derivatives.

work at Monsanto Company. One of the earliest assignments was based on the raw-material approach; the problem was whether monomeric styrene could serve as the source of industrial organic chemicals. With the advent of synthetic rubber, monomeric styrene would be cheap and available in tremendous quantities.

Styrene can be chlorinated:

$$C_6H_5CH{=}CH_2 + Cl_2 \xrightarrow{CCl_4} C_6H_5CHClCH_2Cl + C_6H_5CH{=}CHCl \qquad (30)$$

It can be hypochlorinated:

$$C_6H_5CH{=}CH_2 + HOCl \longrightarrow C_6H_5CHOHCH_2Cl \qquad (31)$$

This led to the interesting chemistry shown in Fig. 5 [7]. Wherever the letter *P* is shown beside a reaction, a patent was obtained (a total of five).

The raw-material approach has been almost completely abandoned. It was a scientific success and a commercial failure. In this case the chemistry was a lot of fun and five patents were obtained easily. But the only product for which a commercial application was known was phenylacetaldehyde, and any process involving chlorine eliminated its sole use, as a perfume ingredient. (A very few chlorine atoms ruin the aroma.) Also, as will be shown shortly, acetophenone can be made much more cheaply in another way.

C. Technical-Field Approach

The third approach to offensive research, which has proved highly viable, was to investigate a specific technical field. This is why petroleum companies study hydrocarbon chemistry and catalysis, and why aluminum companies study new alloys. Much of the success of this method involves the jigsaw puzzle concept of science. If a chemist finds some of the pieces and starts putting them together, other people will find other pieces that fit also.

Perhaps the most successful exploiters of this technical-field approach have been the pharmaceutical companies that synthesize compounds with expected biological activity and couple this synthesis with detailed observations in pharmacology, drug metabolism, biochemistry, and biology. It is this close coordination of the several disciplines involved that has made this research effort so outstandingly successful. In 1933, the year before the author graduated from college, there were no vitamins, no sulfa drugs, no antibiotics, no corticosteroids, no polio, measles, or mumps vaccines, no oral diabetic drugs, no anticoagulant therapy, no antihypertensives, no antihistamines, no tranquilizers, no antidepressants, no agents for gout control, no nonaddicting morphine congeners, no thiazide diuretics, no "pill", no blood plasma, no gamma globulin [8]. All this progress has taken only 40 years.

Very early in his career with Monsanto Company, the author was asked to evaluate the Union Carbide Corp. process for the manufacture of styrene. This opened up the whole field of the liquid-phase oxidation of alkylaromatic hydrocarbons.

The present route to styrene, and the one Monsanto Company was then using, is

$$C_6H_5C_2H_5 \xrightarrow[\substack{600\text{–}660°C, \\ H_2O \text{ diluent}}]{SiO_2 \cdot Al_2O_3} C_6H_5CH{=}CH_2 + H_2 \tag{41}$$

Since the conversion is only about 35%, the purification by fractional distillation is horrendous. Ethylbenzene boils at 136°C and styrene at 145°C, so tremendous fractionating columns are needed. The only reason this separation is feasible is that Dow Chemical Company discovered that sulfur, a styrene polymerization inhibitor, can be added at the top of each distillation tower and flows down through it. Otherwise the styrene would polymerize in the column, a terrible mess.

H_2N–C_6H_4–Cl ← (H) — O_2N–C_6H_4–Cl* — Cl_2 → O_2N–C_6H_3(Cl)–Cl (9,10)

Intermediate | Intermediate

(H) ↓

H_2N–C_6H_3(Cl)–Cl* (11,12, 13,14)

Intermediate

H_2O, OH^- → O_2N–C_6H_4–OH*

CH_3OH, OH^- → [O_2N–C_6H_4–OCH_3] ↓ (H) H_2N–C_6H_4–OCH_3

Intermediate

C_2H_5OH, OH^- → [O_2N–C_6H_4–OC_2H_5] ↓ (H) H_2N–C_6H_4–$OC_2H_5^*$ (15.16)

Intermediate

↓ $(CH_3CO)_2O$

CH_3CONH–C_6H_4–$OC_2H_5^*$ (17.18)

Pharmaceutical

O_2N–C_6H_4–OH* ↓ Na salt + Cl–P(=S*)(OC_2H_5)(OC_2H_5) Intermediate

O_2N–C_6H_4–OP(=S*)(OC_2H_5)(OC_2H_5)

Insecticide
(Parathion)

Na salt

\+

$Cl{-}P(=S^*)(OCH_3)(OCH_3)$

Intermediate

$O_2N{-}C_6H_4{-}OP(=S^*)(OCH_3)(OCH_3)$ (19)

Insecticide
(Methyl Parathion)

Other possible products not marketed by Monsanto Company:

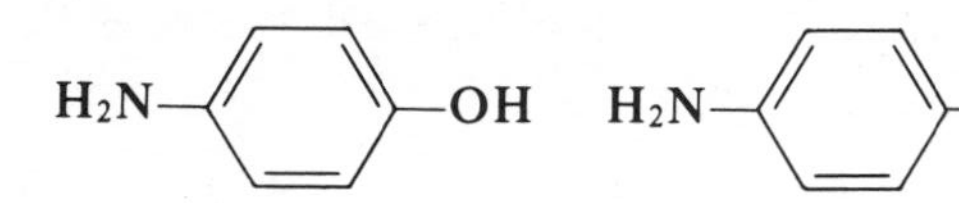

[2]

Intermediates

FIG. 3. *p*-Nitrochlorobenzene derivatives.

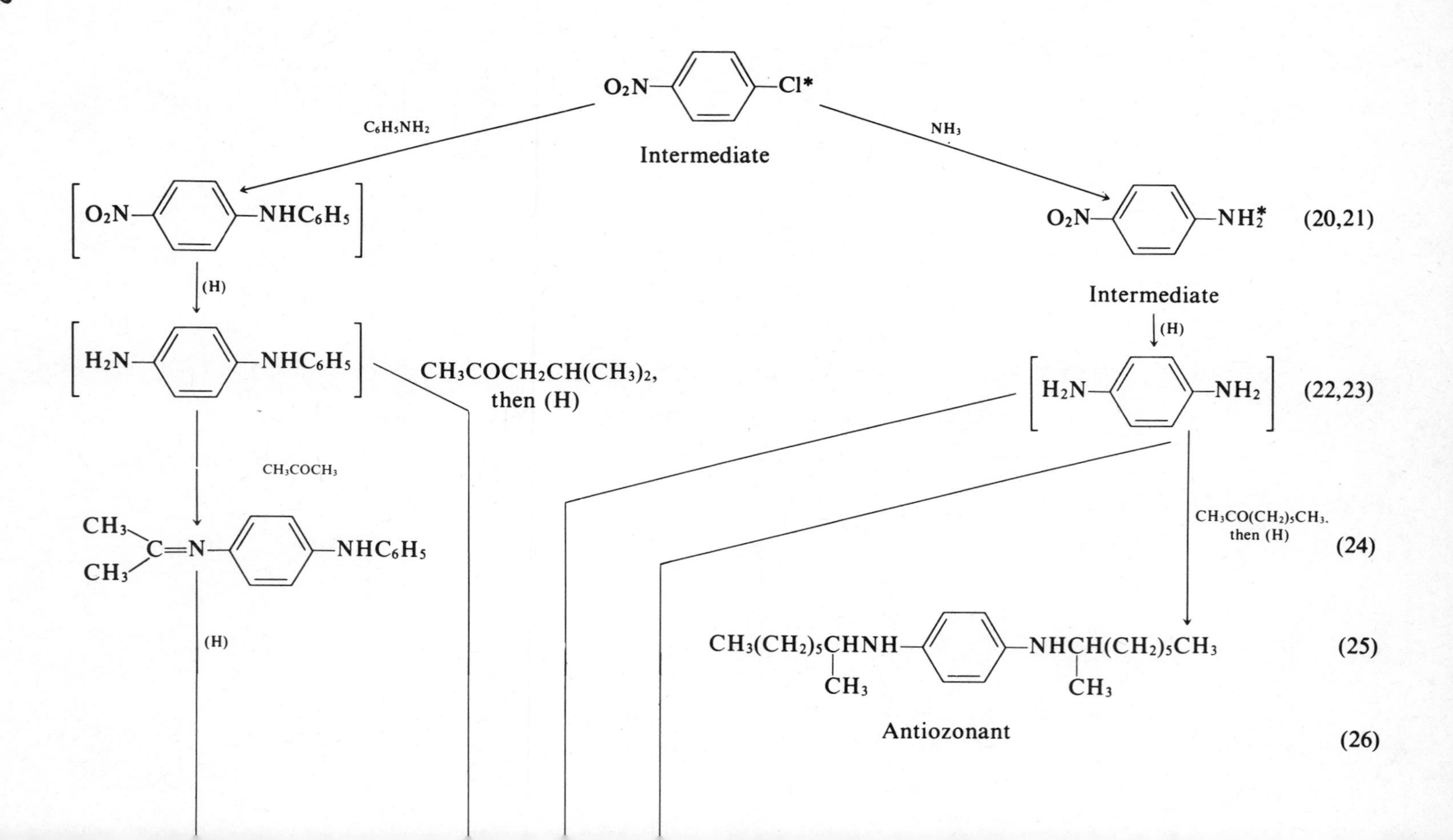

O2N–C6H4–Cl*
Intermediate
C6H5NH2
NH3
O2N–C6H4–NHC6H5
(H)
H2N–C6H4–NHC6H5
CH3COCH2CH(CH3)2,
then (H)
CH3COCH3
(CH3)2C=N–C6H4–NHC6H5
(H)
O2N–C6H4–NH2*
(20,21)
Intermediate
(H)
H2N–C6H4–NH2
(22,23)
CH3CO(CH2)5CH3,
then (H)
(24)
CH3(CH2)5CH(CH3)NH–C6H4–NHCH(CH3)(CH2)5CH3
(25)
Antiozonant
(26)

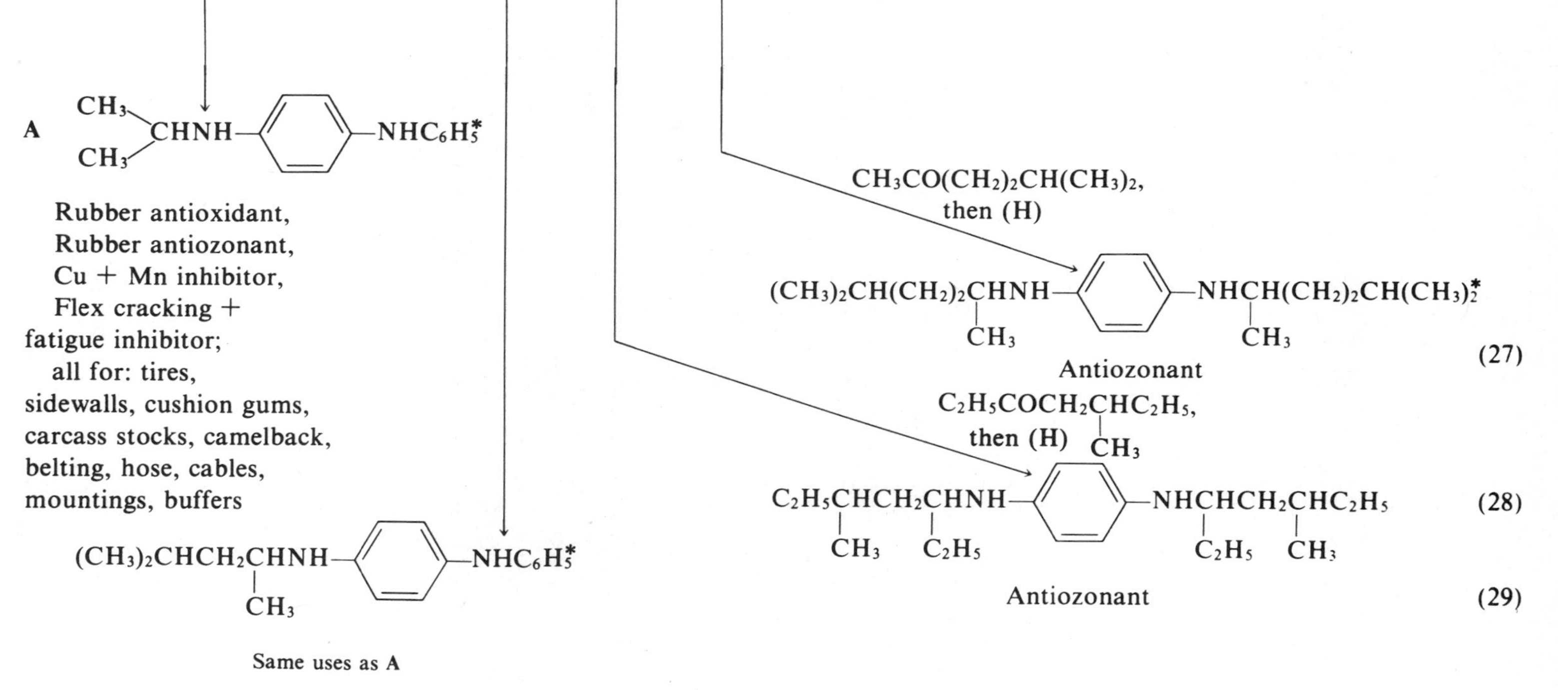

FIG. 4. *p*-Phenylenediamine derivatives.

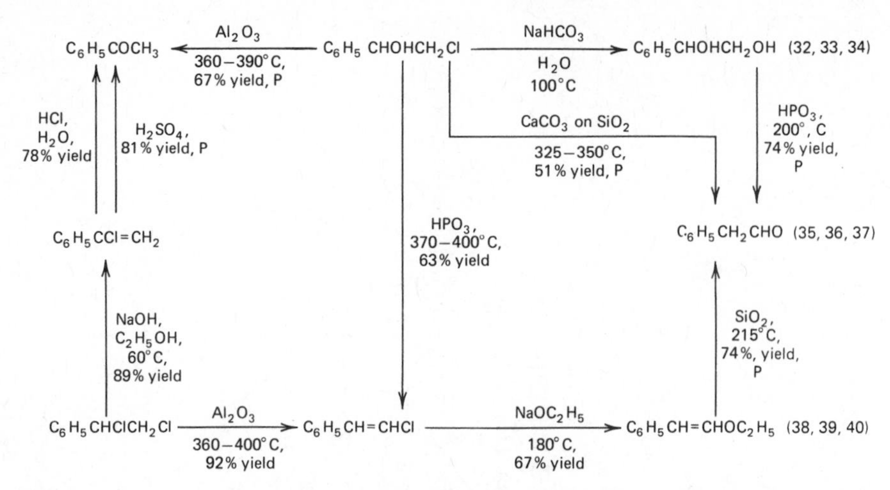

FIG. 5. Chlorinated styrene chemistry. Reprinted with permission from W. S. Emerson and E. P. Agnew, *J. Am. Chem. Soc.* **67,** 1945, p. 518. Copyright 1945 by American Chemical Society.

The Union Carbide Corp. process involved three steps, but none of the separations are difficult, since acetophenone boils at 162°C/12 mm and its hydrogenation is essentially quantitative.

$$C_6H_5C_2H_5 \xrightarrow[\text{liquid phase}]{\text{air}} C_6H_5COCH_3 \xrightarrow[\text{catalyst}]{H_2} C_6H_5CHOHCH_3 \xrightarrow{-H_2O} C_6H_5CH{=}CH_2 \quad (42)$$

Various catalysts can be used for the oxidation, of which the most effective are Cr_2O_3, $Co(OH)_2$, MnO_2, V_2O_5, $Ca(OH)_2$, CuO, and PbO. Chromium oxide produced completely reproducible results just as spooned out of the reagent bottle, but conversions and yields using manganese dioxide varied considerably, depending on its method of preparation.

Although not competitive with the catalytic cracking of ethylbenzene for styrene manufacture, the route shown in Equation 42 is a very convenient one for the preparation of substituted styrenes:

$$C_2H_5{-}C_6H_4{-}OCOCH_3 \xrightarrow[Cr_2O_3]{\text{air}} CH_3CO{-}C_6H_4{-}OCOCH_3$$

24% conversion, 79% yield

$$\xrightarrow{\text{2 steps}} CH_2{=}CH{-}C_6H_4{-}OCOCH_3 \quad \text{[Ref. 9]} \quad (43)$$

$$C_2H_5-C_6H_4-CO_2CH_3 \xrightarrow[Cr_2O_3]{air} CH_3CO-C_6H_4-CO_2CH_3$$

54% conversion, 66% yield

$$\xrightarrow{\text{2 steps}} CH_2{=}CH-C_6H_4-CO_2CH_3 \quad \text{[Ref. 10]} \quad (44)$$

$$C_2H_5-C_6H_4-CH_2OCOCH_3 \xrightarrow[Cr_2O_3]{air} CH_3CO-C_6H_4-CH_2OCOCH_3$$

23% conversion, 55% yield

$$\xrightarrow{\text{2 steps}} CH_2{=}CH-C_6H_4-CH_2OCOCH_3 \quad \text{[Ref. 11]} \quad (45)$$

This research represents a transition from the investigation of a scientific field to the fourth approach, research directed toward a specific use. In this case Monsanto Company was investigating a wide variety of vinyl copolymers in order to improve the properties of plastics and of synthetic rubber.

Although none of the monomers in Equations 43–45 proved to be commercially promising, there was an interesting and attractive by-product of the research. Two other oxidations were effected [12]:

$$o\text{-}NO_2C_6H_4C_2H_5 \xrightarrow[Cr_2O_3]{air} o\text{-}NO_2C_6H_4COCH_3 \quad (46)$$

$$O_2N-C_6H_4-C_2H_5 \xrightarrow[Cr_2O_3]{air} O_2N-C_6H_4-COCH_3 \quad (47)$$

Monsanto Company's patent department decided not to file a patent application, since applications had been filed covering several similar oxidations and these last two were just more of the same. The editor of the *Journal of the American Chemical Society* agreed with Monsanto Company's patent department. Publication of the oxidations shown in Equations 46 and 47 finally was achieved as a short note only because someone in the *Journal of the Chemical Society* (London) had said the reactions would not work [13]. It turned out that *p*-nitroacetophenone was the key intermediate for Chloromycetin, made by Parke, Davis and Company,

$$O_2N{-}C_6H_4{-}CHOHCH(NHCOCHCl_2)CH_2OH \qquad [4]$$

and the publication of this note led to several million dollars' worth of business for Monsanto Company. It can be profitable for a company to publish the results of its research. This is an excellent example of the jigsaw puzzle concept; two pieces discovered in separate laboratories fitted together beautifully.

D. Fulfilling-Need Approach

The fourth approach to selecting offensive research projects, which is most generally in use today, is research directed toward fulfilling a specific need, when it would be profitable businesswise to effect a technological change. The market has to be anticipated, and it is highly desirable to stress proprietary areas in which sales will expand rapidly.

For many years Monsanto Company has served the rubber industry with antioxidants and vulcanization accelerators. When white sidewall tires became fashionable, it was necessary to discover nondiscoloring antiozonants. This apparently was successful.

Success with the oxidation of ethyl groups attached to an aromatic ring led to the question of whether the technique could be applied to methyl groups similarly situated. Here there was a ready market, because at that time Monsanto Company was selling benzoic acid at the price of $0.45 per pound. The company was preparing it by the decarboxylation of phthalic anhydride.

$$C_6H_4(CO)_2O \xrightarrow[\text{catalyst, } H_2O]{\Delta} C_6H_5CO_2H + CO_2 \qquad (48)$$

Air oxidation of toluene would be much cheaper.

$$C_6H_5CH_3 \xrightarrow[\text{catalyst}]{\text{air}} C_6H_5CO_2H + H_2O \qquad (49)$$

The cost calculations are most illuminating and provided a tremendous stimulation to the investigation (Table 1) [14]. The possible economic advantage was obvious. Moreover, since the company already was manufacturing benzoic acid, there was no need to develop specifications and quality-control proce-

TABLE 1. **Benzoic Acid Raw-Material Costs**

Compound	Molecular Weight	Cost/lb
Phthalic anhydride	148	\$0.11
Toluene	92	0.032
Benzoic acid	122	0.45

Phthalic anhydride: 148 × 0.11 = \$16.30
Toluene: 92 × 0.032 = 2.86

Raw-Material Costs for Benzoic Acid in Comparison with Yields

	100% = 122 lb	75% = 92 lb	50% = 61 lb
Phthalic anhydride	16.30/122 = \$0.134	16.30/92 = \$0.177	16.30/61 = \$0.267
Toluene	2.86/122 = 0.024	2.86/92 = 0.031	2.86/61 = 0.047

dures, and purification problems were expected to be minimal. Of the ethyl-group-oxidation catalysts, only $Co(OH)_2$ and PbO were effective. Cobalt was selected over lead since the latter produced a larger amount of tars and "high boilers."

Because of its low boiling point (110°C), the liquid-phase oxidation of toluene must be conducted under pressure. With the two experimental autoclaves available in the laboratory, the surprising and very disconcerting phenomenon was observed that oxidation took place regularly and reproducibly in one autoclave but never in the other. The reason was never discovered. Although the problem was circumvented by the use of a soluble cobalt salt as catalyst, the organic chemicals division of Monsanto Company understandably was slow in picking up the process from the central research department for fear they would build a plant constructed of the same materials as the negative autoclave.

The successful liquid-phase oxidation of toluene to benzoic acid led logically to the question of whether it would be desirable to oxidize other methyl aromatic compounds. The oxidation of the three xylenes to the corresponding toluic acids is even easier than the oxidation of toluene. Developmental quantities of these acids were prepared and offered to the trade. With no applications known, the sole result of this effort was to stock the shelves of most of the repackaging outfits in the country.

At that time polyester fibers were just being commercialized, thus creating a need for terephthalic acid. Its preparation by the liquid-phase oxidation of *p*-xylene was investigated. Unfortunately, at a certain point this oxidation stops (although at a respectable conversion to *p*-toluic acid) since, with this acid melting at 180°C, the oxidation mixture becomes extremely viscous. The Hercules, Inc. chemists (Chapter 3) circumvented this problem by esterifying the *p*-toluic acid with methanol at this point and continuing the oxidation. Even here the fact that terephthalic acid melts at 425°C (under pressure), sublimes at over 300°C, and is not soluble in much of anything necessitates its purification through the much more tractable dimethyl ester.

Two other possibilities were investigated at Monsanto Company. Applying the coal-oxidation technique, it was found that aqueous potassium *p*-toluate can be oxidized to potassium terephthalate in up to 77% conversion and yield. Since the reaction required a temperature of 260–275°C and an applied pressure of 1000 psi of pure oxygen, it was economically unattractive. A liquid-air plant would have to have been part of the manufacturing operation. Also with the Teflon packing then available, the stuffing box problems in the stirred autoclave were horrendous [15].

Another way to burn off methyl groups was found. In the presence of a catalyst comprising cobalt, lead, and manganese acetates, with pure oxygen as the oxidizing agent in boiling butyric acid solution, *p*-xylene easily gave a 56% yield of *p*-toluic acid. Although the technique was unattractive to the laboratory neighbors, it was nonetheless effective and of some general applicability, even though it did not lead directly to terephthalic acid [16].

At this point the situation was evaluated, and it was decided to abandon the program. The technically most promising route to terephthalic acid appeared to be liquid-phase oxidation of *p*-xylene in aliphatic acid solution. Commercially this would have to be done in glacial acetic acid under pressure, for which at that time only tantalum-lined equipment could be used, hardly an economical material of construction. (As discussed in Chapter 3, the present commercial process for oxidizing *p*-xylene to terephthalic acid in acetic acid under pressure had to await the development of metallurgical techniques for forging titanium.) On the business side of the deliberations, there was only one terephthalic acid customer in the United States, and it was known that several other potential suppliers were developing processes for its production. To be one of several suppliers to the only customer is not an attractive business situation. The investigation was dropped.

IV. TRANSITION TO DEVELOPMENT

A brief study in the liquid-phase oxidation field illustrates the transition from research to process development. At one point it appeared that *m*-chlorostyrene might be of commercial interest to Monsanto Company. It ultimately was not, but two possible routes to its manufacture were delineated (Fig. 6) [17].

Although not an article of commerce, *m*-chloroethylbenzene could have been manufactured easily. In the presence of excess aluminum chloride, the orientation in a Friedel-Crafts alkylation becomes meta rather than the ortho-para that results when catalytic quantities of aluminum chloride are used. In this instance the alkylation of chlorobenzene with ethylene could have been commercially attractive if a sufficiently large volume use for *m*-chlorostyrene had developed.

Since no process work was ever conducted on the two possible routes to *m*-chlorostyrene, the potential development problems can be examined from a totally impartial point of view. First come raw-material costs on the arbitrary

C_2H_5 (Cl) —air, Cr_2O_3, 140–155°C→ $COCH_3$ (Cl) —H_2, Cu Chromite, 140°C, 980 psi→ $CHOHCH_3$ (Cl) 94% yield

26% conversion, 76% yield

↓ Cl_2, PCl_3, UV

$CHClCH_3$ (Cl) —$CaSO_4$, 425–475°, 93% yield→ $CH{=}CH_2$ (Cl)

44% conversion, 91% yield

↓ Al_2O_3, 313–340°C, 84% yield

(50,51)

FIG. 6. Routes to *m*-chlorostyrene.

basis of the then-possible cost of $0.25/per pound for *m*-chloroethylbenzene (Table 2). That gives the chlorination route a $0.12/per pound advantage based on yield. In addition, the chlorination route is two steps, whereas the oxidation route is three steps. In both cases purification of a chemical by distillation under reduced pressure probably would be necessary.

In both routes a conversion involving recycling is necessary. The chlorination of *m*-chloroethylbenzene has to be stopped at about 50% conversion to avoid the formation of dichloroethylbenzenes. The oxidation has to be stopped because the rate becomes too slow to be practical and appreciable side reactions occur (76% yield at 26% conversion). Residence time is expensive. In a detailed evaluation of the two routes, this oxidation step is the one that would require the most study in an effort to increase both yield and conversion.

There would be important investment considerations. Hydrogen chloride is the by-product in the chlorination route, and equipment would have to be pro-

TABLE 2. *m*-Chlorostyrene Raw-Material Costs

Compound	Molecular Weight
m-Chloroethylbenzene	140.6
m-Chlorostyrene	138.6

At 100% yield raw materials = 140.6 × 0.25 = $35.15

$\$^{35.15}/_{138.6} = \$0.254/\text{lb}$

Significance of Yields

% Yield	Raw-material cost/lb
60 (Oxidation route)	$^{35.15}/_{83} = \$0.42$
85 (Chlorination route)	$^{35.15}/_{118} = 0.30$

vided to dispose of it. Likewise its presence probably would necessitate the use of expensive alloy steel in both steps.

In the oxidation route, pressure equipment would be required in which to conduct the hydrogenation, as well as a hydrogen compressor, and this would be an appreciable additional investment if hydrogenations were not already being conducted in the plant. The investment in an air compressor for the oxidation probably could be avoided by drawing air through the hot *m*-chloroethylbenzene at slightly reduced pressure.

V. SUMMARY

In this chapter the nature of research was mentioned, and defensive research (and why it is the kind most commonly found in industry) was described. Offensive research in relation to the four successive historical approaches was discussed. Finally, the kind of thinking that goes into translating research results into an industrial process was illustrated.

It has been shown that good chemical research is conducted in industry. Economic considerations have been emphasized as well as the need for utility. If no one can use a chemical, there is no sense in making it. Therefore it is the job of the industrial chemists "to learn how to make new reactions go with commercially attractive yields under conditions that can be established economically to make products for which there is a demand or a potential demand" [18]. An industrial chemist must devise products and processes that will be profitable in the company's field of interest. This is not a restriction but a challenge, just as the challenge in academia is to train students to produce research of scientific merit.

REFERENCES AND NOTES

1. This idea is explored in some detail by D. A. Schon, *Technology and Change*, Delacorte Press, New York, 1967.
2. Chemical Profiles, *Chem. Market. Report.*, **214,** July 1, 1978.
3. Chemical Profiles, *Chem. Market. Report.*, **210,** October 1, 1976.
4. See the Union Carbide Corp. catalogue, 1973–1974 edition.
5. These three tables originally were constructed from the thirty-eighth edition of the Monsanto Company product catalogue. Only those compounds marked by asterisks are listed in the fortieth edition. This is a good illustration of how products are discontinued as the business situation changes.
6. The chemistry shown has been derived by the author based largely on the intermediates offered for sale by Monsanto Company.
7. W. S. Emerson and E. P. Agnew, *J. Am. Chem. Soc.*, **67,** 1945, p. 518.
8. L. A. Battista, *private communication.* Taken from a speech given at the 1974 American Chemical Society meeting in Los Angeles and quoted from a previous paper given by M. Tischler.

9. W. S. Emerson, J. W. Heyd, V. E. Lucas, W. B. Cook, G. R. Owens, and R. W. Shortridge, *J. Am. Chem. Soc.*, **68,** 1946, p. 1665.
10. W. S. Emerson, J. W. Heyd, V. E. Lucas, E. C. Chapin, G. R. Owens, and R. W. Shortridge, *J. Am. Chem. Soc.*, **68,** 1946, p. 674.
11. W. S. Emerson, J. W. Heyd, V. E. Lucas, W. I. Lyness, G. R. Owens, and R. W. Shortridge, *J. Am. Chem. Soc.*, **69,** 1947, p. 1905.
12. W. S. Emerson, J. W. Heyd, V. E. Lucas, J. K. Stevenson, and T. A. Wills, *J. Am. Chem. Soc.*, **69,** 1947, p. 706.
13. A. H. Ford-Moore and H. N. Rydon, *J. Chem. Soc.*, 1946, p. 679.
14. These figures, although taken from 1969 data, are in line with those at the time of the investigation. The $0.45 per pound for benzoic acid was the price at the time of the investigation.
15. W. S. Emerson, T. C. Shafer, and R. A. Heimsch, *J. Org. Chem.*, **16,** 1951, p. 1839.
16. W. S. Emerson, V. E. Lucas, and R. A. Heimsch, *J. Am. Chem. Soc.*, **71,** 1949, p. 1742.
17. W. S. Emerson and V. E. Lucas, *J. Am. Chem. Soc.*, **70,** 1948, p. 1180.
18. W. A. Franta, *Chemtech,* **3,** November, 1973, p. 650.

7

INDUSTRIAL CHEMICAL DEVELOPMENT

I. INTRODUCTION

In the preceding chapter it was shown that industrial chemical research produces scientific knowledge, not products and processes. However, chemical companies are not in the business of producing and selling scientific knowledge, although they may derive a small amount of income from licensing patents. They are in the business of manufacturing and selling chemicals. Therefore, the job of the industrial chemist "is to learn how to make new reactions go with commercially attractive yields under conditions that can be established economically to make products for which there is a demand or a potential demand" [1].

Putting something new and useful into the channels of commerce is called *innovation* [2]. In the chemical business this may be a new product or a new process. The method by which innovation is effected is called *development.* In this chapter the technical side and in Chapter 16 the commercial side of industrial chemical development are described.

Technical development, then, is the use of known scientific knowledge to create new products and new processes. It is objective, conscious, and deliberate. The goal is always obvious and the risks are measurable. Planning is essential. In fact if one does not plan, one can lose one's shirt. Development is measured by achievement rather than by discovery. The emotional involvement is the same as that of a plant manager who has appreciably exceeded his production quota or of a salesman who has just landed a large, new account. However, it must be emphasized that creativity and imagination are just as necessary in development as in research, and in a large development program, as will be shown in the next chapter, it may be necessary to perform some research in order to fill in gaps in scientific knowledge.

In most chemical companies development constitutes much the major portion of the technical budget, often as much as 90%. This is partly because it takes a lot more time and effort to commercialize a chemical discovery than it does to make the discovery in the first place. Thus one creative research chemist with a technician helper can keep a large team of development technologists busy. Secondly, the business executives who pay the bills are partial to programs whose goal is obvious and whose risks and ultimate value are measurable.

II. TYPICAL CHEMICAL DEVELOPMENT EXAMPLE

As a start a fictitious [3], but typical, example of a development project will be examined, and then the general considerations involved in the consecutive phases of a technical development project will be considered in some detail. The example is the development of a new surface-coating resin.

1. In the laboratory an organic chemist found that styrene oxide reacted with phenol to give a hard, clear amber-colored resin. The reaction was reproducible. The resin was tested in a surface-coating formulation and it appeared promising. The company was in the surface-coating-resins busines, so further work was planned immediately. Otherwise the lead would have been dropped then and there as being outside the field of company interest.

This first phase was the *research* involved. It took about two weeks and cost $4000–$5000.

2. The research and development director agreed that further work should be done. The chemist tried such other phenols as *p*-cresol, *p*-*t*-butylphenol, and *p*-phenylphenol, phenols that commonly are used as raw materials for phenol-formaldehyde type surface-coating resins. The resins thus produced were tested in several different formulations, and that prepared from *p*-*t*-butylphenol appeared to be the most promising. The preliminary outline of a commercial process was delineated. A patent search showed that the resins and their surface-coating formulations appeared to be novel. An initial market study showed that there was a place in the surface-coating field for a resin with these properties.

This second phase will be called *definitive development.* It took about two months and cost $15,000–$20,000.

3. At this point the research and development director went to technical management, and a development project was established, involving a much larger team effort. It should be noted that even in phase 2 a synthetic organic chemist, a surface-coatings expert, a patent attorney, and a market-research person were involved.

The patent survey had shown that both the resins and their surface-coating formulations were novel. Therefore, in order to provide patent examples, additional phenols were tested, as well as such substituted styrene oxides as *p*-methyl and *m*-chloro.

The optimum ratio of styrene oxide to *p*-*t*-butylphenol was determined, both for patent purposes and as a basis for product and process development. The partial substitution of such other materials as formaldehyde, benzaldehyde, and furfural for styrene oxide was tested to see if improved properties could be obtained, as well as for additional patent protection.

Larger samples of the experimental resin were prepared in order to complete the formulation studies and to provide material for preliminary sampling to potential customers. Aging tests of wood panels coated with the formulation were conducted in a weatherometer. The details of the intended process for preparing the resin and formulating it were developed.

Since for laboratory purposes it had been most convenient and economical to use the literature method for preparing styrene oxide, namely, by treating styrene bromohydrin with potassium hydroxide, the process study had to include the development of a commercially feasible method for preparing styrene chlorohydrin and dehydrohalogenating it with sodium hydroxide rather than with potassium hydroxide. A special method for purifying styrene oxide had to be devised, since in the laboratory the compound has to be distilled under vacuum (see Section IV.A). Otherwise extensive and unpleasant decomposition can take place. Distillation at reduced pressure is very expensive on a commercial scale.

Finally, the market study was sharpened on the basis of the preliminary sampling.

This third phase will be called *laboratory development*. It took about six months and cost $70,000–$80,000.

4. The next step was a big jump. A pilot plant was designed, constructed, and operated, primarily in order to verify the proposed processes, but incidentally to provide material for extended field sampling. Quality-control methods were developed in conjunction with the pilot-plant operation.

On the basis of the field trials and the pilot-plant data, final product specifications were established. These specifications had to be and were met in the pilot plant. From the results of these field trials, the commercial development department was able to estimate the potential market with reasonable accuracy and therefore could help size the commercial plant.

The weathering tests were extended outdoors to several geographical locations: Southern Arizona—very sunny, hot, and dry; Florida—sunny, hot, and humid; the Seattle area—overcast, moderate temperature, and very humid; North Dakota—moderate humidity and tremendous temperature variations, including very cold. Obviously, the results of these tests were most helpful in accurately defining the potential market, since different surface-coating formulations are used in different geographical locations. Then the several patent applications were filed.

This fourth phase will be called *design development*. It took about two years and cost $450,00–$500,000.

5. Finally, a site was selected for the commercial plant. The plant was designed, constructed, and placed in operation. The usual number of start-up problems were encountered.

Quality-control methods were finalized.

Sale of the product was started.

This fifth phase will be called *final development*. It took about two years and cost about $2.5 million including the plant investment.

MARKET	PHASE	PROCESS ECONOMICS
Request for market survey	1 Decision about company interest Idea seeking | Research	Request for investigation
Estimate of size and nature of market	2 Definitive Development Patent search Applications research Evaluation of properties Product definition | Process definition Commercial process outline	Estimated costs: Order of magnitude +50% to −30% Guiding cost and potential studies
Sampling and follow–up Estimate of specific markets	3 Laboratory Development Product improvement Applications development Large sample preparation Indicated patent studies | Process development Ancillary development (ecology)	Preliminary cost estimate: +30% to −15% Process decision
Establishment of specifications Firm estimate of initial market and rate of growth Field trials	4 Design Development Quality–control methods Filing of patent applications Extension of stability tests on new products Regulatory–agency contacts | Process design and verification Computer engineering Miniplant pilot plant	Definitive project cost estimate ± 10% Scale of commercial operation
Settling of strategy Training sales representatives	5 Final Development Completion and evaluation of use tests Final quality–control methods Introductory sales | Interim production plant	Project decision Plant start–up

FIG. 1. Summary chart of phases of industrial chemical development. Adopted with permission from H. Skolnik and L. F. McBurney, *Chemtech*, February, 1971, p. 82. Copyright 1971 by American Chemical Society.

TABLE 1. Summary of Surface-Coating-Resin Development[a]

Phase	Elapsed Time	Cost
1. Idea seeking (research)	2 weeks	$ 4,000–5,000
2. Definitive development	2 months	15,000–20,000
3. Laboratory development	6 months	70,000–80,000
4. Design development	2 years	350,000–400,000
5. Final development	2 years	~2,500,000

[a] The relative orders of magnitude for the costs and elapsed times of these phases are reasonably typical.

From this example it can be seen that the research, or idea-finding stage of a successful commercial chemical development, is a very small element of the total cost and time involved. And from stage to stage in the development, both cost and time increase exponentially.

In order to visualize this situation more clearly, a pyramid has been constructed in which time increases downward and outlay in dollars increases with the length of the horizontal baseline (Fig. 1) [4]. The pyramid obviously is not to scale.

In this example of the surface-coating resin, the idea came from the research department (phase I). It might equally well have involved purchased technology, such as the licensing of a patent. Companies often purchase technology to save both time and money. Occasionally such purchase is due to lack of nerve.

The rest of the steps in the example of the surface-coating-resin development correspond pretty closely to the pyramidal chart. Table 1 summarizes the process.

III. PHASES OF INDUSTRIAL CHEMICAL DEVELOPMENT

The reasons elapsed times escalate from phase to phase are fairly obvious from the example in Section II. Now costs and therefore functions will be examined in more detail from phase to phase.

A. Phase 1: Idea Seeking—Research

Using the example in Section II, in phase 1 one research chemist and one coatings technologist plus such supporting services as laboratory technicians, routine analyses by the analytical laboratory, and the usual part-time supervision by a research group leader are involved.

B. Phase 2: Definitive Development

In phase 2 we encounter for the first time a *development team* drawn from a variety of technological disciplines. One of the biggest things a chemist in industry has to learn is how to work as a member of such an interdisciplinary team. It may be very difficult for the chemist to realize that a researcher must be able to cooperate and work with so many different kinds of people in order to see a discovery brought to commercialization.

Applications Research. In phase 2 we encountered for the first time *applications research,* a type of scientific endeavor with which many chemists are unfamiliar. Applications research involves finding commercial uses, usually for new materials. In order for a meaningful market survey to be made, as many applications as possible for a new product must be identified, based on a knowledge of the needs of a large number of possible customers. The size of the potential market determines the size of the commercial plant. On the other hand, applications research can even indicate that the potential market is too small to warrant commercialization of the new product.

In applications research the new material is screened by simple but meaningful tests for as many potential uses as possible. Wherever this is relevant, its properties are compared with those of other materials presently in commercial use. In the case of a new polymer, usually many formulations must be examined.

Many chemical companies have screening agreements with other firms in areas in which the chemical companies have no applications expertise, such as agricultural chemicals and pharmaceuticals. The wide variety of fields that can be involved in an applications-research study is shown by an assignment of the author's in which he was asked to find possible uses for about 450 organic compounds. These were research by-products for which no uses had been found. In the study leads were developed for a potential hair-waving agent, a nickel-plating brightener, an antioxidant for polyethylene and polypropylene in the presence of carbon black, two stabilizers for oil-immersion transformer board, a thermal stabilizer for paper and cotton cloth, and a possible size for use in epoxy-bonded fiberglass laminates.

It must be emphasized that applications research continues throughout the life of a product. Nylon originally found use in parachutes and hosiery. Now there are any number of other applications including tire cord, sweaters, carpets, and machine parts.

One outgrowth of applications research is use patents (patents protecting the use of a given compound or material in a specific application). Even if a company holds a composition-of-matter patent on a new chemical or a new polymer, it is highly embarrassing to have one customer control the largest market for same with a use patent.

Philosophically, applications research is intriguing in that it constitutes an answer looking for a problem.

In phase 2 the product is defined (definitive development). For a new polymer this can mean an extensive study. The effect of such factors as molecular weight, molecular-weight distribution, copolymerization, and formulation on such properties as color, stability, hardness, tensile strength, and adhesion must be examined. Obviously these properties have to be related to possible uses.

This study defines the product or products selected for laboratory development in phase 3.

In phase 2 the process is defined. An example showing a few of the factors that can be involved was shown for the preparation of *m*-chlorostyrene in the preceding chapter. Additionally, the cost and availability of possible raw materials and the more obvious environmental problems were examined.

In phase 2 the initial economic evaluation is made. It is order of magnitude only, as shown in Fig. 1. Besides relevance to company interest, it is based on size and type of market, whether or not the market can be served with one product, the competition in the market, and the minimum economic scale of production (which varies from one product and one company to another). To illustrate an extreme case of the last criterion, a petroleum company would hardly be interested in manufacturing a watch oil, no matter how good its properties were.

In phase 2 the initial patent search is made.

For the very general case of a new polymer, what kind of a development team may be involved? Following is a typical list:

1. A project supervisor, part time.
2. One to two chemists plus laboratory technicians defining the product and process, full time.
3. Analysts in the analytical laboratory as needed.
4. Physical testing technicians as needed.
5. Many applications research people looking for new uses, a very small fraction of each one's time.
6. An engineer doing the economic evaluation and looking for possible engineering problems in the process, part time.
7. A market research person getting a preliminary reading on possible uses, their market size and type, part time.
8. A patent attorney or patent searcher, part time.
9. A purchasing agent looking at raw-material sources, part time.

C. Phase 3: Laboratory Development

In phase 3, applications research is continued and extended to include detailed performance evaluations in comparison with materials presently in commercial use if the application in question so warrants. More formulations are examined, and stability and storage data are obtained if called for. The product is prepared from commercial-grade raw materials. An attempt is made to anticipate potential customers' needs. A preliminary cost-versus-performance ratio is determined for as many applications as possible. Continued emphasis is placed on use patents. Consideration of packaging is started.

Commercial development people need firm applications data in order to do their job, and this aspect of development will be reemphasized and expanded in Chapter 16. An advertising brochure containing physical and chemical properties plus lots of literature references is scientifically interesting reading, but of dubious commercial utility. Unless commercial development people can accompany their initial sample distribution to prospective customers by use-test data sufficiently interesting to stimulate additional work on the part of said customers, the only feedback they are likely to receive from the evaluation of the sample is shelf-life studies. They also will stock the shelves of most of the repackaging houses in the country. Such data hardly are helpful in the preliminary estimates of specific markets; such estimates are necessary at this point, and upon them the scale of commercial operation must begin to be estimated.

In phase 3, as a result of the extended applications studies, tentative product specifications are established, based on commercial-grade raw materials. Larger amounts of product are prepared, often in some form of technical laboratory, and samples are distributed by the commercial development department to selected prospective customers.

A detailed evaluation of the chosen process is made in the light of the technical considerations to be discussed. A preliminary sketch of the proposed plant is made. The selection and development of test methods is begun. And consideration is made of possible environmental consequences.

Costs are reviewed in more detail with heavy engineering department involvement to bring the accuracy of the preliminary estimate to approximately +30% to −15%. As a result of the applications research and preliminary sampling results, the commercial-development department begins to develop merchandizing concepts. Technical and economic feasibility is again confirmed.

The personnel involved in phase 3 are much the same as those in phase 2 plus technical-laboratory personnel. There is a much heavier involvement on the part of the engineering, market-research, and commercial-development departments. The quality-control department (one methods-development person) begins to help in the selection and development of quality-control methods.

D. Phase 4: Design Development

In phase 4 the results of the preliminary sampling, conducted in phase 3, are summarized and carefully evaluated. In the light of these results and customers' specifications as known, the tentative product specifications are reviewed and finalized. Information is compiled on techniques that prospective customers should use in applying the product. This is extremely important, because in the author's experience, merely giving a prospective customer a sample to evaluate without proper directions on how to use it, usually results in a negative test with resultant complete loss of interest on the part of the prospective customer.

In phase 4, the commercial-development department reviews the schedule of extended field trials with the marketing and the sales departments. On the basis of this review, the amount of material needed is ordered from the semiworks or pilot-plant operation. The commercial-development department, with backup from the department of research and development, conducts the field trials. The results of these trials lead to reevaluation of the merchandising ideas and confirm the market approach. A firm estimate can now be made of the size and nature of the initial market and of its probable rate of growth.

At this point safety considerations, shipping problems, and government regulations have to be considered. Packaging plans are firmed up. Preliminary work is started on technical bulletins and on labels. Advertising programs are planned. Much more extended stability tests are conducted on new products if such are indicated.

It is now obvious that as a development project progresses the responsibility for product use moves gradually from the department of research and development (of which applications research is a part) to the department of commercial development. As a corollary it means that in the chemical industry commercial-development people must have a high technical capability.

In phase 4, scale-up of the proposed process is begun with an attendant large increase in expense. This means the design, construction, and operation of a pilot plant. It must be emphasized that even here as many engineering data as possible are collected on a bench scale in metal. Even a continuous bench-scale operation usually requires only the engineer or chemist in charge plus an operator on each shift, and many operations can be so instrumented that they can run overnight without checking, so only one technician on the day shift is required.

A pilot plant means one or more operators plus an engineer in charge on each shift. Much larger quantities of raw materials are required compared with laboratory-scale operation, and a pilot plant represents an appreciable capital investment. Therefore, for economic reasons, the trend is to pilot only those steps in a process in which appreciable scale-up is necessary for an accurate technical evaluation. In a pilot plant, special techniques and equipment, not readily available in the laboratory, are evaluated. Materials of construction are evaluated as well as such process variables as temperature control and control of the flow of liquids and gases. The use of vacuum or pressure also is checked.

Many chemical companies have semiworks facilities for the preparation of field-trial samples. A pilot plant is used to obtain process data for designing the commercial plant. With its high labor costs, it is uneconomical to use a pilot plant for the preparation of field-sample material.

In the pilot plant the ease of meeting product specifications is established. There may be purity problems, since by-products are produced in many processes, often on such a small scale percentagewise that they are of little significance in a laboratory bench-scale operation. Their buildup, isolation, and disposal must be determined in a pilot plant. Also in a pilot plant such ecological problems as the handling of noxious fumes and the whole matter of waste disposal become significant and have to be evaluated.

Not only must the pilot-plant design and operation be coordinated closely with the department of research and development, but it also must be coordinated closely with the manufacturing department. A pilot plant is a scaled-down plant, not a scaled-up laboratory operation. Perhaps the process can be integrated with present commercial production facilities, a tremendous economic advantage if partially idle equipment can be utilized.

In phase 4, plant engineers and operators are trained, and a preliminary draft of an operating manual is drawn up.

Safety considerations take on great importance in phase 4. For instance, in a Parathion plant all operating personnel have to have a blood test once a month even when there are no accidental spills. The Occupational Safety and Health Administration (OSHA) is directing increasing attention and publicity to these considerations.

In phase 4, process and product quality-control methods are firmed up. With processes, as much reliance as possible is placed on instrumental control with its instant readout capability. For product analysis and control, methods are developed that a technician can use quickly and repeatedly with reproducible accuracy.

Finally, on the basis of these studies, a definitive cost estimate is made. Its accuracy is of the order of ±10%. This accuracy is needed because phase 5 is the design and construction of the commercial plant, a big investment. Obviously all the results of the analyses just described are summarized in report form for this final decision. Fixed- and working-capital estimates are made, as well as flow diagrams and expected operating costs for the plant.

The final marketing plan is presented. All possible patent applications are filed—product, process, and use—and their impact on the project is evaluated.

Following is a summary of the personnel involved at this stage of development, although it is even more difficult to estimate their exact number in each category from project to project:

1. A project supervisor, full time.
2. Engineers for the design, construction, and operation of the pilot plant, for evaluating the data from the pilot plant, for helping decide on the final process, for cost evaluation, and for liaison with the manufacturing department.
3. Operators for operating the pilot plant.
4. Commercial-development people for planning, running, and evaluating the field trials, for helping decide the marketing approach, for helping set the product specifications, and for writing the technical bulletins.
5. Chemists for handling all the chemical problems, for developing the quality-control methods, for helping with the field trials, for anticipating possible customer technical problems, for developing method of application, for helping evaluate the field trials, for helping set the product specifications, for helping evaluate the pilot-plant data, for consultation on pilot-plant operation, and for helping with the final process.
6. Manufacturing personnel for liaison with the project and for helping adapt the process to present plant facilities and equipment.

7. Personnel people for helping plan for and procure the necessary staff for commercial-plant operation.
8. Salespeople for helping to decide the marketing approach and for helping with the technical bulletins.
9. Patent attorneys for preparing and filing the patent applications.
10. Purchasing people for lining up the necessary raw materials.

E. Phase 5: Final Development

Phase 5 involves the commercialization of the product and the process.

On the marketing side there is a final review with the sales department of the market, costs, pricing, expected profitability, competition, raw-material availability, and method of product distribution. This review is imperative. A colleague of the author's tells a horror story of a project in which when the plant was constructed, it was found that the raw material was no longer available [5]. Also an adverse patent may have been issued or unforseen technical difficulties may have been encountered. The final estimate of the market and its rate of growth are essential for sizing the commercial facility.

With some products, further market testing may be necessary. These tests are planned, conducted, and reviewed as in the case of the field trials in phase 4.

In phase 5 the merchandizing and marketing program is firmed up and a preliminary budget is prepared. Negotiation of sales agreements and contracts is started. Salespeople are trained, perhaps with the aid of introductory sales. Advertising material is prepared. A technical-service department is organized and established. All this leads to the acceptance of and assumption of responsibility for the product by the sales department.

On the technical side, the site for the plant is selected. Market location is an important factor. In the case of sulfuric acid plants, as an extreme example, freight rates are the determining factor in plant location. American Potash and Chemical Corp. located its second chlorate plant in Mississippi in order to be near the southern paper industry.

Access to raw materials can be equally important. American Potash and Chemical Corp. located its ethylene dibromide plant next to a refinery, since ethylene could be piped directly over the fence. Thus transportation, whether by rail, barge, pipeline, or truck, is important. In many cases access to deep water or to a railroad spur is imperative. The author knows of a custom drug manufacturer who considered building a facility in Puerto Rico because of other very attractive economic considerations. This was ruled out because the business depended upon rapid, certain, and cheap product delivery, which could be uncertain from the island location.

Other obvious siting factors include taxes, zoning, and waste-disposal problems. Even a trace amount of a smelly chemical in a residential sewer can lead to all kinds of continuing problems with the local inhabitants. Labor availability and cost are very important. This was a big attraction of the Puerto Rico location for the drug manufacturer. Specialized staff must be available—engineers to operate the plant, chemists to operate the quality-control laboratory, even accountants and purchasing agents to conduct the everyday business of the facility. And obviously the cost and availability of land are extremely important.

After the site is selected, the plant is designed and constructed based on the process studies conducted in phase 4. Both supervisors and operators are trained, often in the pilot plant. The final operating manual is written. Final quality-control methods are established, and the laboratory is constructed and staffed. A plant technical-service group is organized. Assistance in breaking in the plant at the startup of production by the appropriate project people leads to acceptance of production responsibility by the manufacturing department.

There is no reason to redetail the personnel involved in phase 5. They are the same as those in phase 4 with increased responsibility taken by line operating people in the manufacturing and the sales departments.

From this discussion it can be seen that industrial chemical development is a very complex process involving many different kinds of people working together as a team. Obviously costs increase tremendously as the project progresses. Therefore, there is every reason to expedite the project and move as rapidly as possible. Although somewhat arbitrarily divided into phases in the previous discussion, with readings taken at the end of each phase, the process actually is a continuous one. And the estimates of the costs of building and running the plant become more and more accurate as additional data are acquired at each step.

Every time a reading is taken, a long list of items must be rechecked. Since the description given here should give a good idea of the complexity of the process, these lists are omitted, but the reader who so desires can readily refer to them elsewhere [6].

IV. COMPARISON OF LABORATORY WITH PLANT

Laboratory research in industry is conducted in very much the same manner as it is in academia. However, it is conducted under very strict government (such as OSHA) regulations, from which academia is largely exempt [7]. Therefore, the differences between these two types of endeavors are mentioned in this section, although the author believes most academic laboratories try to follow OSHA guidelines.

Chemical manufacture is totally different from anything in a university. Technical development, which has just been outlined in detail, is the translation of a chemical process from the laboratory to the plant.

In this section such items as chemicals, equipment, process operations, and so forth are compared. Graduate chemical engineers, particularly if they have attended a practice school, probably have some idea of these comparisons. Undergraduate chemical engineers probably have some inkling, but unless they have worked in industry, chemists probably do not.

A. Chemicals

Item	Laboratory	Plant
Raw materials	Pure (reagent grade)	Industrial grade
	Cost no real factor	As cheap as possible

The chemist in the laboratory uses pure reagents. In a plant, commercial-grade chemicals must be used. This can have an appreciable effect on the reaction and product in question. For instance, carbon monoxide in hydrogen prepared from the reforming of natural gas can adversely affect a hydrodesulfurization catalyst [8]. A liquefied petroleum gas from a plant stream in which only hydrogen sulfide had been identified was found to be corrosive to copper, so carbonyl sulfide also had to be removed before the product could be marketed [8]. This necessitated additional equipment. In the case of condensation polymers with which exact stoichiometry is essential, extensive raw-material purification may be necessary on a commercial scale.

In the laboratory the chemist's time is the biggest expense. When the author was examining the chemistry of styrene oxide, he prepared it from the bromohydrin, since a good method was described in the literature.

$$C_6H_5CH{=}CH_2 + HOBr \longrightarrow C_6H_5CHOHCH_2Br \xrightarrow{OH^-} C_6H_5\underset{\diagdown O \diagup}{CH{-}CH_2} \qquad (1)$$

~85% yield

A process, perhaps through the much cheaper chlorohydrin, would have had to have been developed had the oxide been commercialized at that time. Also, in Chapter 3 examples were shown of many industrial chemical processes that have been superseded by ones with lower-cost raw materials.

Item	Laboratory	Plant
Conversion	If reasonable, O.K.	Recovery and recycling of raw material essential
	Unreacted starting material may not have to be recovered and recycled	Recycling and recovery are expensive

In the laboratory a reasonable conversion is all that is needed. A small amount of impure recovered starting material can usually be discarded. Usually in a plant all unreacted starting material must be recovered, purified, and recycled.

Item	Laboratory	Plant
By-products	No immediate problems	Must be collected and disposed of

In the laboratory, by-products can easily be discarded. In a plant they have to be collected and their disposal can cause considerable trouble and expense. In continuous processes special steps have to be taken to continuously remove by-products to prevent their buildup.

Item	Laboratory	Plant
Storage	Bottles and cans	Tanks, drums, sacks and bins
	Dangerous chemicals usually in hoods	Special precautions for dangerous chemicals

In the storage area, both laboratory and plant, the government is very strict. Caution labels must be on all containers holding such toxic chemicals as benzene, cyanides, and carcinogens. Gas cylinders (laboratory) must be properly chained in a completely secure position.

OSHA defines most organic liquids in various inflammable and combustible classes based on their flash point (defined in Chapter 2). On this basis OSHA determines the sizes and materials of construction for containers and defines the maximum amount of a liquid that can be stored in each type of container in an industrial laboratory. Container types include glass or approved-plastic bottles, metal cans, safety cans, metal drums, and approved portable tanks. Thus no more than a pint of *n*-pentane may be stored in glass, whereas up to a quart of ethanol may be [7].

A safety can is a metal can, usually painted red, of no more than 5-gallon capacity. It may have a spring-closing lid and must have a spring-loaded and self-sealing spout cover. The pressure applied by the spring on the spout cover is so designed as to safely relieve internal pressure on heating (as from a fire). There are flame-arrester screens in the spout to prevent flashback from any combustion.

Static electricity is a problem in both industrial laboratories and plants. It can be created by a liquid's passing through a hose or pipe. In order to minimize its danger (sparks and then fire), bonding is employed. When a chemist withdraws an inflammable liquid from a drum, the tank nozzle and the receiving container first are bonded electrically, to make the entire operation an electrical unit. The storage tank of course has been grounded.

These same types of rules apply to storage cabinets, bottle sizes, amounts of inflammables, and so forth.

The rules for storage rooms within a building are much the same. In addition, the floor of such storage rooms must either be four inches lower than the surrounding area or else protected by a four-inch dike. Storage rooms must be protected by fire doors as well as by proper sprinkling systems, and so forth.

The storage of innocuous chemicals varies but little from laboratory to plant except for the size and material of construction of the containers. In the laboratory corrosive and toxic chemicals can be placed in a hood. Highly inflammable solvents such as ether and carbon disulfide also can be stored in a hood, and their amounts strictly limited in accordance with the regulations described above. In a plant the storage of all such noxious materials requires all kinds of special precautions.

Item	Laboratory	Plant
Solvents	Essentially unlimited supply	Amounts must be kept at a minimum

In the laboratory essentially unlimited amounts of solvents are available for reactions, extractions, and so forth. In a plant the amounts must be kept to a minimum, and all must be recovered for reuse or disposal, often at some trouble and expense.

B. Equipment

Item	Laboratory	Plant
Materials of construction	Glass or porcelain	Metal

Most laboratory equipment is made of glass and in a few instances of porcelain. Setups are easily assembled and dismantled. Laboratory equipment is easily cleaned, by hot cleaning solution if necessary. There are no corrosion problems except when concentrated alkali is used. The chemist can observe what is happening in a glass flask. High temperatures cannot be achieved.

Plant equipment is made of metal. It takes time to assemble and it can be extremely difficult to dismantle. It can be very difficult to clean, as anyone who has had to unplug a clogged pipe can testify. Corrosion always is a major factor. Mild steel is preferred, since it is relatively inexpensive. Stainless steel of any type adds appreciably to the capital investment, and glass-lined equipment or such special materials as titanium are even more expensive. Cost is a major factor. The chemist cannot watch what is going on in the kettle, but has to rely on instruments. There usually is no practical limit to the temperatures that can be achieved.

Corrosion has just been emphasized. The chemist in industry will very soon become aware of it. If an acid is to be used in the process being developed, the engineers automatically will ask if the hydrochloric acid that was used in the laboratory can be replaced by sulfuric or phosphoric. (Steel can be made passive to sulfuric acid.) The chemist will be asked to run corrosion tests. These are quite simple. Carefully weighed samples of the metals that might be used in the plant are suspended in the media under the conditions to which they would be exposed in production. After a suitable period of time, the samples are reweighed, and the loss in weight is translated in terms of a dimension that the author (as a chemist) has always found intriguing, namely, corrosion in inches per year.

Item	Laboratory	Plant
Weighing	Laboratory balances	Scales
Reactor	Glass flask	Kettle or other specially designed reactors

Laboratory reactions standardly are conducted in a three-necked flask, which can be equipped with such usual appurtenances as a stirrer, thermometer, reflux condenser, dropping funnel for liquid addition, or inlet tube for gas addition. A plant kettle can be similarly equipped. In addition, plant reactors can be designed in an infinite variety of shapes and sizes with all kinds of modes of access and egress.

Item	Laboratory	Plant
Stirrer	Propeller on a shaft, or magnetic	Usually propeller type

In the laboratory, stirring usually is effected by a propeller at the end of a shaft or by a magnetic stirrer. The magnetic type eliminates the need for a stuffing box, which can be as simple as a lubricated piece of rubber tubing located where the glass stirrer shaft enters the piece of glass tubing that holds it in place at the entrance to the flask.

In a plant a propeller-type stirrer at the end of a metal shaft usually is employed. Stuffing boxes, packed with a variety of what essentially constitute washers, are used to control and lubricate the revolving shaft and prevent the escape of chemicals at this point. In reactions conducted under pressure, stuffing boxes can be quite elaborate. When the author was conducting oxidations at 260°C in aqueous solution under an applied pure-oxygen pressure of 1000 psi, the only packing material available was Teflon, and the deterioration was horrendous [9]. The stuffing box had to be taken apart and repacked nearly every week. In a plant, stirring can have a far greater effect on reaction rate and on heat transfer than it does in the laboratory.

Item	Laboratory	Plant
Liquid addition	Dropping funnel	From a tank with valve and flow meter

Liquid addition in the laboratory is very simple. A dropping funnel is easy to manipulate, and if a very slow rate of addition is indicated, the drops can be counted. In the plant, a tank can be emptied through a valve and the rate of addition controlled by observing a carefully calibrated flow meter. Continuous addition can be made and controlled by a metering pump.

Item	Laboratory	Plant
Solid addition	Spatula or pour from beaker	Shovel or hopper

Here the difference is only in scale.

Item	Laboratory	Plant
Temperature	Thermometer	Thermocouple
Condenser	Yes	Yes
Filtration	Büchner funnel or centrifuge	Much the same

Most laboratory filtrations are conducted through a Büchner funnel or by means of a centrifuge. In a plant the Nutsche filter is just a large Büchner funnel, and centrifuges are standard equipment. Many industrial processes also use the plate and frame press, which is described in more detail in Chapter 10.

Item	Laboratory	Plant
Extraction	Separatory funnel	Various methods

Extractions in the laboratory are carried out in separatory funnels, except in the few cases in which much more elaborate apparatus is needed for continuous work. In a plant, the two liquid phases can be stirred in a tank and then allowed to stand and separate, or else a variety of countercurrent equipment can be used.

Item	Laboratory	Plant
Distillation	Fractionating columns	Same

Fractionating columns of a wide variety of types and efficiencies are available for laboratory distillations. For safety a metal catch pan usually is placed under the setup. Plate columns usually are used in plants and many such are in operation. However, they are expensive and require lots of energy to operate. Wherever possible they are replaced by membranes, extraction, and so forth. Rather than distilling high-boiling plasticizers, chemical engineers often clean up these materials and decolorize them by stirring with such reagents as potassium permanganate.

Item	Laboratory	Plant
Plant size	Whatever is convenient	Varies, but size extremely important

Glass laboratory equipment comes in various convenient sizes, and it is easy to select apparatus that is suitable for the experiment in question. A plant is essentially a permanent fixture and represents a large capital investment. All process equipment must be in balance, and the plant must be in reasonable balance with the market. It is uneconomical to run a plant at appreciably below design capacity, and it is much worse to have to build a new plant because the first one was too small to satisfy the market.

C. Process Operations

Item	Laboratory	Plant
Process time	Whatever is convenient	As short as possible

In the laboratory a reaction time often is unimportant. If the car pool leaves at 5:05 PM, the reaction can sometimes be allowed to run overnight, when actually only an hour or two may be necessary [10]. A very bad example is that of the preparation of a pesticide wherein the yield was quantitative in 10 minutes, but only 65% when the reaction mixture was refluxed overnight [10].

In a plant, every time you save a few minutes, you save dollars in getting maximum use from equipment and in conserving energy, such as that used in heating a reactor or running a motor.

Item	Laboratory	Plant
Heating	Gas flame, electric mantle, oil or metal bath, furnace	Steam, oil bath (Dowtherm, Aroclor type), flame, electricity

A wide variety of methods is available for heating laboratory reaction vessels, the only limitation being the temperature that the apparatus will withstand. In heating organics, open flames are avoided wherever possible. The same is true in a plant, although here high temperatures often are used because the equipment is made of metal rather than glass. However, in a plant, energy costs become very important, and there can be a big time lag in bringing large volumes of chemicals to the proper reaction temperature.

Item	Laboratory	Plant
Cooling	Various baths, reflux condenser	Various methods

Often in the laboratory a reflux condenser imparts sufficient cooling to a boiling solvent to keep the reaction under control. Various cooling baths also are available: water, ice, dry ice suspended in a solvent, or even liquid nitrogen. In a plant, refrigeration is very expensive, so river water at ambient temperature is used whenever possible. A reflux condenser can be used in a plant as well as in the laboratory, plus such other techniques as vapor or liquid injection or the use of an inert gas in vapor-phase reactions, techniques not readily available in the laboratory. In the case of very exothermic reactions, the plant reactor can be so designed that only small amounts of material react continuously, as in a pipe.

In the laboratory the investigator seldom worries very much about the heat of a reaction, except in extreme cases. If reasonably agile, the chemist usually can get an ice bath under the reaction flask in time to prevent disaster if the reaction takes off. If necessary the investigator can even throw the setup out the window on very short notice. These procedures are not possible in a plant. You simply cannot suspend a 1000-gallon vessel from a derrick and dunk it in a swimming pool, much less project it into a vacant lot if a reaction takes off. For commercial production, heats of reaction have to be calculated carefully and adequate cooling planned therefor. Moreover, in a plant the cooling surface that can be provided per unit volume of reaction mixture is exponentially less than that which is available with laboratory apparatus. Also, a reaction that may appear to be endothermic in the laboratory because of the cooling provided by the air around the flask may actually be exothermic and must be treated accordingly in the plant.

Item	Laboratory	Plant
Vacuum	Water aspirator, oil pump, mercury vapor pump	Steam jet

Almost any desired vacuum can be achieved in the laboratory. As many water aspirators as the local water pressure can support are easily installed in every laboratory. Any chemist who needs one has an oil pump. In the plant a steam jet can go down to about 20 mm. Pressures below that are very difficult if not impossible to obtain. This is another reason that high-boiling plasticizers are not purified by distillation.

Item	Laboratory	Plant
Pressure	Reactions usually at 15 psi Various apparatuses available for higher pressures	Various methods

Most laboratory reactions are conducted at atmospheric pressure, although in development work it should be remembered that an increase of 5 psi often increases the boiling point of a liquid sufficiently to double the speed of a reaction. For higher pressures in the laboratory, all sorts of equipment is available, all the way from Coca Cola bottles to very heavy steel autoclaves and their appurtenances. If a gas is needed at high pressure, it usually can be bled from a cylinder.

In a plant the key is capital investment. Small increments above atmospheric pressure offer no problem in suitably gasketed metal equipment. High pressures require very expensive compressors and reactors plus other expensive auxiliary equipment. In both laboratory and plant operations, adequate shielding is extremely important.

Item	Laboratory	Plant
Vapor phase	Hot tube	Wide variety of methods

In the laboratory, vapor-phase reactions are conducted in special setups involving glass, fused-quartz, or steel tubes, jacketed by some kind of heating bath or else directly wound with heating wire. Temperatures and fluid flows are difficult to control within narrow ranges because of the small quantities of materials involved.

In a plant an infinite variety of reactors can be designed. Temperatures and flows are much easier to control, particularly after equilibrium has been achieved.

Item	Laboratory	Plant
Continuous operation	Very difficult	Easy and widely used

In the laboratory it is very difficult to keep reaction temperatures and flows of small amounts of materials in balance. On the much larger scale of plant operation, this is relatively easy, and continuous processes are in operation everywhere.

Item	Laboratory	Plant
Fluid bed	Difficult and inconvenient	Easy and widely used

Again, handling a fluid bed in the laboratory is difficult because of the small quantities of materials involved. The catalyst usually has to be regenerated separately, batchwise. Fluid-bed catalytic processes are used widely in industry on a large, continuous scale.

Item	Laboratory	Plant
Analyses	Any time, of any nature	Can be problems

In a well-equipped analytical research laboratory, whether academic or industrial, almost any kind of analysis can be obtained, the only problem being priorities. If something very unusual and beyond the skills of the laboratory in question is required, the particular phase of the research work involved can be halted temporarily and the samples sent elsewhere.

In a plant all necessary analyses must be anticipated so that the requisite personnel and equipment will be on hand. Analyses usually are needed quickly, because the response of a process to controls may be slow. One simply cannot shut down a plant to wait for an analysis that has to be sent somewhere else.

D. Wastes

Item	Laboratory	Plant
Soluble salts	Sink and sewer problems	Problems

In the laboratory, waste soluble salts, except for a few toxic items such as cyanides, altogether too often may be flushed down the drain. In a plant all such salts have to be separated, as they should be in the laboratory, often by a special process, and then taken to a special disposal site. Usually the sewer cannot handle plant-scale volumes, and the salts may well interfere with the process of disposing of other wastes in a sanitary sewer system. The problem of disposing of waste soluble salts is a big one, as the current uproar over phosphates indicates.

Item	Laboratory	Plant
Insoluble salts	Ash can and disposal site	Disposal site

The disposal of waste insoluble salts is essentially the same in both cases, although in the case of a plant the trucking expense can be appreciable and the dumping site had better not be an eyesore nor drain into a water supply.

Item	Laboratory	Plant
Acids and alkalis	Sink and sewer	Problems

In the laboratory, acids and alkalis seldom are considered to offer a special treatment problem. There usually is enough dilution in the sewer to take care of normal amounts of such wastes, although they should be taken to a special disposal site. In a plant, the volumes usually are so large that waste acids and alkalis have to be neutralized and the resulting salts disposed of as above.

Item	Laboratory	Plant
Solvents	Waste-solvent can and disposal site Some recovery	Recover if possible, special disposal

In the laboratory, waste solvents usually are collected in a waste-solvent can similar to the safety can described previously and then transported to a special disposal site. In a plant, waste solvents are recovered if at all possible. If the producer finds this recovery uneconomical, there are people who recover waste solvents on a custom basis (chemical junk people). Also, higher-boiling waste solvents, and other organics, often can be added to the plant fuel and burned. A special disposal site may be required, like Monsanto Company's terphenyl lake.

Item	Laboratory	Plant
Heavy-liquid organics	Same as insoluble salts	Same as solvents
Solid organics	Same as insoluble salts	Same as solvents

In a plant, heavy-liquid and solid organics often can be added to the plant fuel and burned.

E. Personnel

Item	Laboratory	Plant
Chemist	Skilled chemist	Operator

This is one area in which the transition from academic or industrial laboratory to industrial plants perhaps means the greatest change for a graduate chemist. In the laboratory all but routine manipulations are carried out by skilled scientists. In a plant all operations are conducted by operators with minimum supervision on the part of the engineers in charge of the shift. Although usually intelligent, these operators seldom have had any chemical training. If recruited in a city, they may never even have used a wrench or tightened a bolt [11]. Although they do what they are told, directions for them must be spelled out extremely clearly, and they are very limited by their own skills (or more often lack of them) and by plant equipment. If told to crack a valve, they may even break it with a hammer [11]. They will not recognize the unusual. The story is told of an operator who suffered considerable discomfort in adding formic rather than phosphoric acid to a dye batch, with resultant ruination of said batch [11].

At one time the author had occasion to repeat an ostensibly simple preparation in the chemical literature, the chloromethylation of ethylbenzene [12]:

$$C_6H_5C_2H_5 + CH_2O + HCl \xrightarrow{ZnCl_2} H_5C_2-C_6H_4-CH_2Cl + H_2O \quad (2)$$

The reaction starts at 66°C, and above 69°C everything turns to tar. An experienced investigator can hold a small-scale reaction within a 3°C temperature range in the laboratory, but no operator using plant-scale equipment could possibly do the same with a large batch.

Another example of this type of problem involved instructions to a pilot-plant operator to add a gelling agent to a liquid with vigorous, high-speed stirring. The operator dumped the gelling agent into a tank, added the liquid, and turned on the stirring motor. The propeller got tied up immediately in the very viscous gelling agent, the agitator seized, and the motor immediately burned out. The operator became the proud owner of a mess of goop in the bottom of a tank, plus a burned out stirring motor [13].

Item	Laboratory	Plant
Analyst	Skilled chemist	Technician

This is a special case of the previous. In a research laboratory the investigator can call on the services of an analytical chemist with essentially any special skill needed. Plant and quality-control laboratories are staffed primarily with technicians with very limited chemical training. This is why pharmaceutical quality-control departments have methods-development groups whose job it is to simplify and streamline research analytical procedures so that technicians can conduct them accurately and quickly. Plant analyses are based on the skills of the people and the equipment available. Numbers are obtained in reply to specific questions with no interpretation and no extraneous observations that could be extremely helpful.

F. Safety

Item	Laboratory	Plant
Work place	Can be very sloppy	Rigidly monitored

In industry, OSHA specifies that work places, both laboratory and plant, must be clean, orderly, and sanitary. Aisles must be marked and kept clean. There must be plenty of exits, and they must be so labeled. Nonexit doors and passages also must be so labeled. Eyewash fountains, safety showers, and safety blankets must be present in adequate numbers and be readily available, in good condition, and of the right type.

In a plant, explosionproof wiring often is required. Spills must be catchable and must be cleaned up quickly. Ventilation must be adequate. Fire hoses as well as fire extinguishers must be readily available.

Item	Laboratory	Plant
Clothing	Uniforms, shoes, glasses	Same plus hats

This is where the contrast between academic and industrial laboratories unfortunately also can be sharp. Many schools are requiring safety glasses in the laboratory, but other protective clothing often is minimal. In industrial chemical laboratories, safety glasses almost always are required. Usually some sort of protective clothing, such as a rubber apron or a white cotton coat (usually preferred in pharmaceutical companies, since it appears more professional) also is required.

For many years Monsanto Company made available to its research personnel lightweight cotton shirts and slacks for use in the laboratory. Street clothes were left in the individual's locker. Although this was somewhat disconcerting to the egos of fresh PhDs, and a few did not wear them, these clothes, essentially mechanics' uniforms, were extremely serviceable and prevented a tremendous amount of damage to street clothes. Sitting inadvertently on a little dilute sulfuric acid can render the investment in a fashionable pair of slacks highly unprofitable.

Although safety shoes usually are not required in an industrial chemical research laboratory, they often are recommended. Years ago the author found that glass laboratory apparatus bounces off one's toe without damage, but the same is not true when the glass hits a concrete floor. Safety shoes are almost a must for anyone who has developed such reflexes and who suddenly starts working with metal equipment.

In a plant, uniforms, safety glasses, safety shoes, and hard hats almost always are required. Gloves may also be required. In pharmaceutical plants, in which sanitation is preeminent, hard hats often are replaced by paper hats, and women are required to wear hairnets. Hard hats and other types of headgear always are required even of visitors.

Item	Laboratory	Plant
Fumes	Hood	Elaborate precautions

A laboratory reaction that evolves noxious fumes is conducted in a hood, which takes care of the situation. In a plant elaborate disposal precautions must be taken, particularly if there are neighbors in the vicinity with sensitive noses and vivid imaginations. Also, OSHA sets exposure limits for certain toxic chemicals. This may mean that personnel are required to wear respirators for their own protection.

Item	Laboratory	Plant
Dangerous reactions	Shield, usually in hood	Elaborate and expensive precautions

In the laboratory, reactions that are liable to get out of control are run behind a safety shield, usually in a hood. Work with explosive materials is conducted on as small a scale as possible, to minimize risk. Even so, a colleague of the author's was deaf in one ear from working with C_3N_{12} (cyanuric triazide).

In a plant, work with dangerous materials and reactions means extremely elaborate precautions, such as running the process behind reinforced concrete under remote control. Thus in a munitions plant, small units are operated separately in a wide area under remote control, with facilities for immediately dumping everything if something seems to be going amiss.

Item	Laboratory	Plant
High pressure	Special laboratory equipment	Same as dangerous reactions

Although high-pressure reactions in academia usually are conducted in a separate laboratory in the research building, it may only be a different room with no other safety precautions. The author has been told of a situation involving a hydrogenation bomb in an academic laboratory, in which the reaction got out of control, and the bomb went out the window and bounced around on a city street [14]. Fortunately no one was in the way.

Industrial laboratories usually conduct high-pressure reactions under remote control in a separate building behind reinforced concrete; the building is equipped with a flimsy roof or back wall that will blow out in case of an explosion.

In a plant, high-pressure reactions are pretty much treated as dangerous reactions. The precautions for dangerous reactions have been outlined, and it must be remembered that these plants represent a very high capital investment.

G. Legal

Item	Laboratory	Plant
Patents	Part of the chemical literature	Can be very restrictive

In academia, patents are part of the chemical literature. In industry their claims can be very restrictive, and an extensive investigation may be necessary to avoid infringing the claims of adversely held patents. Licensing may even be necessary.

REFERENCES AND NOTES

1. W. A. Franta, *Chemtech,* **3,** November, 1973, p. 650.
2. M. A. Glaser, *Chemtech,* **6,** March, 1976, p. 182.
3. The first phases of this example are true, only the project did not work out that way.
4. This pyramid has been adapted by the author from the one originally published by H. Skolnik and L. F. McBurney, *Chemtech,* **1,** February, 1971, p. 82. The accuracy of the cost estimates at the different stages was adapted from S. Katell, *Chemtech,* **1,** November, 1971, p. 648.
5. L. W. Bass, private communication.

6. L. W. Bass, *The Management of Technical Programs,* Praeger, New York, 1965, Chapter 9. J. H. Saunders, *Careers in Industrial Research and Development,* Dekker, New York, 1974, pp. 106–113.
7. For OSHA regulations the author found S. Sichak, *The Laboratory Safety Deskbook, Chemical Bulletin Supplement,* **66,** No. 2, February, 1979, most helpful.
8. A. L. Conn, *Chemtech,* **5,** March, 1975, p. 154.
9. W. S. Emerson, T. C., Shafer, and R. A. Heimsch, *J. Org. Chem.,* **16,** 1951, p. 1839.
10. B. J. Luberoff, *Chemtech,* **5,** April, 1975, p. 193.
11. Several of these examples are taken from M. L. Nadler, *Chemtech,* **2,** May, 1972, p. 318.
12. W. S. Emerson, J. W. Heyd, V. E. Lucas, W. I. Lyness, G. R. Owens, and R. W. Shortridge, *J. Am. Chem. Soc.,* **70,** 1948, p. 1180.
13. R. A. Nash, *Chemtech,* **6,** April, 1976, p. 241.
14. R. C. Elderfield, private communication.

8

DEVELOPMENT EXAMPLES

I. INTRODUCTION

In this chapter a few examples of industrial chemical developments, all taken from the recent literature, are considered. Since the details of some of the processes were probably published in order to promote the sale or licensing of the process in question, it is quite likely that some of these details were presented in the best possible light. The author is not promoting any of the processes. They are simply being used as modern examples to illustrate important points in technical chemical development.

II. CHLOROPRENE

The first case history is of the Petro-Tex Chemical Corp. process for preparing chloroprene from butadiene [1]. It is based on the Distillers, Ltd., process.

Traditionally chloroprene has been prepared by dimerizing acetylene and then adding hydrogen chloride to the vinylacetylene thus produced:

$$2HC{\equiv}CH \xrightarrow{CuCl} HC{\equiv}CCH{=}CH_2 \xrightarrow{HCl} CH_2{=}CClCH{=}CH_2 \qquad (1)$$

Although this process looks very simple on paper, it involves the handling of acetylene, which is both difficult and dangerous, and the by-product in the first step, divinylacetylene, is extremely sensitive. In fact it is the author's understanding that no one in the E. I. DuPont de Nemours and Company is allowed to work with it [2].

On a commercial scale, operating personnel can be reasonably well protected by conducting the reactions behind reinforced concrete, but it still is highly inconvenient to have one's plant disappear into thin air every so often, particularly when the capital investment happens to be quite high.

An alternate route starts with butadiene, which is available commercially in tonnage quantities*:

1. $CH_2{=}CHCH{=}CH_2 + Cl_2 \xrightarrow[\text{phase}]{\text{vapor}}$

$$CH_2ClCHClCH{=}CH_2 + CH_2ClCH{=}CHCH_2Cl \quad (2)$$

2. $CH_2ClCH{=}CHCH_2Cl \xrightarrow{\text{metal salt}} CH_2ClCHClCH{=}CH_2$ (3)

3. $CH_2ClCHClCH{=}CH_2 \xrightarrow{-HCl} CH_2{=}CClCH{=}CH_2$ (4)

The optimization of this known chemistry to produce a viable industrial process is a typical example of industrial chemical development.

When the development began, it was known that liquid-phase chlorination of butadiene gives high boilers. On the other hand, at 65–75°C in the vapor phase, 75–77% dichlorobutenes was produced, the remainder being high-boiling polychlorinated materials.

An investigation of higher temperatures gave the results listed in Table 1. Thus a reaction temperature of 290–330°C was found to be optimum. By-products that had to be separated were as given in Table 2.

Since the addition of only one molecule of chlorine was desired, the butadiene had to be present in excess. The optimum butadiene-to-chlorine ratio proved to be 3:1 to 6:1. Although higher ratios did not have too much effect on yield, the amount of vinylcyclohexene (butadiene dimer) did increase. At the 3:1 to 6:1 ratio, the excess butadiene helped to control the temperature. This use of

TABLE 1. Butadiene Chlorination[a]

Maximum Temperature (°C)	Yield of Dichlorobutenes (%)
250	76.8
270	86.9
300	88.5
320	90.5
340	86.1

[a] Reprinted with permission from A. J. Besozzi, W. H. Taylor and C. W. Capp, Preprints, "The Commercial Production of Chloroprene Via Butadiene," Division of Petroleum Chemistry, Inc., New York, August 27–September 1, 1972, p. E15. Copyright 1972 American Chemical Society.

* Equations 2–4 reprinted with permission from A. J. Besozzi, W. H. Taylor, and C. W. Capp, Preprints, "The Commercial Production of Chloroprene Via Butadiene," Division of Petroleum Chemistry, Inc., New York, New York, August 27–September 1, 1972, p. E15. Copyright 1972 American Chemical Society.

TABLE 2. **Butadiene Chlorination By-Products**[a]

1-Chlorobutadiene	Hydrogen chloride
Chloroprene	Tars
Dichlorobutanes	Carbon oxides
Tetrachlorobutane	Carbonaceous materials
Trichlorobutadiene	High boilers (Telomers)
Vinylcyclohexene	

[a] Reprinted with permission from A. J. Besozzi, W. H. Taylor and C. W. Capp, Preprints, "The Commercial Production of Chloroprene Via Butadiene," Division of Petroleum Chemistry, Inc., New York, August 27–September 1, 1972, p. E15. Copyright 1972 American Chemical Society.

an inert gas to help control the temperature of a vapor-phase reaction is a standard industrial chemical technique.

The investigators worried whether or not by-product hydrogen chloride, produced by substitution side reactions, might be deleterious. It was not. The hydrogen chloride did not add to the butadiene under the conditions in the reactor, and it helped absorb heat as another inert diluent. It could be bled off as it accumulated.

It was known that a low surface-to-volume ratio in the reactor was desirable. Thus a small tubular reactor proved best. It provided good mixing and a fast temperature rise, which gave a quicker reaction. The optimum residence time was 12 seconds. When a packed tubular reactor was used, the yields were lower and more tars were produced.

The process that proved successful has just been outlined. Several alternatives were thought of, tested, and found wanting.

Obviously, direct chlorination of butadiene to chloroprene would be highly desirable. It was found that at high temperatures, yields were low, with carbonaceous materials becoming major by-products. At 150°C in a solvent, more 1-chlorobutadiene was produced than chloroprene.

Another alternative was to use as raw material a crude C_4 stream containing both butenes and butadiene. (Petro-Tex Chemical Corp. would have had this available since it was in the butadiene business.) At 260–300°C in the presence of a cupric chloride on pumice catalyst in a fluid bed, butadiene indeed was chlorinated selectively in high yield, but the capital cost of the equipment needed to handle the large gas volumes was too high and ruled out this alternative.

Oxychlorination of butadiene with hydrogen chloride and oxygen did not appear promising, although a recent patent suggests that this is indeed feasible [3].

Just from this one example it can be seen that in the development of an industrial chemical process, not only chemistry (every conceivable chemical route must be examined), but also engineering and economics are involved. Obviously, the industrial chemist must be thoroughly aware of the latter two considerations, even though not involved in their detailed evaluation.

For the second step (Equation 3), it was known that in the presence of a catalyst the two dichlorobutenes are in equilibrium with each other. At this point it was necessary to do some research (discover new scientific facts) in the course of a development project. In the presence of cupric chloride the following equilibria obtain:

$$\begin{array}{ccc} & \textit{trans}\text{-}CH_2ClCH{=}CHCH_2Cl\ (A) & \\ \swarrow\!\!\nearrow & CuCl_2 & \searrow\!\!\nwarrow \\ \textit{cis}\text{—}CH_2ClCH{=}CHCH_2Cl\ (B) & \rightleftharpoons & CH_2ClCHClCH{=}CH_2\ (C) \end{array} \qquad (5)$$

For reactor design it was necessary to measure the kinetics at three different temperatures and establish the equilibria. It was found that at 120°C the equilibrium mixture comprised 68.5% A, 7.5% B, and 27% C.

For the purposes of this process it was fortunate that 1,2-dichlorobutene boils lower than either of the 1,4 isomers. Therefore, the reactor comprised a pot containing the cupric chloride catalyst topped by a distillation column through which the desired 1,2-dichlorobutene could be separated as it was formed and the 1,4-dichlorobutenes could be returned to the pot for further isomerization. The optimum pot temperature was found to be 120–160°C. Obviously, by-products had to be bled from the pot periodically, and the dichlorobutenes they contained had to be recovered and recycled. Likewise the 1,2-dichlorobutene overhead stream contained some chlorobutadiene, which also had to be separated.

As a bonus, it was found that certain additives increased the effectiveness of the cupric chloride catalyst–adiponitrile, heptaldoxime, *p*-nitroaniline, and so forth.

This step again emphasizes the engineering inputs that are necessary for the successful development of an industrial chemical process. It also shows the necessity for very careful physical chemical measurements (in this case kinetics) for accurate engineering design of a plant.

For the third step (Equation 4), there were three possibilities. The first was to use aqueous sodium hydroxide as the dehydrochlorinating agent:

$$CH_2ClCHClCH{=}CH_2 \xrightarrow[H_2O]{NaOH} CH_2{=}CClCH{=}CH_2 \qquad (6)$$

The desired chloroprene product was obtained in excellent yield. A small amount of 1-chlorobutadiene was produced as a by-product, but it could be separated from the chloroprene by fractional distillation.

In industry, vapor-phase pyrolysis always is a viable and often attractive alternative. In this case direct pyrolysis led to the wrong isomer:

$$CH_2ClCHClCH{=}CH_2 \xrightarrow{\Delta} CHCl{=}CHCH{=}CH_2 + HCl \quad (7)$$

The same results were obtained in the liquid phase in the presence of a metal halide catalyst.

Of course, catalysts can be used in the vapor phase. In the presence of an acid catalyst, the major pyrolysis product again was 1-chlorobutadiene. A basic catalyst worked very well to give chloroprene, but unfortunately it was deactivated rapidly and could not be regenerated.

On the basis of these studies, the sodium hydroxide route was chosen, even though sodium hydroxide was needed in molar quantity and the chlorine values were lost as sodium chloride. Again in this step some research was necessary. Since the kinetics proved to be very difficult to measure in this two-phase sys-

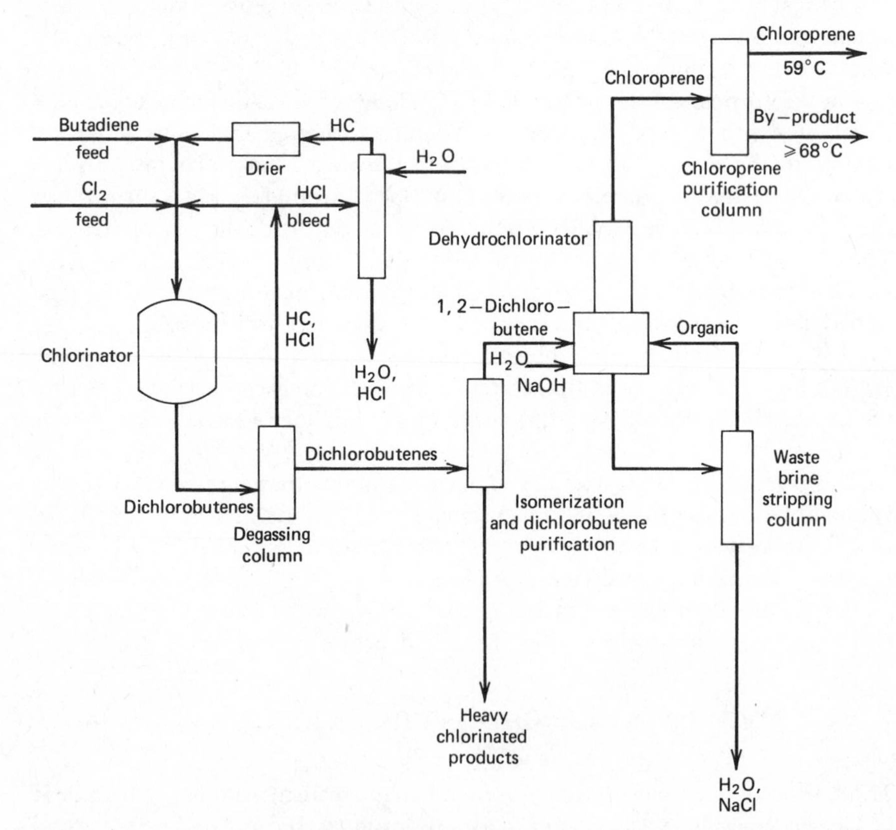

FIG. 1. Chloroprene process flow sheet. Reprinted with permission from A. W. Besozzi, W. H. Taylor and C. W. Capp, Preprints, Division of Petroleum Chemistry, Inc., New York, August 27–September 1, 1972, p. E16. Copyright 1972 by American Chemical Society.

tem, they were only determined qualitatively. Stirring of the reaction mixture was very important because of the two phases.

Chloroprene reacts with oxygen to give a peroxide that initiates polymerization. Therefore, oxygen had to be excluded in the third step. Since the standard polymerization inhibitors such as hydroquinone and *t*-butylcatechol are less effective with chloroprene, compounds such as potassium thiocyanate and sodium sulfide nonahydrate had to be employed, not only to stabilize the final product, but in the reactor as well.

In the third step, the chemistry and not the engineering determined the choice of routes. And for the second time in the project, some new scientific facts had to be determined.

The plant layout is shown in Fig. 1.

Schematic diagrams of this type are used in industry to show what equipment is to be used and the relationships between the several pieces. It is not necessary to draw a pretty picture of each reactor or still, but it is essential that all streams be included, so that the engineers can size the equipment and connect it properly.

It should be noted again that the isomerization (Equation 3) is carried out in a still. The dichlorobutenes from the degasser are fed onto one of the lower plates of the column, so that the desired lower-boiling 1,2-dichlorobutene can be distilled off directly and the higher-boiling 1,4-dichlorobutene can drop into the pot and be isomerized by the cupric chloride.

From this example it can be seen that the differences between research and development are not always as clear-cut as is indicated by the definitions in Chapters 6 and 7, in that in this case facts new to science had to be discovered in order for the development project to succeed. However, the approach and the risks involved in research and in development are totally different.

III. ANALOGY TO THE AGE OF DISCOVERY

Perhaps it would be helpful to draw an analogy with the Age of Discovery. A country (the company) wants to explore an unknown coast, either because it appears promising to the rulers of the country or because an explorer has sold them on the idea. (You can select many of your own research problems in industry.) A vessel is fitted out and sent on a six-month voyage of discovery (*research*). The cost of the expedition can be estimated easily (exploratory work by one man for six months), but what the expedition will discover is unknown. The risk is high, and there is an excellent chance nothing of promise will be found.

At the end of the six months the results of the voyage are reviewed. It may be that a promising harbor has been found and that the natives appear friendly and disposed to trade. Exploitation is indicated (*development*). On the other hand, there may be indications that a longer voyage (more *research*) is worth the risk. Remember that the captain of the good ship *Research* will always recommend the longer voyage, since discovery is his way of life. He finds it hard to learn that the country's (company's) resources are limited. If he has discovered a promising harbor, he may be asked to defer the next voyage of discovery (*research*) and help explore the harbor (mostly *development*).

It may be that the results of the six-month voyage of discovery are very unpromising, in which case the country, although dedicated to discovery (*research*), may feel that exploration of another part of the world (another area of technology) is a more fruitful area in which to invest its limited resources.

On the other hand, a promising harbor must be explored in some detail (*definitive development*). Soundings have to be taken, the shore has to be explored for a suitable trading post site—firm ground near deep water with an adequate supply of fresh water—and more information is needed about the number and affluence of the natives (*preliminary market research*). The risks at this stage are low, and the cost and duration of the exploration can be estimated reasonably accurately from previous experience and from what the discoverer saw of the harbor.

Now the results of this exploration (*definitive development*) are reviewed (another checkpoint), and a decision is made whether or not to proceed further. Buoys probably will be needed at the harbor entrance and along the channel, and some dredging may be necessary. Although a suitable site for the trading post (plant) may have been found, considerable difficulty may be encountered in clearing and leveling the land. Based on experience, the cost of overcoming these obstacles again can be estimated with reasonable accuracy along with the amount of time that will be required (cost and duration of this phase of the project) (*laboratory* and *design development*).

If the decision is made to proceed, exploration still must be continued. Perhaps the natives are found to be less plentiful and less affluent than was at first supposed (the market is not as big as was first thought). Or else the channel silts up very rapidly (there are unexpected corrosion problems that will make the cost of plant maintenance unusually high). On the other hand, other, more numerous and more affluent natives may have been contacted, so the post must be much larger than originally planned (*applications research* has uncovered other uses, so a much bigger plant will be needed).

Then the trading post has to be built, staffed, and placed in operation (*final development*).

There is one more analogy. There is a well-known harbor where several other countries (companies) have trading posts (plants) and are conducting a profitable trade with the natives. Your country (company) decides to establish a trading post there also (become another producer). The exploratory (technical) risk is low (the technology can be purchased), so all that is needed is to make certain there is enough business to support another post (a good market survey would show this).

IV. TEREPHTHALIC ACID

The next development example to be outlined is that of the Lummus Company process for manufacturing terephthalic acid [4].

Originally, polyethylene terephthalate (Terylene, Dacron) was prepared by the reaction between dimethyl terephthalate and ethylene glycol, because the first process for the air oxidation of *p*-xylene went through methyl *p*-toluate as an intermediate, and the final product, dimethyl terephthalate, is readily purified. Terephthalic acid could offer two advantages over its dimethyl ester. It could be cheaper, particularly since at the same price per pound, 17% less of the acid is needed per pound of polyester. Use of the free acid enables ethylene oxide to be substituted for the glycol. Ethylene oxide is the raw material for manufacturing ethylene glycol, and its use in this application cuts in half the amount of volatile material (water or methanol) that must be removed in the final polymerization.

However, ultra high purity acid is necessary, since in a condensation polymerization the molar amounts of the two components must be balanced exactly, and there can be no monofunctional impurities to reduce the molecular weight of the polymer by ending the chains.

Any new process for manufacturing a commodity chemical had better be lower cost, preferably in terms of both capital and manufacturing, or else the present producers will lower the price of the product and leave the new "competitor" with a brand-new plant it cannot afford to run. And at the present time, environmental considerations also are becoming increasingly important.

There were several reasons why the Lummus Company felt there was an opportunity for a new terephthalic acid process. In Chapter 3 it was mentioned that the air oxidation of *p*-xylene in acetic acid solution under pressure involves such horrendous corrosion problems that the reactor must be made of titanium, not a minimal plant investment. In addition, heavy metal ions from the catalyst are hard to remove, and they create undesirable color bodies. Also, special steps at considerable expense are necessary to remove the aldehyde (structure **1**), an obvious chain stopper, whose concentration in the final product must be held below 50 ppm. Also, formulation problems are encountered because of the morphology of the terephthalic acid particles produced by this process.

$$OHC-C_6H_4-COOH \qquad [1]$$

As in the first case (chloroprene) a new route was envisioned for development based on known chemistry:

$$H_3C-C_6H_4-CH_3 \xrightarrow[(O)]{NH_3} NC-C_6H_4-CN \xrightarrow{H_2O} HOCO-C_6H_4-CO_2H \qquad (8)$$

In a conventional ammoxidation (first step), the amount of hydrocarbon in the feed is kept below the lower limits of the explosive concentration. This leads to low space-time yields and high costs for gas circulation and recovery. In the new process a metal oxide was used both as the "catalyst" and as the oxygen source. The mixed *p*-xylene and ammonia vapors were passed over this catalyst in a fluid bed. Needless to say, the identity of the "catalyst" was not divulged by the authors of the article. The use of this catalyst eliminated the need to worry about explosive limits and markedly speeded up space-time yields. Selectivity was 90% to mononitrile and dinitrile. If the initial product was largely mononitrile (not stated by the authors), there could be appreciable recycling problems connected with the process.

The concept of explosive limits is something the academic chemist or chemical engineer is probably not aware of, but it is very important in industry. For every organic chemical vapor, there are upper and lower concentration limits within which admixture with oxygen produces a vapor that can be detonated. These concentration limits are determined experimentally. They usually are reasonably close to the oxygen balance for complete combustion to gases of all the atoms in the organic compound in question. For obvious reasons, chemical companies prefer to operate their processes on either the chemical-rich side or the oxygen-rich side of the explosive mixtures, outside the explosive limits.

In the new Lummus Company process, the product from the first step in Equation 8 was filtered to remove heavy metals. This was primarily a physical separation, since metal ions do not form salts with nitriles as they do with acids (see the direct air oxidation of *p*-xylene in acetic acid). The product was then cooled and separated into gaseous, liquid, and solid phases.

The solid terephthalonitrile (TPN*) was washed with fresh *p*-xylene (part of the *p*-xylene feed) to remove *p*-tolunitrile and was centrifuged. The product then was again washed, this time with water, and again centrifuged. The liquid fraction comprising unreacted *p*-xylene and *p*-tolunitrile plus the *p*-xylene wash was recycled to the reactor. The gaseous phase was sent to the ammonia recovery system, where the ammonia was recovered and the carbon dioxide, presumably ammonia free, was vented to the atmosphere.

Since the various streams and their selective fates are beginning to become pretty involved, the process flow sheet should be in view (Fig. 2).

The spent catalyst in the reactor was removed, steam stripped to recover organics, and regenerated (presumably with air or oxygen) in a fluid bed. The vapors were exhausted through a cyclone to remove fine solid particles and then vented.

The main-stream hydrolysis obviously takes place in many steps, some of which are shown in the following equation:

---

* The author hates the use of nicknames and abbreviations such as TPN because they are extremely confusing in reports, particularly to those not in immediate touch with the project. However, they become more or less necessary in flow diagrams because of the paucity of space available.

$$NC{-}C_6H_4{-}CN \xrightarrow{H_2O} NC{-}C_6H_4{-}CONH_2 \xrightarrow{H_2O}$$

$$H_2NOC{-}C_6H_4{-}CONH_2 \xrightarrow{H_2O} H_2NOC{-}C_6H_4{-}CO_2NH_4 \xrightarrow{H_2O}$$

$$NH_4OCO{-}C_6H_4{-}CO_2NH_4 \xrightarrow{H_2O} NH_4OCO{-}C_6H_4{-}CO_2H \xrightarrow{H_2O}$$

$$HO_2C{-}C_6H_4{-}CO_2H \quad (9)$$

It was found best to conduct this series of reactions in three stages.

In the first stage the wet dinitrile cake and the mother liquor from the end of the stage were blended and fed to the top of a staged countercurrent reactor, where hydrolysis and steam stripping both were effected. The bottoms were cooled and centrifuged to separate the solid monoammonium terephthalate:

$$HO_2C{-}C_6H_4{-}CO_2NH_4 \quad [2]$$

The mother liquor from the centrifuge was recycled as indicated in Fig. 2. The steam and ammonia were sent to the ammonia recovery still so that the ammonia could be separated and recycled.

In this first stage it was found that the rate-controlling step was the hydrolysis of the amide to the ammonium salt. The rate was found to increase with pH, although it was not possible to use the stoichiometric amount of alkali without increased by-product formation. The mother liquor was found to contain both the ammonium salt of the monoamide and the diammonium salt of the acid:

$$H_2NOC{-}C_6H_4{-}CO_2NH_4 \qquad NH_4OCO{-}C_6H_4{-}CO_2NH_4 \quad [3]$$

In all this hydrolysis the amount of amide had to be kept low (<10 ppm because this grouping introduced nitrogen and therefore color into the final product as well as serving as a molecular-weight-limiting chain stopper.

In the second stage the solid, dry monoammonium terephthalate was steam stripped in a heated rotary decomposer to give impure terephthalic acid. This second stage was noncatalytic. The ammonia-steam mixture from the stripping was passed through a cyclone to recover fine solid particles, and then was cooled. The ammonia-rich vapor went to the ammonia still, and the dilute aque-

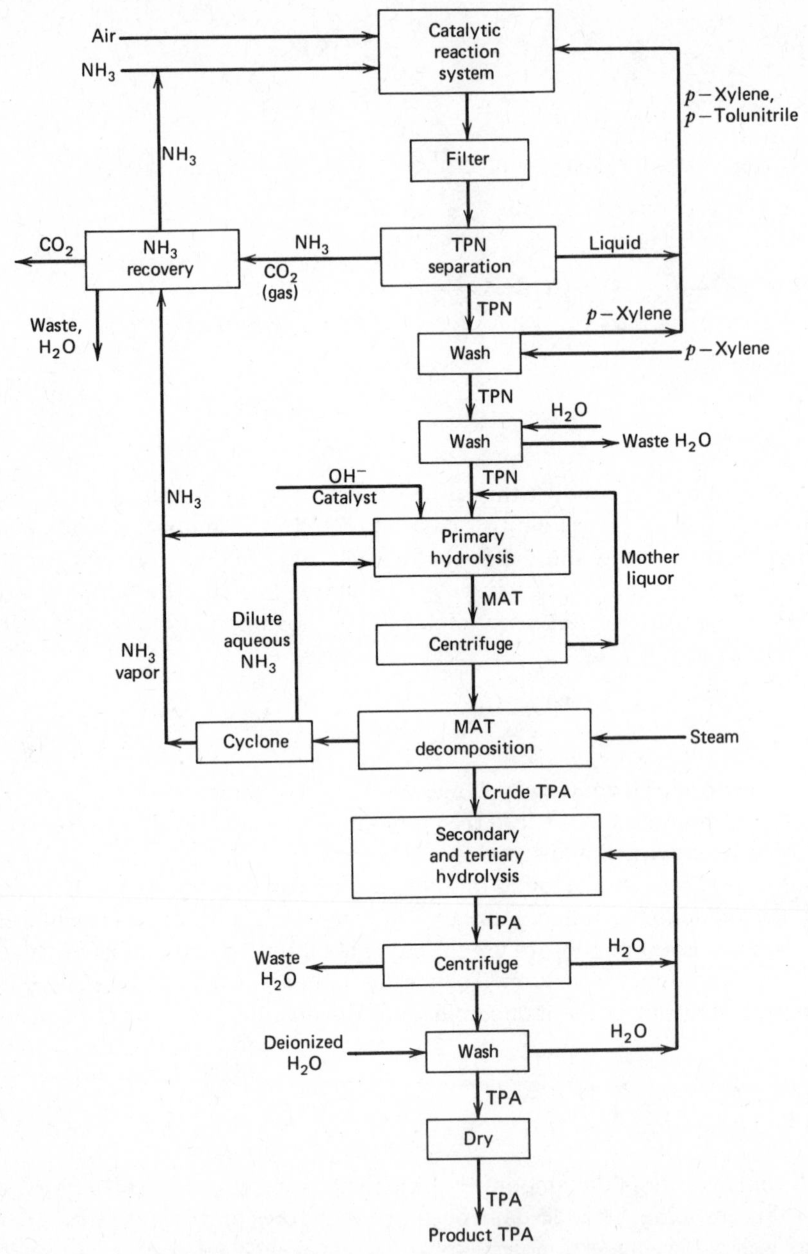

FIG. 2. Terephthalic acid process flow sheet. Terephthalonitrile has been abbreviated to TPN, monoammonium terephthalate to MAT, and terephthalic acid to TPA. Prepared by the author from the article by A. P. Gelbein, M. C. Sze, and R. T. Whitehead, *Chemtech,* August 1973, p. 479.

ous ammonia was fed to the dinitrile-slurry vessel feeding the primary hydrolyzer.

$$HO_2C-C_6H_4-CONH_2 \quad [4]$$

In the third stage the impure terephthalic acid containing the monoamide (structure **4**) was slurried with recycled water to bring the nitrogen content below 10 ppm. In order to keep the terephthalic acid in solution, a high temperature had to be employed (presumably with the system under pressure). Cooling was effected so as to produce the product in the desired crystalline form. This product was separated on a centrifuge, washed with deionized water, and dried. The wash water was recycled.

Next, the concern is with the wastewater streams. Here again it will be clearer if the flow sheet is in view (Fig. 3). The wastewater stream from the final terephthalic acid wash started as deionized water containing no inorganic salts. When it left this part of the process, it contained some organics. It was concentrated by means of waste heat from the main process unit, and the organics were separated by filtration before the water went to the settler.

The wastewater streams from earlier steps in the process were found to comprise hot water plus liquid and solid organic contaminants. These streams were cooled to 25°C in the settler, and the solids at the bottom were separated by filtration. These solids and those obtained above were slurried in fuel oil and burned. The liquid organics (oils) were separated in the decanter and also burned. The water was returned to the settler.

The clear water from the settler, containing some soluble organics, was fed to the top of the cooling towers. Since the organics separated as the water evaporated and cooled, the cooling-tower slats were constructed of plastic rather

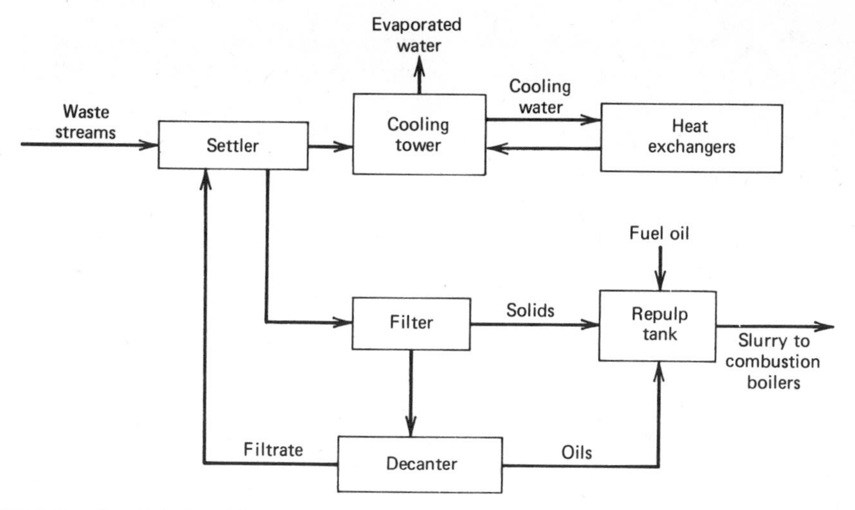

FIG. 3. Terephthalic acid process waste streams. Prepared by the author from the article by A. P. Gelbein, M. C. Sze, and R. T. Whitehead, *Chemtech,* August 1973, p. 479.

than wood so that the sludge could be scraped off periodically and returned to the main settler for separation and burning. The clear water that did not evaporate was sent to the process coolers.

Not all the details of the process were given in the original article. Here and there obvious gaps have been filled in by guesswork, but plenty of loose ends remain.

Again, in this example, some research had to be conducted in a development program—on the "catalyst" that provided the oxygen for the ammoxidation, and in determining the kinetics and yields in the various steps in the hydrolysis. This example places much more emphasis on chemical engineering than the first one does, and therefore reemphasizes the necessity for the chemists and the engineers to work very closely together and for the chemists to be familiar with plant equipment. The example also shows that a process that can appear to be quite simple chemically may require a fairly complex plant for its commercialization.

In industrial chemistry, yield is very important. This is illustrated in this example by the recycling of just about everything. Even water can be an expense, so maximum use is made of the water employed. In industry, energy is very expensive. In this process what might otherwise be waste heat is utilized wherever possible.

One of the selling points made for this process is ecology. In theory at least there are no liquid or solid wastes, only gases that have been scrubbed in a cyclone—water vapor and carbon dioxide, hardly atmospheric contaminants. As environmental regulations become stricter, processes of this type will find more extensive utilization. Buttoning up a process as thoroughly as this one poses a real challenge to the chemist and to the chemical engineer.

V. DIMETHYLFORMAMIDE

The next example is a Japanese process for preparing dimethylformamide, a widely used solvent in polyacrylonitrile-fiber manufacture and for butadiene extraction [5].

The present process for manufacturing this chemical is the reaction between carbon monoxide and dimethylamine:

$$CO + HN(CH_3)_2 \xrightarrow{\text{catalyst}} HCON(CH_3)_2 \qquad (10)$$

Like the older chloroprene process mentioned at the start of this chapter, this process looks great on paper. However, in the manufacture of dimethylamine from ammonia and methanol, even though stoichiometric quantities of the two reactants are used, appreciable amounts of both methylamine and trimethylamine are also produced (Equation 11), and the separation of the three products by fractional distillation is both costly and difficult. In addition the byproducts have to be gotten rid of.

$$NH_3 + CH_3OH \longrightarrow CH_3NH_2 + (CH_3)_2NH + (CH_3)_3N \qquad (11)$$

The Japanese tried combining the two steps in the old synthesis, but the yield of dimethylformamide was low.

$$CO + NH_3 + 2CH_3OH \longrightarrow HCON(CH_3)_2 + 2H_2O \qquad (12)$$

The literature revealed that methanol reacts with acetonitrile to give dimethylacetamide. Why not substitute hydrogen cyanide for acetonitrile in this known chemical reaction?

$$HCN + 2CH_3OH \xrightarrow{\text{catalyst}} HCON(CH_3)_2 + H_2O \qquad (13)$$

Hydrogen cyanide is readily available as about a 10% by-product from acrylonitrile manufacture.

The first problem was to find the right catalyst, partly research and partly development. Some 200 different potential catalysts were screened batchwise in glass or titanium equipment. Tubes were used for soluble catalysts and a stirred reactor for insoluble ones. Hydrogen cyanide was charged as a methanol solution. Qualitatively, analysis for identification purposes utilized the mass spectrometer with gas chromatograph attachment, the infrared spectrophotometer, and nuclear magnetic esonance. Quantitative analysis was by gas chromatography.

Of the catalysts tried, the most effective proved to be the chlorides, oxyhalides, and hydroxides of titanium, zirconium, selenium, and tellurium. The principal by-products were the other two amides plus the expected gases and high boilers.

$$HCONHCH_3 \qquad HCONH_2 \qquad [5]$$

Recycling the other two amides increased the yield of dimethylformamide. The presence of water at the start of the reaction proved beneficial in that the yields of the three amides were increased and there was less color in the product and fewer high-boiling by-products.

The optimum reaction conditions were shown to be temperatures of 240–260°C, retention times of 20–40 minutes, and pressures of 60–100 kg/cm^2. For homogenous catalyst (titanium tetrachloride), 1–2% based on hydrogen cyanide was best; for a heterogenous catalyst (titanium hydroxide), 5–10% based on hydrogen cyanide was needed.

As time went on, an exhaustive analysis was made of the various by-products, which are shown in Table 3.

TABLE 3. Hydrogen Cyanide Methylation By-Products[a]

CO	$(CH_3)_2NH$	$CH_3CON(CH_3)_2$
CO_2	$(CH_3)_3N$	$CH_3CONHCH_3$
NH_3	HCO_2CH_3	CH_3CONH_2
CH_3NH_2	HCO_2H	CH_3CN

[a] Adapted with permission from Y. Fukuoka and N. Kominami, *Chemtech*, November, 1972, p. 670. Copyright 1972 American Chemical Society.

This process deals with catalysis, an obscure subject at best, in which, as in this case, an Edisonian approach in research provides useful results even though theory and logic may not have contributed a great deal to the planning of the experiments.

In this process, materials of reactor construction had a major effect on product yield. Table 4 summarizes experiments conducted in glass at 240°C for two hours. The yield is based on mole percent of hydrogen cyanide consumed. Heavy corrosion was observed with stainless steel and Hastelloy. There was no titanium corrosion in 500 hours.

The titanium tetrachloride catalyst as residue from the vacuum distillation separation of the various products was recycled effectively four times.

A 10 kg/day pilot-plant study was conducted in titanium equipment using a titanium tetrachloride catalyst. Yields were 80–85% based on hydrogen cyanide and 95% based on methanol. Conversions were 100% based on hydrogen cyanide and 87% based on methanol.

A diagram of the proposed process is shown in Fig. 4.

Any unreacted hydrogen cyanide presumably went off with the other gases from the top of the stripper. This accounts for the 100% conversion (no recovery) and some of the 80–85% yield. Not shown (Fig. 4) is a scrubber to remove the hydrogen cyanide before venting the gases to the atmosphere, a piece of equipment that would have to be part of the final plant. Although it might be possible to vent this much hydrogen cyanide in parts of Siberia, it would not be possible in countries like Japan and the United States.

Presumably most of the unreacted methanol came off the top of the water separator and was recycled. This accounts for much of the 87% conversion and 95% yield.

Acetamide and methylacetamide boil closely enough to the corresponding formamides that they doubtless were recycled with them and underwent further alkylation to dimethylacetamide. Since dimethylacetamide boils only about 10°C above dimethylformamide, it must be present as an impurity in the final product. Otherwise the purification tower would be prohibitively expensive.

TABLE 4. Effect of Reactor Material Construction[a]

Material Tested	Total (all 3) Amides mol %	Gases mol %	Residue mol %
Glass	92	2	6
Stainless steel (SUS-32)	68	9	26
Titanium	90	3	8
Hastelloy	72	7	20
Zirconium	89	3	7

[a] Adapted with permission from Y. Fukuoka and N. Kominami, *Chemtech*, November, 1972, p. 670. Copyright 1972 American Chemical Society.

As a logical by-product of this research and development project, a dimethylacetamide process also was developed:

$$CH_3CN + 2CH_3OH \xrightarrow{\text{catalyst}} CH_3CON(CH_3)_2 + H_2O \qquad (14)$$

In this case a stainless steel reactor could be used (SUS-32). The catalyst was zinc chloride. At 350°C and 150 kg/cm^2 applied pressure and with a 30-minute residence time, the conversion based on acetonitrile was 91% and the yield on the same basis was 95%. The yield based on methanol was 96%. Recycled by-products included

$CH_3CONHCH_3$ CH_3NH_2
CH_3CONH_2 $(CH_3)_3N$
CH_3CO_2H

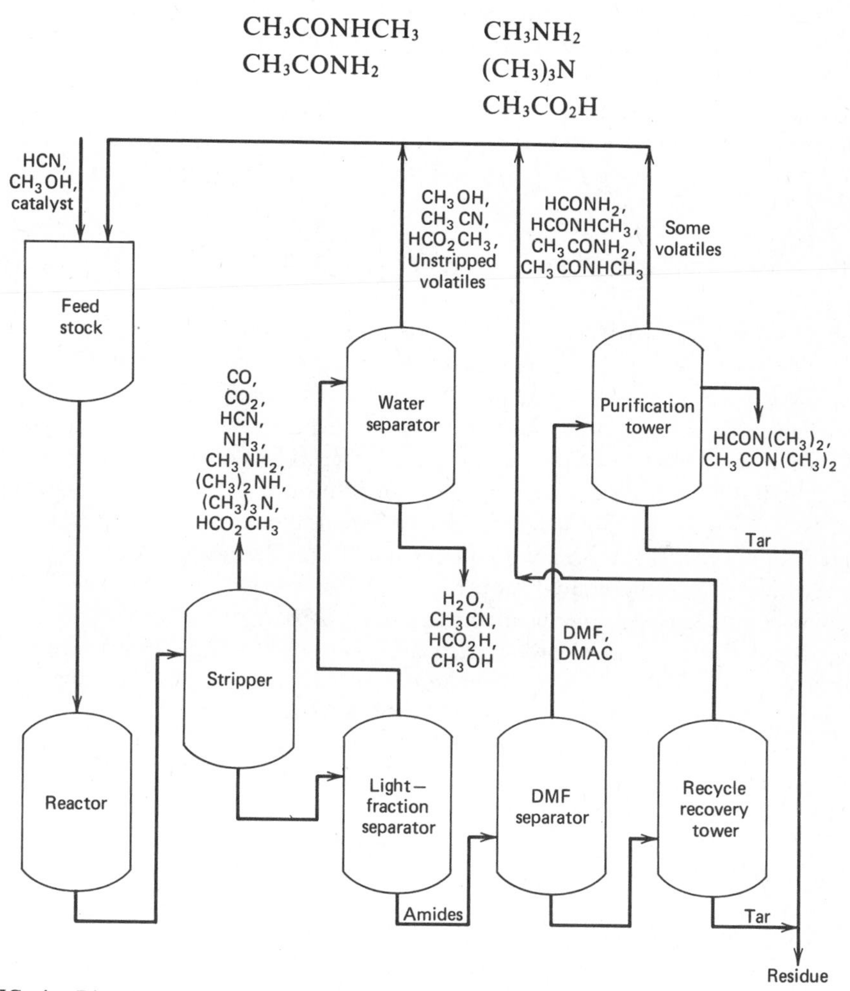

FIG. 4. Dimethylformamide process flow sheet. The streams have been labeled for a better understanding of what transpired. Dimethylformamide has been abbreviated to DMF and dimethyl acetamide to DMAC. Adapted with permission from Y. Fukuoka and N. Kominami, *Chemtech*, November, 1972, p. 670. Copyright 1972 by American Chemical Society.

As in the two previous examples, research again is encountered in the middle of a development program. Likewise, as much material as possible was recycled to insure maximum yield and minimal raw-material costs. Again, this is a process outline without every step described in detail. Of interest is the emphasis on the use of the best modern analytical equipment in conducting the laboratory work.

In this last process a big economic standoff was the capital cost of a plant made of titanium compared with the dimethylamine-manufacture problem. A company in the business of manufacturing and selling all three methylamines obviously has an appreciable advantage in using the older process.

Today operator safety in chemical plants is often discussed. Over the years this must have been a major concern of industrial chemical management and of their technologists. Just in this chapter processes have been described in which such toxic and unpleasant chemicals as hydrogen cyanide, carbon monoxide, chlorine, acetylene, ammonia, and formic acid are involved. Safety is a very major factor in the successful engineering of a chemical plant.

VI. A NEW GOLF BALL

The last example involves the development of a product rather than a process [6]. Here again chemists were members of an interdisciplinary team. This is the development of the new United States Rubber Company golf ball.

The desire for new processes comes from the manufacturing department because of the technical and economic unsatisfactoriness of older processes. Examples have just been shown in the cases of chloroprene, terephthalic acid, and dimethylformamide. The desire for new products comes from the sales department in order to have something better to offer in the marketplace.

In the case of the United States Rubber Company golf ball, what the national commodity manager wanted was a new golf ball, completely certified and ready for manufacture in one year. It had to meet all USGA specifications; it had to look different from existing models; it had to outperform competition; and it had to be sold competitively at a profit. These are the kinds of new-product specifications a development team encounters all the time.

What do some of these marketing specifications really mean? First, what about the USGA requirements? They are as follows: the ball cannot be less than 1.680 inches in diameter; (2) the ball must weigh no more than 1.62 ounces; and (3) the ball must attain no more than a maximum initial velocity of 255 ft/sec on a standard testing machine.

Within these technical limits a ball has to meet the specifications (ideals) of the average golfer: (1) it must go a mile when driven, even if the fairway is waterlogged by a month of continuous rain; (2) it must drop dead on the green, even if there has been no rain for a month and the ground is like concrete; (3) it must putt straight and true no matter how poor a putter the golfer in question may be; (4) it must be white and stay clean in spite of landing in the mud or having the grass of the rough bashed against it; (5) it must never be cut no matter how viciously the golfer in question may top it with his or her niblick; and (6) it must sound and feel pleasant when hit even though heading in a wide arc for the trees out of bounds on the right.

Facing these various and divergent requirements, how does a development team tackle such a problem? As scientists, what they did was to make a fundamental study of why a golf ball acts the way it does. The more one knows of fundamental scientific principles, the easier it is to make a better product.

A golf ball is not a simple projectile with its path in flight determined by weight, launch angle, and velocity at start. Backspin can be as high as 10,000 rpm. Spin and airstream determine the aerodynamics. The path of flight is determined by both aerodynamics and gravity.

Aerodynamics has two components: (1) the lift perpendicular to the line of flight and (2) the drag along the line of flight, which opposes forward motion. Many factors affect lift and drag. Most important are the shape, size, depth, and arrangement of the markings on the ball. These factors were studied in a wind tunnel.

Then came the field trials. These were made photographically on a range with a ball half black and half white to show spin. Under the eyes of several cameras, all balls were machine-hit at night over a carefully surveyed range with reference markers. Both the markers and the ball were floodlighted, and a computer was used to interpret the complete record of spin, velocity, and position of the golf ball.

When these field trials were made, certain requirements had to be met in order to do this research in applied physics. A still, dark night was needed with no moon, no wind, and no surface lights to interfere with the photography. Florida was selected as the location. Unfortunately, altogether too often the nighttime relative humidity exceeded 100%, so the camera plates became fogged. Likewise, on Florida nights both mosquitos and snakes are at large, neither of which are pleasant company in the dark.

What did the results of this fundamental physics study demand of the chemical technologist? Better surface markings had to be designed. It was a matter of personal satisfaction to the organic chemist involved that the optimum design proved to be one of hexagons.

The basic structure of a golf ball consists of the following (Fig. 5): *A*, a hard core center to give weight; *B*, what used to be a long winding of fine rubber ribbon to give resilience (the author has not taken a golf ball apart in a long time); and *C*, a surface coating (cover) to preserve the ball and combat the less accurate strokes of the more inexperienced golfers.

The detailed study had to involve the center, the windings, and the cover of the ball, with regard to both dimensions and materials used. The polymers found to be successful in the center had to be processed differently than rubber. They froze at too low a temperature and were tough for rubber core winding. Ionomers were found to be much better than balata for the cover, but they have to be molded at such high temperatures that if the ball manufacture was performed in the usual way, the rubber inside was ruined. New paints had to be found, since the standard golf ball paints do not stick to ionomers.

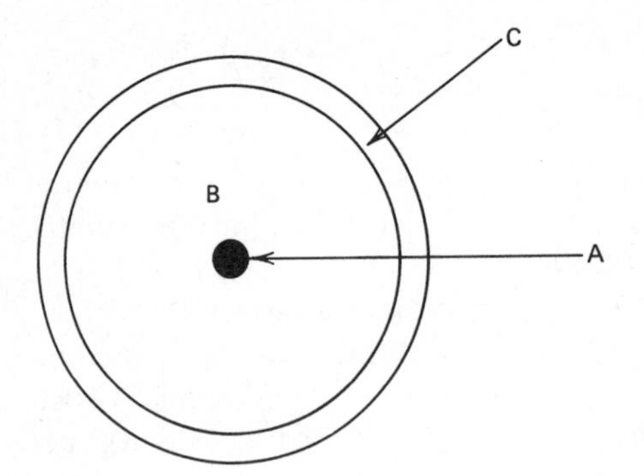

FIG. 5. Diagram of a golf ball.

The scientific and technological problems were solved, but for obvious commercial reasons, the technical details have not been revealed.

This product-development project illustrates first the necessity for some kind of marketing input to help guide the development. Once the goals were set and agreed upon, the development had to be entirely technological. The sales department hardly directed the technical program, although there had to be excellent rapport between the research, the development, and the sales departments for the project to be successful.

Examples of development projects similar to the four given in this chapter can be added almost *ad infinitum*. However, the author hopes that the four given here will be sufficient to give the reader some idea of the kinds of problems he or she may be called upon to solve in industry. Often a very fundamental scientific approach is needed.

The technical depth of the problem as originally presented can be very deceiving. When the author was on the staff of Arthur D. Little, Inc., the company was asked to undertake a technical assignment for a burial-vault company. There are certain religious sects in this country who, when they move, disinter the more recently departed, bring them along, and rebury them near the new home. Moving a concrete burial vault is a major engineering undertaking. This should not be the case with a light plastic vault.

At first glance the solution to this problem might appear to require little technical depth. However, very sophisticated accelerated-aging tests had to be developed in order to be sure that the final plastic formulation would withstand 50–100 years in the ground.

REFERENCES

1. A. J. Besozzi, W. H. Taylor, and C. W. Capp, *Preprints*, Division of Petroleum Chemistry, Inc., New York, August 27–September 1, 1972, p. E15.
2. C. S. Marvel, private communication.
3. M. Ogawa and Y. Yoshimaga, Jap. Pat. ('74) 00806; *C.A.*, **81**, 25040y (1974).
4. A. P. Gelbein, M. C. Sze, and R. T. Whitehead, *Chemtech*, **3**, August, 1973, p. 479.
5. Y. Fukuoka and N. Kominami, *Chemtech*, **2**, November, 1972, p. 670.
6. F. Martin, *Chemtech*, **2**, July 1972, p. 448.

9

PATENTS

I. INTRODUCTION

The subject of patents is one to which most beginning chemists and chemical engineers have had very limited exposure at best. Yet this area is a very important one in industrial life and has been since the founding of this country. Thomas Jefferson was the first patent examiner, and George Washington signed the first patent [1].

In the following discussion the author will use certain standard patent terms. They should be self-defining.

II. NATURE OF A PATENT

What is a patent? A patent is a legal document that can be said to define a piece of property. It gives to the owner thereof the right to exclude others from using the specific piece of property for a period of 17 years. By so doing it is designed to encourage invention. It also is designed to encourage the publication of inventions, so that ultimately they can be used by everybody. Presumably during the 17 years in which the patent is in force, the inventor will not only recover his research costs, but also will make sufficient profit that his investment in time and money during the inventing process will have been worthwhile. Although originally used mainly by individual inventors, the principle is just as applicable to large corporations that invest many thousands of research dollars and wish to establish a patent position. The chemist should remember that one reason research professionals are hired by industry is to invent, and they usually are re-

---

The chapter on patents in J. H. Saunders, *Careers in Industrial Research & Development,* Dekker, New York, 1974, pp. 180–201, is excellent. His description of the determination of inventorship is particularly good.

quired to assign their inventions to their employer as part of their employment contract.

III. STRUCTURE OF A PATENT

Patents pretty clearly are divided into two parts—the *disclosure,* including the *examples,* and the *claims.*

The disclosure is the first part of a patent. It defines the scope of the invention. It may review the prior art, which means pertinent literature of which the inventor is aware. The disclosure also establishes a defensive position. Since one cannot patent anything that has been disclosed in the literature, even if no one has tried it, the disclosure of an issued patent may anticipate work that a competitor otherwise might think of and patent.

The disclosure also is a selling document. It tries to show how unexpected the nature of the experimental results were (even if the experiments were carefully thought out and planned with just these results in the inventor's mind) and how unobvious they were to anyone skilled in the art. Likewise the invention must have utility, which also is shown in the disclosure. One cannot patent an idea.

Next come the examples, which illustrate the scope of the invention and back up the claims. Examples must be repeatable by someone skilled in the art. Otherwise the claims based thereon can be declared invalid in court. However, as anyone who has tried to repeat patent examples knows, the exact procedure to be followed can be quite obscure.

Legally a patent working example need never have been tried. It simply has to produce the desired result when someone skilled in the art follows the directions given. However, chemists by and large treat patent examples as publications of laboratory experiments, and it hardly is considered to be ethical to publish as actual experiments laboratory work that one has never performed. However, patent working examples are not considered publications of actual physical work in many fields other than chemistry. This is an excellent example of something that can be perfectly legal, but which many people consider to be unethical, at least when actual experimental work is implied.

The second part of a patent is the claims. These define the legal property from which others are excluded and are based on the disclosure and reasonable extensions thereof. Claims always are one sentence. Even the German language, which is noted for the complexity of its sentence structure, is a marvel of simplicity compared with the claims covering a complicated mechanical device. As is obvious from an examination of almost any patent, a series of claims is written in diminishing scope. This is so that if the broadest claim is declared invalid, at least some part of the property can be salvaged. It is difficult to obtain claims that cover an area much larger than that disclosed in the examples. For instance, if one is claiming the esters of a new organic acid and gives as examples the methyl, butyl, and octyl esters, claims above decyl or dodecyl are not likely to be allowed.

IV. TYPES OF PATENTS

Before the procedure one goes through in order to obtain a patent is examined in some detail, several types of chemical patents need to be examined broadly.

1. *Product.* A product patent covers a specific composition of matter—a new chemical compound or group of chemical compounds, or a new formulation or range of formulations. It is the most valuable type of patent because it is the most easily policed. All one has to do is analyze a competitor's product to find out if one's patent is being infringed.

2. *Process.* A process patent covers a specific process or a group of closely related processes. It is nowhere nearly as valuable as a product patent. In order to ascertain infringement one really would need a spy in the competitor's plant.

3. *Use,* A use patent covers the use of a chemical or group of chemicals in a specific application. It is about as valuable as a process patent. One pretty much would need to have a spy in the user's plant to determine infringment, unless the use is performed in public, as in the case of an agricultural pesticide.

V. PROCEDURE FOR OBTAINING A PATENT

The possibility of filing a patent application starts with an idea, which is duly recorded in a chemist's notebook. As is the case with all laboratory-notebook pages, the page or pages describing the idea should be signed by the chemist or chemical engineer and witnessed. This is the date of conception, which is very important in determining inventorship in the case of an interference. (Interferences are discussed subsequently in this chapter.) Presumably the inventor is sufficiently familiar with the literature that he or she has every reason to believe the idea is novel.

Many U.S. and foreign patents carry the names of two or three inventors. At one time the U.S. Patent Office felt that only one person could have the "flash of genius" that would lead to a patentable invention. However, in a chemical research group, particularly if the idea is the group leader's, those working on the idea at the bench usually contribute enough to be included as inventors. This is excellent from a morale point of view, because a group leader who hogs all the patents as sole inventor usually does not have a very happy group.

The next step, of course, is to test the idea in the laboratory. The description of the successful experiment is again recorded and the notebook page or pages signed and witnessed. This is the date of reduction to practice. Actually the patent position is stronger if someone other than the inventor reduces the idea to practice. The time between the dates of conception and reduction to practice should not be unreasonably long, although *unreasonable* is hard to define. If this time is held to be unreasonably long, then the inventor has not been exerting due diligence, and it will be held against him in an interference.

After an idea has been reduced to practice, the experiment usually is checked, and perhaps, several related experiments are performed in order to better define the scope of the invention. At this point a company patent disclosure form usually is filled out and submitted to the company patent department or to a law firm if the company employs outside counsel. Most companies require something of this nature. Either before or after the preparation of the disclosure form, the chemist sits down with the patent attorney and discusses the scope of the invention in the light of the known prior art. The attorney probably will suggest a few additional experiments to better define the scope and obtain broader claims.

The patent in which the author and Dr. Tracy M. Patrick, Jr., were coinventors is reproduced in Fig. 1. Examples 1, 2, 3, and 4 in this patent describe basically the same method for preparing thiophenealdehydes from thiophene, 2-methylthiophene, and 2-chlorothiophene. The broadest claim covers the preparation of thiophenealdehyde, monoalkylthiophenealdehydes, and monohalogenated thiophenealdehydes by that method. Much broader coverage (from broader scope) could have been obtained if there had been examples using as starting materials di- and trialkylthiophenes, di- and trihalogenated thiophenes, arylthiophenes, heterocyclically substituted thiophenes, acylthiophenes, thiophenecarboxylic acids and their derivatives, nitrothiophenes, acylaminothiophenes, and so forth. However, although Monsanto Company did not obtain patent coverage on the preparation of all these different types of compounds, it would have been rather difficult for anyone else to do so subsequently using the same method, since it would have been hard to say that it was unobvious to anyone skilled in the art that these compounds might not undergo the same reaction.

Although the chemist presumably is familiar with the literature, a thorough search of the patent literature is nevertheless made. Usually this is conducted in the Patent Office in Washington. Most patent law firms and company patent departments have correspondents in Washington who conduct such searches for them. It is no mean assignment. However, it is facilitated by the fact that the now over 4 million U.S. patents that have been issued are divided by subject matter into some 70,000 official subclasses [1]. There also are over 6 million cross-reference patents and over 10 million foreign patents.

During the course of the procedure that has been outlined so far, after the first reduction to practice, a decision is reached concerning whether to file a patent application. This decision usually is made by the research director and the patent attorney with inputs from the chemist and his or her group leader. It is based on the perceived future value of the patent to the company and the closeness of the prior art. (The latter is a measure of the difficulty and probable expense of obtaining a patent.) These people may also be restricted in their decision if the company has a budgetary limitation on the amount of money to be spent on patents each year.

If the decision is made to file, then the patent attorney prepares the application in the form just outlined. The chemist reviews it. The chemist signs an oath of affirmation attesting to his or her inventorship, and the attorney files the application in the Patent Office and pays the filing fee. The date of filing is most important, since it gives priority in interference proceedings. (The first to file is the senior party, and the burden of proof is on the junior party.) A patent application must be filed within one year after the research results have been published. Patent applications are secret, both while pending and when abandoned.

The first step on the part of the Patent Office is a review by a government security office, which may refer the application to various government agencies, for example, NASA, AEC, DOD, and so forth. A government security office must issue a license if the application is to be filed abroad within six months of the date of the U.S. filing, and the application may be held secret if the security of the United States is involved.

Patent applications are reviewed by *patent examiners.* Many of these are young, embryonic patent professionals gaining experience. There are some 1200 of them divided into three main technical areas—chemical, electrical, and mechanical. Of the more than $50 million budget of the Patent Office, more than half is paid for by the applicants' fees [1].

In general, patent examiners try to make the inventor, or rather the inventor's attorney, prove patentability. It has been stated that fewer than 20% of all applications are allowed the first time around [1]. The author feels this is a very generous figure. When rejecting the claims in an application, the examiner may cite a variety of references alone and in combination; the relevance of the references is not always obvious to the inventor and the attorney. The attorney has to act within specified time periods in replying to the examiner.

The first action starts what can be a long period of negotiations. The inventor's attorney may file amendments to meet the objections of the examiner. If the invention is a valid one, the examiner will allow claims as arguments and evidence are presented. Finally, after obtaining all the claims he or she thinks will be given, the inventor's attorney pays the final fee and allows the patent to issue with the allowed claims. However, allowance can be withdrawn for reason right until the time of issuance. Also, as of July 1, 1981, patents may be reexamined if adequate reasons therefore can be presented [2].

Research scientists are interested in publishing their work. Publication policies vary from company to company. Many companies withhold permission to publish until the corresponding patents are ready to issue. In others the attorneys follow the very reasonable policy of waiting until they have received the first action from the Patent Office before permitting publication, so that they know what art they are up against.

The inventor and attorney are not always satisfied with the claims allowed by the examiner as evidenced by a notice of final rejection. The first appeal is an informal one to the examiner's supervisor. The next appeal is a formal one to the Board of Appeals. An appeal fee is required. It has been reported that the Board of Appeals reverses the examiner in about 30% of the cases [1]. Recourse can then be had to the U.S. District Court for the District of Columbia or to the U.S. Court of Customs and Patent Appeals. Final recourse is to the Supreme Court of the United States.

Once a patent issues, the claims are in force for a period of 17 years from the date of issuance. If a company has a big stake in an area, as E. I. Du Pont de Nemours and Company had in Nylon, then it files improvement patents. Although dominated by the original, broad basic patent, these serve to provide protection after the original patent has expired.

VI. INTERFERENCES

Anyone who has been in research for any length of time knows that there is a definite timeliness to it. The same discovery often is made at much the same time in two or more different laboratories. This can lead to two parties' applying for patent coverage on the same invention at the same time.

If a patent examiner is confronted with two essentially identical patent applications, he can declare an interference. Usually he does not do this if there is a three months' difference in filing dates between the two applications, and he must obtain the approval of the Commissioner of Patents if the difference is more than six months [1]. If a patent issues and an inventor has a similar application in the Patent Office, he can force an interference by copying the claims of the issued patent. In this case the first patent cannot be canceled, but it becomes unenforceable if the second one issues.

In an interference, testimony and depositions of the inventor and of persons who worked on the invention are taken, and their notebooks are examined. The inventor's work must be corroborated, and this is why all laboratory-notebook pages must be signed and witnessed. The patent is awarded to the original inventor, the one who had the earliest conception date. However, the inventor must have exercised due diligence in reducing the invention to practice, so the date of reduction to practice is important.

An excellent example of how close two patents can be and not be in interference is the discovery that 2-thiophenealdehyde can be prepared in good yield by treating thiophene with phosphorus oxychloride and an *N,N*-disubstituted formamide.

$$\text{C}_4\text{H}_4\text{S} + POCl_3 + C_6H_5N(CH_3)CHO \longrightarrow \text{C}_4\text{H}_3\text{S-CHO} \qquad (1)$$

This discovery was made at much the same time in the laboratories of Fordham University [3], Monsanto Company [4,5], and Abbott Laboratories [6].

Patented Jan. 1, 1952 **2,581,009**

UNITED STATES PATENT OFFICE

2,581,009

METHOD OF PREPARING THIOPHENE-ALDEHYDES

William S. Emerson and Tracy M. Patrick, Jr., Dayton, Ohio, assignors to Monsanto Chemical Company, St. Louis, Mo., a corporation of Delaware

No Drawing. Application August 24, 1948, Serial No. 45,935

5 Claims. (Cl. 260—332.3)

1

This invention relates to new methods of preparing thiophenealdehydes and substituted thiophenealdehydes. More particularly, this invention relates to a new method of preparing an intermediate useful in the synthesis of a wide variety of thiophene derivatives.

Thiophenealdehyde has been prepared by a variety of methods, but most of them are tedious or otherwise objectionable. The most practicable methods involve either the oxidation of 2-acetothienone to 2-thienylglyoxylic acid, followed by decarboxylation, or the reaction of 2-thenyl chloride with hexamethylene tetramine, but these methods are objectionable in that the low yields usually are obtained.

The purpose of this invention is to provide a new and useful method of preparing thiophenealdehyde and the aldehydes of various substituted thiophenes. A further purpose of this invention is to provide a novel method of preparing high yields of the various thiophenealdehydes.

In accordance with this invention, the thiophenealdehydes are prepared by reacting thiophene, or a substituted thiophene with phosphorus oxychloride and an N,N-disubstituted formamide, for example diethyl formamide, dibutyl formamide, formpiperidide, N-ethyl formanilide, or N-methyl-p-formtoluidide. The reaction is preferably conducted under anhydrous conditions in solution in a suitable organic solvent, for example toluene, benzene or xylene. After the reagents are mixed, the resulting solution is heated to reflux temperatures for sufficient time to complete the reaction, at which time the reaction mass is poured into cold water or ice, and separated by suitable distillation. An effective separation may be achieved by steam distillation, whereby the major portion of the thiophenealdehyde is collected in the nonaqueous layer of the distillate. A portion of the thiophenealdehyde is dissolved in the aqueous layer and improved yields are achieved by extracting the water solution with toluene, or another water immiscible solvent, for the thiophenealdehyde. The nonaqueous layer and the extracts may then be distilled to separate the thiophenealdehyde.

By this method, thiophenealdehyde may be prepared in unusually pure state and in very substantial yields. In addition to thiophenealdehyde, various substituted thiophenealdehydes may also be prepared. The useful thiophenealdehydes prepared in accordance with this invention are represented by the following structural formula:

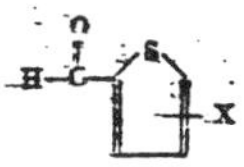

O
‖
H—C— (thiophene ring, S) —X

FIG. 1.

2

wherein X is a radical of the group consisting of hydrogen, halogen and alkyl radicals. Suitable halogen substituents include iodine, and fluorine, although bromine and chlorine are the most important. Although any alkyl substituted thiophenealdehyde may be prepared, those having from one to four carbon atoms in the alkyl group are the most useful, for example methyl, ethyl, isopropyl, n-propyl, isobutyl, n-butyl and t-butyl.

Further details of the preparation of thiophenealdehyde and the new substituted thiophenealdehydes are set forth with respect to the following specific examples.

Example 1

A 1000 ml., 3-necked flask was provided with an efficient stirrer, a thermometer and a reflux condenser sealed by means of a calcium chloride tube. The flask was charged with 135 grams of N-methyl formanilide and then, while cooling on a water bath, 153.4 grams of phosphorus oxychloride was added. The water bath was then removed and the mixture stirred for one hour. A solution of 126 grams of thiophene in 300 cc. of anhydrous toluene was then added and the entire mass boiled at reflux temperatures for one and one-half hours. The reaction mass was then cooled and poured into an equal volume of crushed ice. The mixture was then steam distilled and the aqueous portion of the condensate extracted three times with toluene. The extracts were then combined with the nonaqueous layer of condensate. The product distilled at 85° to 86° C. at 16 mm. total pressure and was identified as 2-thiophenealdehyde.

Example 2

The procedure of Example 1, was repeated except that dimethylformamide was used in place of the N-methyl formanilide. The compound formed was also identified as 2-thiophenealdehyde.

Example 3

The procedure of Example 1 was repeated and 67.5 grams of N-methyl formanilide, 76.7 grams of phosphorus oxychloride and 49 grams of 2-methylthiophene dissolved in 150 cc. of anhydrous benzene were reacted. The resulting product was identified as 5-methyl-2-thiophenealdehyde.

Example 4

Using a procedure similar to that described in Example 1, 67.5 grams of N-methyl formanilide, 76.7 grams of phosphorus oxychloride and a solution of 59.3 grams of 2-chlorothiophene in 100 cc. of anhydrous benzene were reacted. In this reaction, the reagents were combined without cooling and were boiled at reflux temperature for two hours. The final fractionation produced the compound identified as 5-chloro-2-thiophenealdehyde.

The various thiophenealdehydes produced in accordance with the method of this invention are useful intermediates in the preparation of a wide variety of organic compounds. The following examples describe in detail various types of reactions by which new and useful chemical compounds are prepared.

Example 5

A solution of 27.5 grams of p-aminophenol in 200 cc. of ethanol and 28.3 grams of 2-thiophenealdehyde were charged to a reaction flask provided with a stirring mechanism and a reflux condenser. The mixture was heated at reflux temperature for fifteen minutes while being vigorously stirred. The heating was then discontinued and the mixture was stirred for an additional fifteen minutes. The flask was then cooled and a crystalline precipitate was obtained. After filtration, the crystalline material was washed with ethanol and dried in a vacuum desiccator over potassium hydroxide. Three successive crystallizations from ethanol produced a crystalline material identified as 2-thenal-p-aminophenol.

Example 6

A 2-liter, 3-necked flask was fitted with a stirring device, a thermometer and a reflux condenser. The flask was charged with 40.5 grams of 2-thenal-p-aminophenol, 18 grams of magnesium chips and 750 cc. of anhydrous methanol. The mixture was vigorously stirred and the temperature gradually rose to 32° C. after which it was cooled periodically by immersion in an ice bath. After twenty minutes, the reaction was accelerated and it was necessary to maintain the flask in an ice bath to keep the temperature below 32° C. After one hour, all of the magnesium had dissolved and the reaction mass was then heated and 500 cc. of methanol was evaporated. The residue was then poured in to 500 cc. of 5 percent aqueous sodium hyroxide while vigorously stirring the reaction mass. During this addition, the flask was immersed in an ice bath. The reaction mass was then diluted with 250 cc. of 5 percent sodium hydroxide and filtered. The gelatinous mass so separated was mixed with an additional 150 cc. of 5 percent sodium hydroxide and filtered. The combined filtrations and washing were then treated with carbon dioxide and filtered. The filtrate was then cooled and a solid material obtained, which after four successive crystallizations from methanol, was found to have a melting point of 107 to 108° C., and was identified as 2-thenyl-p-aminophenol.

Example 7

A reaction flask charged with a solution of 13.2 grams of sodium hydroxide in 120 cc. of water and 60 cc. of ethanol. While maintaining the solution at 12° C. by means of an ice bath, 31.2 grams of acetophenone was added with vigorous stirring. Thereafter, 29.3 grams of 2-thiophenealdehyde was added and the mixture stirred at 25° C. for three hours. Upon standing, a crystalline material was separated, washed free of alkali with water, and then with 100 cc. of 50 percent aqueous ethyl alcohol. After four successive crystallizations from ethanol, a crystalline material having a melting

point of 59° C. was obtained and identified as 2-thenalacetophenone.

Example 8

Using the procedure described in the preceding example, except that p-methoxyacetophenone was used in place of the acetophenone, the solid compound so obtained was identified as 2-thenal-p-methoxyacetophenone.

Example 9

A reaction flask was charged with 11.2 grams of thiophenealdehyde, 22.7 grams of 2,4,6-trinitrotoluene, 1 cc. of piperidine and 100 cc. of xylene. The reaction mass was heated at reflux for fifteen minutes and 1.9 cc. of water was removed from the condensate in a Dean and Stark trap. Upon cooling, the reaction mass crystallized and was separated by filtration after diluting with 25 cc. of benzene. An additional crop of crystals was obtained by diluting the filtrate with hexane and cooling. The combined precipitates were dissolved in 200 cc. of benzene, decolorized by heating with activated charcoal, filtered while hot, diluted with an equal volume of hexane, and cooled. An orange colored crystalline solid, thereby obtained, was separated by filtration and dried. The resulting product was identified as a mixture of the cis- and trans-isomers of 1-(2'-thienyl)-2-(2,4,6-trinitrophenyl) ethylene.

Example 10

A mixture of 12.3 grams of rhodanine, 10.3 grams of 2-thiophenaldehyde, 20 grams of anhydrous sodium acetate, and 100 cc. of glacial acetic acid was charged to a reaction flask and boiled at reflux temperance for twenty minutes. Upon cooling, 250 cc. of water was added and the precipitate, so obtained, separated by filtration, washed with water, and dried in a vacuum desiccator in the present of potassium hydroxide. The resulting solid substance, after recrystallization from glacial acetic acid solution, was identified as 5-(2'-thenal)-2-thio-2,4-(1,3)-thiazolinedione.

Example 11

A reaction flask was charged with 31.3 grams of o-aminothiophenol and 75 cc. of pyridine. The flask was then charged gradually, while vigorously stirring, with 28 grams of 2-thiophenealdehyde. The mixture was then stirred for one-half hour at 85 to 95° C., cooled in an ice bath, and acidified with dilute hydrochloric acid. An oily substance separated and soon solidified. The solid matter was separated by filtration, washed with water and dried over potassium hydroxide. Three successive crystallizations of the product, from an ethanol solution in an atmosphere of carbon dioxide, produced a substantial yield of a compound identified as 2-(2'-thienyl)-benzothiazoline.

Example 12

A solution of 31.5 grams of the compound, prepared in accordance with the preceding example, in 200 cc. of warm ethanol was charged to a reaction flask, and then a solution of 26 grams of ferric chloride in 50 cc. of ethanol was gradually added over a period of one-half hour. The mixture was cautiously warmed for one-half hour, diluted with 100 cc. of water, and cooled. The thick slurry so obtained was diluted with a 300 cc. portion of water and filtered. The residue was washed with water and dried in vacuum

FIG. 1. (*Continued*)

2,581,009

5

over potassium hydroxide and sulfuric acid. After recrystallizing the product twice from ethanol, it was found to have a melting point of 98 to 99° C. and was identified as 2-(2'-thienyl)-benzothiazole.

The invention is defined by the following claims.

We claim:

1. A method of preparing a thiophenealdehyde, which comprises reacting a thiophene having the structural formula:

wherein X is a radical of the group consisting of hydrogen, halogen and alkyl radicals, with an N,N-disubstituted formamide and phosphorus oxychloride, in an anhydrous medium and separating the resulting product.

2. A method of preparing thiophenealdehyde, which comprises reacting thiophene with an N,N-disubstituted formamide and phosphorus oxychloride, in an anhydrous solution and separating the resulting product.

3. A method of preparing thiophenealdehyde, which comprises reacting thiophene, in an anhydrous solution, with N-methyl formanilide and phosphorus oxychloride, and separating the resulting product.

4. A method of preparing 5-methyl-2-thiophenealdehyde, which comprises reacting in an anhydrous solution, 2-methylthiophene with N-methyl formanilide and phosphorus oxychloride, and separating the resulting product.

5. A method of preparing 5-chloro-2-thiophenealdehyde, which comprises reacting in an

6

anhydrous solution, 2-chlorothiophene with N-methyl formanilide and phosphorus oxychloride, and separating the resulting product.

WILLIAM S. EMERSON.
TRACY M. PATRICK, Jr.

REFERENCES CITED

The following references are of record in the file of this patent:

FOREIGN PATENTS

Number	Country	Date
514,415	Germany	Dec. 13, 1930
519,444	Germany	Feb. 28, 1931

OTHER REFERENCES

Bernthsen and Sudborough: Organic Chemistry, page 549, Van Nostrand, N. Y., 1922 edition.

Whitmore: Organic Chemistry, 884, 893, Van Nostrand, N. Y., 1937.

Richter: Organic Chemistry, pages 649–650, Wiley, N. Y., 1938.

Ailes: J. Pharm. Exp. Ther. 72, 265 (1941).

Seemann: Canadian Journal of Research, vol. 19, sec. B, page 291 (1941).

Steinkopf: Die Chemie des Thiophens, page 21, Steinkopf, Dresden, 1941, Edwards Lithoprint 1944.

Powers, Advancing Fronts in Chemistry, vol. II, page 33, Reinhold Pub. Co., N. Y., 1946.

Williams, Detoxication Mechanisms, pages 194, 197, 198, Wiley, N. Y., 1947.

Caesar and Sachanen: Ind. Eng. Chem. 40, 922 (1948, May).

King and Nord: J. Organic Chem. 13, 635–640 (1948).

Ex parte Bywater and Coleman 83 U. S. P. Q. 4.

FIG. 1. (*Continued*)

The two patents that issued [4,6] are reproduced here (Figs. 1 and 2). Table 1 shows the dates.

The Weston (Abbott Laboratories) patent was a continuation-in-part of an earlier application, which means Weston submitted additional experimental material after filing his original application. The Emerson and Patrick paper

TABLE 1. Publication and Patent-Applied-for Dates

Article/Patent	Submitted/Applied for	Published/Issued
King and Nord article (Fordham University)	3/16/48	9/48
Weston patent (Abbott Laboratories)	9/25/47 and 5/12/49	6/24/52
Emerson and Patrick patent (Monsanto Company)	8/24/48	1/1/52
Emerson and Patrick paper (Monsanto Company)	3/28/49	9/49

Patented June 24, 1952 **2,601,479**

UNITED STATES PATENT OFFICE

2,601,479

THIOPHENE AND HALOTHIOPHENE-CARBOXALDEHYDES

Arthur W. Weston, Waukegan, Ill., assignor to Abbott Laboratories, North Chicago, Ill., a corporation of Illinois

No Drawing. Application May 12, 1949, Serial No. 92,961

4 Claims. (Cl. 260—329)

1

This application is a continuation-in-part of my prior application Serial No. 776,155, filed September 25, 1947, now abandoned.

The present invention relates to new compositions of matter and more particularly to certain heterocyclic compounds containing the thiopene ring and the preparation thereof.

The invention comprises the preparation of novel thiophene-carboxaldehyde and derivatives thereof. The general formula of the novel compounds is:

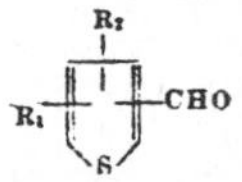

$$\text{(thiophene ring with } R_2 \text{ at position 3, } R_1 \text{ at position 5, CHO at position 2; S in ring)}$$

wherein R_1 and R_2 are hydrogen, alkyl groups containing from one to four carbon atoms inclusive, or halogen atoms.

The novel thiophenecarboxaldehydes are valuable intermediates in the preparation of organic compounds, pharmaceutical compounds and the like, some of which are disclosed in the above mentioned application.

Generally, the compounds of the invention may be prepared by reacting a thiophene with an N-lower alkyl substituted formanilide in the presence of phosphorus oxychloride, phosphorus oxybromide, thionyl chloride, and like condensing agents.

The invention may be illustrated in more specific detail by the following examples:

EXAMPLE I

2-thiophenecarboxaldehyde

A mixture of 0.5 mole of N-methylformanilide and 0.5 mol. of phosphorus oxychloride is placed in a flask and allowed to stand for about one half hour. About 0.5 mole of thiophene is then added to the mixture while the temperature of the reaction mixture is maintained at about 25–30° C. After the initial exothermic reaction subsides, the reaction mixture is allowed to stand for about 16 hours at room temperature. The mixture is then poured into ice water with vigorous stirring. The oily layer is separated from the aqueous mixture and is combined with the subsequent ether extracts of the aqueous phase. The ether solution is then washed with a dilute sodium bicarbonate solution until the washings are neutral, and then dried over sodium sulfate. The ether is distilled off and further distillation of the residual oil gives the 2-thiophenecarboxaldehyde, B. P. 91–92° C. at 25 mm., n_D^{25} 1.5888.

2

EXAMPLE II

5-bromothiophene-2-carboxaldehyde

To a mixture of 0.3 mole phosphorus oxybromide dissolved in a small amount of chlorobenzene and 0.3 mole of N-methylformanilide is added dropwise about 0.36 mole of 2-bromothiophene while maintaining the temperature at about 50° C. The reaction mixture is stirred at room temperature for about 18 hours, and then hydrolyzed in ice water and isolated as per Example I. Distillation of the oily residue gives 5-bromothiophene-2-carboxaldehyde, B. P. 118-121° C. at 15 mm. The aldehyde is further purified by conversion to the sodium bisulfite derivative, regenerated with sodium carbonate solution, and fractionated to give the pure aldehyde, B. P. 114–115° C. at 14 mm., n_D^{25} 1.6328.

By employing phosphorus oxychloride in place of phosphorus oxybromide the aldehyde is also obtained.

EXAMPLE III

5-chloro-2-thiophenecarboxaldehyde

Following the procedure of Example I, about one mole of 2-chlorothiophene, one mole of N-methylformanilide and about one mole of phosphorus oxychloride are reacted to give a crude 5 - chloro - 2 - thiophene - carboxaldehyde. The crude product from the ether extracts is shaken with an equal volume of saturated sodium bisulfite solution giving a precipitate of the addition complex, which is filtered, washed with ether and dissolved in water. The aldehyde is released by adding an excess of sodium carbonate solution, which is extracted with ether. The ether extracts are dried, concentrated and the residue is distilled to give the pure aldehyde, B. P. 91–92° C. at 13 mm., n_D^{25} 1.6017.

EXAMPLE IV

3-methyl-2-thiophenecarboxaldehyde

By following the procedure described in Example I, 0.3 mole of 3-methylthiophene, 0.3 mole of N-methylformanilide and about 0.3 mole of phosphorus oxychloride are reacted to produce the 3-methyl-2-thiophenecarboxaldehyde, B. P. 113-114° C. at 25 mm., n_D^{25} 1.5833.

EXAMPLE V

5-methyl-2-thiophenecarboxaldehyde

By heating a mixture of about 0.3 mole of N-ethylformanilide and 0.3 mole of phosphorus oxychloride with about 0.35 mole of 2-methyl-

FIG. 2.

2,601,479

3

thiophene, and isolating the product as described in Example I, 5-methyl-2-thiophenecarboxaldehyde is prepared, B. P. 113–114° C. at 25 mm., n_D^{29} 1.5782.

Following the procedure of Example I and using the desired substituted thiophene, the following compounds may be prepared. 5-tertiary butyl-2-thiophenecarboxaldehyde, B. P. 135–136° C. at 25 mm., n_D^{26} 1.5428, and 2,5-dimethyl-3-thiophenecarboxaldehyde, B. P. 116–117° C. at 25 mm., n_D^{25} 1.5599. Other alkyl or halogen substituted thiophenecarboxaldehydes may be prepared by the use of the appropriate alkyl or halogen substituted thiophene.

I prefer to carry out the reaction of the invention at about room temperature. Higher temperatures will work but the yield is usually decreased. For instance, if the reaction temperature in preparing 5-chloro-2-thiophenecarboxaldehyde is raised by carying out the reaction on steam-bath, the yield is about 20% lower than carrying out the reaction at about room temperature.

In the preferred process equi-molar amounts of the reactants or a slight excess of the thiophene compound gives the best results. Using an excess of the other reactants gives lower yields.

Others may readily adapt the invention for use under various conditions of service, by employing one or more of the novel features disclosed or equivalents thereof. As at present advised with respect to the scope of my invention, I desire to claim the following subject matter.

I claim:

1. The process of producing a compound of the formula

R–⟨S⟩–CHO

where R is a member selected from the class consisting of hydrogen and halogen, which comprises reacting a compound of the formula

R–⟨S⟩

4

where R is a member selected from the class consisting of hydrogen and halogen, with N-lower alkyl formanilide in the presence of a condensing agent.

2. The process which comprises reacting 2-chlorothiophene and N-lower alkyl formanilide in the presence of phosphorus oxychloride to produce 5-chlorothiophene-2-carboxaldehyde.

3. The process which comprises reacting 2-bromothiophene and N-lower alkyl formanilide in the presence of phosphorus oxybromide to produce 5-bromothiophene-2-carboxaldehyde.

4. The process which comprises reacting thiophene and N-methyl-formanilide in the presence of phosphorus oxychloride to produce 2-thiophenecarboxaldehyde.

ARTHUR W. WESTON.

REFERENCES CITED

The following references are of record in the file of this patent:

UNITED STATES PATENTS

Number	Name	Date
1,717,567	Kalischer	June 18, 1929
1,807,693	Kalischer	June 2, 1931

OTHER REFERENCES

King: J. Org. Chem., 13, 635–640 (1948).

Steinkopf: Die Chemie des Thiophens, pp. 21, 25 and 67, Steinkopf, Dresden, 1941 (Edwards Lithoprint, 1944).

Bernthsen and Sudborough, Org. Chem., p. 549, van Nostrand, N. Y., 1925.

Richter: Org. Chem., pp. 649–650, Wiley, N. Y., 1938.

Whitmore: Org. Chem., p. 893, van Nostrand, N. Y., 1937.

Lands: Proc. Soc. Exp. Bio. Med. 57, 55–56 (1944).

Alles: J. Pharm. Exp. Ther. 72, 265 (1941).

Powers: Advancing Fronts in Chemistry, vol. II, p. 33, Reinhold Pub. Co., N. Y., 1946.

Williams: Detoxication Mechanisms, p. 194, Wiley, N. Y., 1947.

Viaud: Produits Pharm. 2, 58, Feb. 1947.

Caesar and Sachanen: Ind. Eng. Chem. 40, 922 (1948).

FIG. 2. (*Continued*)

was submitted seven months after their patent was applied for, which was after the first action was received from the Patent Office. In those days it took about three years to process a patent application.

Analysis of the two patents shows that the preparations of 2-thiophenealdehyde, 5-methyl-2-thiophenealdehyde, and 5-chloro-2-thiophenealdehyde using *N*-methylformanilide and phosphorus oxychloride are essentially identical in both patents. Weston also gave examples using phosphorus oxybromide with 2-bromothiophene and one using 3-methylthiophene. Emerson and Patrick (Monsanto Company) gave an example using *N*-methylformamide. Weston also said that 2-*t*-butylthiophene and 2,5-dimethyliophene *may* undergo the same reaction, which means he probably did not try them, but quite rightfully had every good reason to believe they would react as expected. In addition, both

TABLE 2. Differences in Claims

Weston	Emerson and Patrick
No anhydrous limitation	Limited to anhydrous medium
Lower alkyl formanilide	*N,N*-Disubstituted formamide
H- and 2-Halogen only, no alkyl (even though two examples were included)	*H*, Halogen, and alkyl, no 2-position limitation
Condensing agent	Phosphorus oxychloride only

applications disclosed alkylthiophenes with up to four carbon atoms in the alkyl group, all four halogen-substituted thiophenes, and substitution in both the 2 and 3 positions in the thiophene nucleus. Weston also disclosed thionyl chloride as a condensing agent.

Although essentially the same, the claims did differ slightly (Table 2).

The references cited by the examiners against both patents were essentially identical.

VII. VALUE OF A PATENT

The entire process that has just been described costs money. How much? Estimates vary, but in 1973 the author received an estimate that the patent-application work of a corporate patent department cost $3000–$6000 per issued patent [7]. This figure includes work on abandoned applications. A 1975 estimate was $1700 for a simple application and $5000 for a complex one [8].

Then why get a patent? Why not keep the invention as a trade secret? The advantages and disadvantages of both courses of action have been summarized admirably by J. P. Kennedy [9] (Tables 3 and 4). These are some of the considerations that have to be weighed by the research director and the patent attorney when deciding whether or not to try to patent a given invention.

TABLE 3. Patent Advantages and Disadvantages[a]

Advantages	Disadvantages
A monopoly that may be limited by dominating patents for 17 years	Protection lost after 17 years
Basis for obtaining outside financing	Expensive to obtain and defend, and to maintain abroad
Stimulation to one's own research	Incentive and intelligence provided to competition
Possibility of licensing and obtaining royalties	Prosecution lengthy
Ease of selling the invention thus covered	False sense of security if the patent subsequently is held invalid
Benefit to society	

[a] Adapted with permission from J. P. Kennedy, *Chemtech*, March, 1974, p. 156. Copyright 1974 American Chemical Society.

TABLE 4. Trade-Secret Advantages and Disadvantages[a]

Advantages	Disadvantages
No cost	Protection easily lost, tough to police
Possibility for protection much longer than 17 years	Infringement not self-policing
	Later patenting may be impossible
	Later inventor may get a valid patent covering the same invention
	Sale of invention difficult
	Disadvantageous for society

[a]Adapted with permission from J. P. Kennedy, *Chemtech,* March, 1974, p. 156. Copyright 1974 American Chemical Society.

There are two other advantages to a patent position that Professor Kennedy did not list. From a patent position one can pick one's competition in issuing licenses. Secondly, a company's patents do not always cover a whole field. Thus two companies can cross-license each other and thereby share the benefits of the total field they have jointly discovered.

Patents can give an excellent picture of a company—its technical position, market thrust, and acquisition value [10]. However, a patent position in commodities may be of little value, since coverage would be in the process area, particularly if other companies are competing successfully without it. A strong patent position in a specialty field can be very important and worth a lot of money in an acquisition. If a company's patent holdings are growing, it usually means aggressive research and development efforts.

Patents can be a good source of technical intelligence in two ways. If one keeps a patent file of one's competitors by company, the direction and scope of their technical efforts often can be ascertained. The Patent Office publishes the Official Gazette, which gives a very brief summary of each patent as it issues. From this, one can order, and for a small fee obtain, copies of those patents in which one is interested. Emphasis always should be placed on filing dates, not on issue dates. In this country most patents now issue in 20–30 months from the date of filing, although the range can be 18 months to five or six years, or even longer. In the United States the contents of a patent are disclosed only when the patent issues. In such foreign countries as Belgium, Denmark, West Germany, Holland, Norway, Sweden, and Japan, the contents are disclosed 18 months after filing or 18 months after the original U.S. filing. Thus one has to keep track of foreign filings as well as those in this country.

In the same way as has just been described in the case of companies, one can keep up with the technical advances in a given field by means of patents. This involves obtaining all the patents in a given Patent Office subclass or subclasses as they issue. Also what usually results from such a search is the knowledge of what companies technically dominate the field in question.

VIII. FOREIGN PATENTS

Obviously patent laws vary from country to country. France, Italy, and many others grant registration patents in which no search is made by the patent office, and validity is determined by the courts. West Germany, the Netherlands, and Japan examine in much the same way the United States does. Although Great Britain, Austria, and Switzerland are also examining countries, their standards are believed to be lower [11].

In Great Britain an inventor files a preliminary disclosure and then has a year in which to file the complete application. This gives inventors an earlier filing date than would have been the case if they had had to determine the entire scope of the invention before filing, as is true in this country.

Some countries publish a patent application a specific time (usually 18 months) after it has been filed and before the patent has been granted. Under this procedure competitors undesirous of seeing the application issue as a patent can institute opposition proceedings in the patent office in question. Invalidation of an issued patent is effected by suit in court, as is the case in the United States.

Since opposition proceedings are not a part of U.S. patent practice, they are described here [12]. They are standard practice in most European countries.

The grounds for instituting opposition proceedings are the same as those used against applications by patent examiners in this country: (1) the subject matter is not patentable over the prior art; (2) the disclosure is not sufficiently complete that one skilled in the art can practice the invention, and (3) the claims are broader than the subject matter disclosed. This opposition must be supported by such evidence as prior publications, the declarations of experts, and so forth. The applicant can argue the case and can amend the description, claims, and drawings. The overall procedure is similar to that in the U.S. Patent Office except that a third party is involved and the examiner is more of a judge. He may reject the application, accept it as amended, or reject the opposition altogether and allow the application as submitted to issue as a patent.

This opposition procedure has two advantages. Most of the prior art available to patent examiners consists of patents. Opponents cite art from the literature, which may not be available to the examiner. Secondly, an opposition to a patent application is much cheaper than a lawsuit to invalidate an issued patent or settle an infringement. In a large corporation the decision to file an opposition is usually made at the same administrative level as that to file a patent application, not at a much higher level. Also, it is always possible to make a deal with an opponent, such as a royaltyfree license, and still have a subsequently issued patent as protection against other possible infringers.

Some countries such as Italy, Spain, Yugoslavia, and Japan do not allow the patenting of pharmaceuticals and food.

At the present time a patent attorney has two choices of routes to follow in seeking foreign patent coverage. A third may be available in 1984 or 1985.

The first option is to file a separate application in each country in which patent protection is desired. The amount of prosecution and its nature vary from country to country. Translation into the language of each country is necessary. A local patent attorney almost certainly must be employed to help. There is a filing fee, of course. At present, estimated costs for obtaining a patent in a non-English-speaking country average about $2000 and can be as high as $4000 in Japan and Sweden [13]. Many countries charge an annual fee to keep a patent in force. Usually this starts a few years from the date of issue (so that there has been time to commercialize the invention) and increases a little bit each year (as the patent presumably becomes more valuable) until the patent expires. This fee varies from zero in Canada to about $4000–$5000 in West Germany [7]. It is now zero in the United States, but this is expected to change shortly. Appeals as to patentability and infringement procedures are made in accordance with the laws of each country in question.

The second option, primarily for European patent coverage, is to file under the European Patent Convention (EPC). The countries that have signed this convention are Austria, Belgium, France, West Germany, Italy, Liechtenstein, Luxemburg, the Netherlands, Sweden, Switzerland, and the United Kingdom.

The application may be filed at a central location, such as the European Patent Office in Munich, or the International Patent Institute in The Hague, or else in one of the signatory countries. It is given a filing date, and a preliminary search of the prior art is made. The application, an abstract thereof, and the search are then published, promptly 18 months from the filing date or the priority date if earlier. The applicant must request an examination.

Patentability is determined by the European Patent Office. Although the preliminary search may be entrusted to a national patent office, the final decision is made by three examiners in the European Patent Office [14]. (Appeals also are handled by three examiners, and a fourth, legal one, if necessary.) The granted patent is again published for purposes of opposition. It is then forwarded to those countries in which the applicant desires patent protection.

Oppositions are handled by a special section of the European Patent Office. Opposition occurs after the patent is granted, so instead of rejecting a patent application, in this case the board of three examiners may actually revoke an issued patent or allow an issued patent to be amended.

What is obtained is a bundle of national patents. However, all patents are identical and result from one case instead of from separate cases in each country [15]. During prosecution the application can be converted to a bundle of separate national applications [14]. Since the patents obtained under EPC are national patents, their interpretation, infringement procedures, and so forth, vary from country to country. Invalidation also takes place in the courts of each country, and invalidation in one country does not affect a patent's standing in the others. A court is being set up to handle this situation within the European Patent Office.

In the United States, patent protection is for 17 years from the date of issue of the patent. Under the EPC it is 20 years from the date of filing. Under the EPC, opposition proceedings may start after the patent is granted, not during the application stage as in some countries. Under this arrangement an opponent cannot reduce the time of protection by opposition during the application stage. Opposition proceedings must start within nine months of the issuance of the patent.

Patentability requirements are the same as in the United States—the invention must be new, unobvious, and useful. The German requirement of a technical advance as well has been dropped under the EPC. In addition one does not have to show the preferred way of carrying out the invention as in the United States [15].

In the United States if inventors publish their inventions or place them on sale, they have a year in which to file a patent application. This is not true under EPC. Publication or sale anywhere invalidates the application. This is the concept of absolute novelty [13].

In the United States, secret use for more than one year forfeits patent rights. Under EPC, secret use is all right as long as it really is secret [13].

Again under EPC there is a whole raft of translations and fees [14].

The third option, which should be available in 1984 or 1985 will be to file under the Community Patent Convention (CPC).

In general the basic organization of the European Patent Office and its procedures under CPC will be the same as under EPC. However, a single supranational patent will be obtained, valid in all the contracting countries. Although interpretation will be on a national basis, infringement and opposition (post grant) procedures as well as invalidation will be centralized in the European Patent Office, rather than in each individual country.

The U.S.S.R. patent system is totally different from anything in the free world [16]. In the Soviet Union a patent primarily is a means for technology transfer. Patents are for foreigners. Russians get *inventor's certificates.*

Both patents and inventor's certificates are issued by the State Committee for Inventions and Discoveries. This committee can influence national planning and can authorize the use of new technology. In an inventor's certificate all rights are assigned to the state, and they can be used in any Soviet enterprise. The inventor may be paid up to $28,000 and may receive a larger pension and preferential treatment in employment and housing.

A U.S.S.R. patent is valid for 15 years from the date of filing. Its subject matter is the same as in this country, although new chemicals, pharmaceuticals, and medical processes are excluded. (They are not in the case of inventor's certificates.) Criteria for patentability are the same as in this country. In the Soviet Union priority is given to the first to file, in the United States to the first to invent. The Soviet Union charges maintenance fees.

In the Soviet Union an inventor must apply for a patent through the U.S.S.R. Chamber of Commerce. Pleading is before the Committee for Inventions and Discoveries. Although this has not yet been done, the Soviet Union can require the compulsory licensing of a patent to the state. (Great Britain has the same law.) Validity is challenged before the Committee for Inventions and Discoveries.

Obviously, policing patents in the Soviet Union is essentially impossible. Travel is limited, particularly for foreigners, and a foreigner is not allowed in a U.S.S.R. plant.

REFERENCES

1. A. P. Kent, *Chemtech,* **2,** October, 1972, p. 599.
2. Anon., *Chem. Eng. News,* **59,** July 15, 1981, p. 5.
3. W. J. King and F. F. Nord, *J. Org. Chem.,* **13,** 1948, p. 635.
4. W. S. Emerson and T. M. Patrick, Jr., U.S. Pat. 2,581,009.
5. W. S. Emerson and T. M. Patrick, Jr., *J. Org. Chem.,* **14,** 1949, p. 790.
6. A. W. Weston, U.S. Pat. 2,601,479.
7. F. M. Murdock, private communication.
8. Anon., *Chem. Eng. News,* **53,** May 12, 1975, p. 14.
9. J. P. Kennedy, *Chemtech,* **4,** March, 1974, p. 156.
10. Much of this is taken from M. A. Greenfield, *Chemtech,* **5,** April, 1975, p. 219.
11. G. Cramer, *Chemtech,* **4,** November, 1974, p. 670.
12. M. N. Meller, *J. Pat. Off. Soc.,* **61,** 1979, p. 1.
13. E. P. Winner, *J. Pat. Off. Soc.,* **62,** 1980, p. 419.
14. J. B. Van Benthem, *Pat. L.A.Q.J.,* **5,** 1977, p. 258.
15. E. Zurrer, *Pat. L.A.Q.J.,* **5,** 1977, p. 272.
16. B. Ancker-Johnson, *Chemtech,* **5,** July, 1975, p. 409.

10

ENGINEERING AND UNIT OPERATIONS

I. INTRODUCTION

This chapter is primarily for the benefit of chemists. However, the chemical engineers may find the first little bit helpful.

When a chemist enters industry, the first representative of another profession with whom he or she will have to work probably will be a chemical engineer. What are chemical engineers? Most of them hold BS or MS degrees and have extensive practical experience. Usually they have had as good undergraduate training in chemistry as a chemistry major and, of course, have had extensive engineering training as well. However, as will be seen later, their thinking is very different from that of chemists.

Where are chemical engineers found in a company? If a chemist initially enters a research group, chemical engineers may well be among the group's membership. Likewise, particularly in a large company, there may very well be chemical engineering research groups as such. Chemical engineers may also be involved in product development and they will most certainly be involved in process development. They design and run pilot plants, so after a chemist has finished the laboratory phases of a project, further development probably will be taken over by a chemical engineer. Chemical engineers design manufacturing plants, build them, and run them. Not only the plant engineering group, but also most of the technical supervision in a plant probably will be made up of chemical engineers. Chemical engineers do a great deal of technoeconomic evaluation, not only in groups devoted entirely to this function, but also in connection with the regular tasks of process development; pilot-plant design, construction, and operation; and manufacturing-plant design, construction, and operation. Thus the opportunities for chemical engineers in the chemical industry are both varied and numerous.

How do chemical engineers think? Although they must understand the chemistry of the reactions with which they are dealing and be familiar with the physical and chemical properties of the compounds involved, they think in terms of unit operations, such as the movement of tons of a solid chemical, the kinetics and thermodynamics of a particular reaction, and various separations, such as a solid from a liquid. The scope of each of these operations can be quite broad. Thus a solid may be separated from a liquid by settling and decantation, by filtration, by centrifugation, or by evaporation of the liquid (drying).

Secondly, chemical engineers think in terms of continuous operation. When larger quantities of a chemical are needed in the laboratory, the chemist usually uses a larger flask and scales everything else up proportionately. At the much higher levels of chemical manufacture, not only may there be a feasible limit to the amount of scale-up possible, but it also is extremely expensive to start and stop a batch plant. Thus continuous operation usually is much more economical on a large scale.

This leads directly to the third aspect of chemical engineering thinking—economics. Chemical engineers are constantly weighing process, and plant design and operation choices against costs. This means that except for a few chemical engineers who are in chemical engineering research, chemical engineers are technologists, not scientists. They use knowledge; they do not create it. This is a fact of life chemists must recognize when working with chemical engineers.

II. UNIT OPERATIONS

Before considering what chemical engineers do in the aggregate, it will be helpful to examine some representative details of unit operations [1]. The mathematics will be omitted and principles only will be examined. Some idea of equipment and of measuring devices will be presented also.

A logical approach is to start with raw materials. Although chemical engineers designing a plant have to worry whether their raw materials are coming by pipeline, railroad, truck, or barge, this discussion will be limited to how these materials are moved once they arrive at the plant location.

A. Transport of Liquids and Gases

In general, gases and liquids are transported in pipes. Pipe sizes vary depending on the amount of fluid to be transported. The walls vary in thickness depending on the internal pressure needed. Materials of construction vary depending on the corrosiveness of the fluid being transported. Since transportation distances may be appreciable, the effect of bends, corners, fittings, and valves must be considered. In piping appreciable distances, the thermal expansion of the pipe has to be taken into account. Another consideration is how the pipe must be supported. All these factors have to be balanced and optimized in order to reduce capital costs to a minimum and still have piping that will stand up under continued use. A broken pipe or connection can shut down a plant and cause the loss of many dollars' worth of production time. There are industrial standards and codes for piping, that take into account wall thickness, materials of construction, seams, welds, and joints.

The chemical engineer has to know whether flow is going to be laminar or turbulent. In piping two-phase systems, turbulent flow is highly desirable, in order to ensure good mixing of the two phases, so that transportation of the phases is equal. Separation of the phases, such as a solid settling out of a gas or liquid, is to be avoided.

Unless the pipe in question has an appreciable downhill pitch, fluids do not move through the pipe of their own accord. They have to be pumped. In selecting and sizing pumps, the chemical engineer must consider the distance the fluid must travel as well as how high it must be lifted. The density of the fluid, its viscosity, and its coefficient of friction with the pipe wall must be known. The velocity at which the fluid must travel also must be known. The calculation of these factors can become quite complicated in the case of solutions and slurries.

The velocity of fluids in pipes can be measured in a number of ways. In this chapter, enough typical examples of measuring instruments and of processing equipment will be described to enable a chemist to understand how a chemical engineer works. Again principles will be emphasized and mathematics avoided by saying "From such and such measurements the desired values can be calculated."

One method of measuring fluid velocity in a pipe is by means of a pitot tube with sidewall static tap. In the diagram shown in Fig. 1, *A* is the sidewall static tap, which measures the pressure against the sidewall, and *B* is the pitot tube with its opening directly into the fluid stream in order to measure the impact pressure. The velocity in the pipe can be calculated from the difference between the two pressures as shown. There are several modifications of pitot tubes; that shown in Fig. 1 is the one that just has been discussed.

The velocity of gases can be measured by a variety of anemometers. The velocity measurement depends on the cooling of a hot wire or thermocouple by the velocity of the flowing gas.

There are a number of head meters, of which the venturi tube shown in Fig. 2 is perhaps the best known. All head meters are based on the principle that if a constriction is placed in a pipe carrying a fluid, there is an increase in the fluid velocity and a lowering of the pressure at the constriction as a result of the increased kinetic energy. The rate of discharge from the constriction can be calcu-

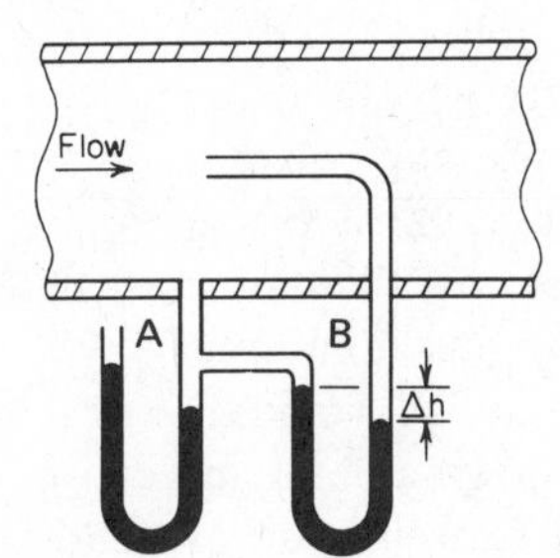

FIG. 1. Pitot tube with sidewall static tap. From *Chemical Engineer's Handbook.* Fifth Edition, p. 5–8, by R. H. Perry and C. H. Chilton. Copyright 1973 by McGraw-Hill, Inc. Used with permission of McGraw-Hill Book Company.

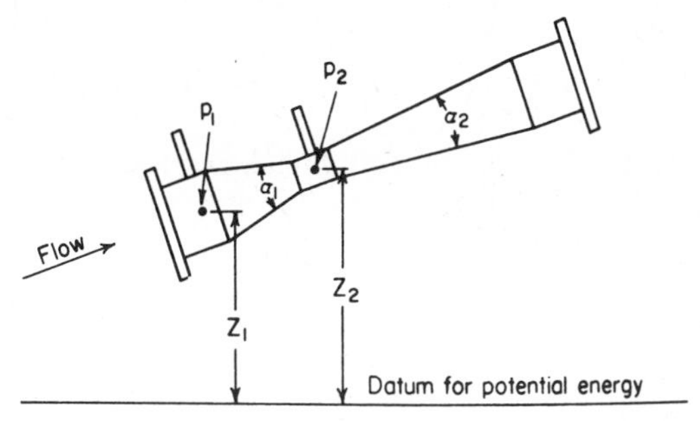

FIG. 2. Herschel-type venturi tube. From *Chemical Engineer's Handbook*, p. 5-10, by R. H. Perry and C. H. Chilton. Copyright 1973 by McGraw-Hill, Inc. Used with permission of McGraw-Hill Book Company.

lated from the pressure drop, the area for flow at the constriction, and the density of the fluid. Nozzles and orifices that are inserted into pipes function in much the same way.

There are, of course, other ways of measuring fluid flow, but these examples should give the chemist some idea of how the chemical engineer approaches the problem.

As mentioned previously, fluids do not go through pipes of their own accord. They have to be pumped. For moving liquids in a chemical plant, the centrifugal pump probably is the most widely used. In such a pump a power-driven shaft rotates a bladed impeller inside a fixed casing. As each blade passes the inlet, a slight vacuum is produced, which draws the liquid into the pump. The liquid slides tangentially along the blades with increasing velocity under the influence of centrifugal force and is expelled from the pump into the pipe. There are many designs of centrifugal pumps.

Another type is the reciprocating pump. Here a piston or similar device moves back and forth in a chamber. The chamber is so valved that as the piston is withdrawn, the liquid is drawn into the chamber. Then as the piston proceeds forward, the inlet valve closes and the liquid is expelled (pumped) through the outlet valve. Here again there are many designs of reciprocating pumps. This same priniciple is employed in rotary pumps, in which the liquid is drawn in between the vanes of a rotating impeller and then forced out the casing exit.

The pumping of gases is similar to the pumping of liquids. When large volumes of gases have to be moved at low pressures, as in ventilation or in supplying air for drying, fans are used. There is no need to detail them further here. Everyone uses them at home.

At higher pressures, turboblowers may be used. Essentially they are centrifugal pumps, that have to handle a compressible gas rather than an essentially noncompressible liquid. Reciprocal compressors essentially are the same as reciprocating pumps, except that when one needs to achieve fairly high pressures, the compressors are used in stages, so one works up to the final presure stepwise. Again, needless to say, there are all kinds of models of turboblowers and reciprocal compressors.

In dealing with the transfer of fluids, chemical engineers have to run overall energy and cost balances. Not only do they have to be sure that the fluid arrives where it is supposed to on time and in the correct quantity, but they also have to be sure it has arrived by the most economical method. It takes energy to run pumps and compressors, and joints and valves enter into capital and operating costs.

B. Transport of Solids

In a small chemical-intermediate-manufacturing or dye-manufacturing plant, it may be easiest and cheapest for an operator to shovel the requisite amount of a solid chemical out of a bin and carry it in a bucket or wheelbarrow to the batch reactor. On a large scale this method of moving solids becomes inordinately costly and inefficient.

A few typical methods for moving solids will be mentioned. Screw conveyors (Fig. 3) are old and quite widely used. They operate on the principle of a worm mounted on a shaft, which forces the solid along a trough or through a pipe. Screw conveyors can be used to mix solids in the process of moving them.

Belt conveyors also are quite widely used because their speed and capacity can be varied tremendously. They are simply a belt, either flat or in the form of a trough, with the weight supported by rollers. They can be used to move solids horizontally or up modest inclines. If a right-angled corner must be turned, it usually is easiest to transfer the material (by dumping) to another belt.

For lifting solids vertically, bucket elevators, of which there are several designs, usually are used. They comprise a series of buckets attached to a belt or chain (Fig. 4).

Pneumatic conveyors likewise are widely used. Here the solid, in the form of a dust or as fine particles, is carried through a pipe by a current of air, often under pressure or vacuum.

The design and selection of the proper conveyor depends on such factors as the particle size of the material, its flowability, its abrasiveness, its weight per cubic foot, its corrosiveness, whether or not it is oily or hygroscopic, and so forth. All these factors enter into the selection of the materials of which the conveyor is to be constructed, the size of the conveyor, and the speed at which it will be operated. The chemist should have some idea of the kinds of questions the chemical engineer is likely to ask about the materials used in his laboratory experiments. The many devices for loading and emptying conveyors have not even been touched on.

Solids are measured by weighing, which today is essentially all automatic. The major assumption is that the solid is uniform in composition and density.

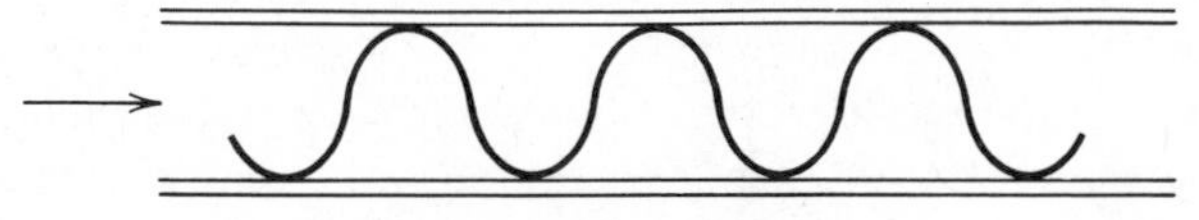

FIG. 3. Representation of screw conveyor.

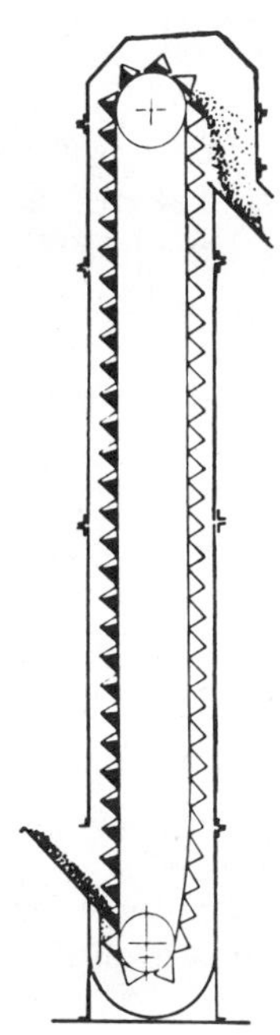

FIG. 4. Bucket elevator. From Stephens-Adamson Manufacturing Company (division of Borg-Warner). Used with their permission.

Batch weighing with electronically controlled discharge into the receiver is used for such tasks as weighing charges for packaging machines or weighing fixed amounts of raw materials to be charged to reactors. In the case of large amounts of materials, continuous weighing is used. The material on a belt is passed over the scales and the weight is recorded and totaled continuously.

There is no need to go into much detail about packaging and storage here, since the subject was mentioned in Chapter 7. In the laboratory, gases are stored in cylinders, liquids in bottles or metal cans, and solids in bottles. In a plant, gases and liquids are stored in tanks, solids in bins, silos, bags, drums, or even boxes. In a warehouse, bags and drums can be stored on pallets and moved around by fork trucks.

C. Reactors

Reactors can be of almost infinite design. However, for small-scale manufacture, batch-reaction kettles are used. They are often called Pfaudlers because so many are designed and manufactured by the Pfaudler Company. These kettles are nothing but large metal versions of laboratory three- or four-necked flasks.

Most reactions have to be stirred. The paddles or their equivalent are designed to best suit the particular reaction in question. Sealing and stuffing boxes were described in Chapter 7.

Larger-scale industrial chemical reactions are run continuously. The reactor may be a single tank or kettle, or two or three in series, or a tube through which the reactants pass continuously. If a catalyst is involved, it may constitute a fixed bed in the reactor tube, or a fluidized bed may be used. Fluidized beds are particularly useful in vapor-phase catalytic reactions in that the fine catalyst particles are suspended in the vapor stream. The catalyst is fed in continuously and withdrawn continuously for regeneration in another reactor. Actually there is a tremendous variety in reactor design, since many reactions involve the mixing of two phases. Some typical equipment will be seen in the section on separations, since such equipment is often used for both purposes.

In order to properly design and size a chemical reactor, a chemical engineer has to know the kinetics of the chemical reaction in question. The mechanism of the particular reaction must be well enough understood that the rate-determining step can be recognized. Thus the reaction may be controlled (as in the laboratory) by the gradual addition of one reactant. The chemical engineer also has to run a complete material balance. Minor by-products, which are ignored in the laboratory, can build up in a plant and cause all kinds of trouble.

Likewise the chemical engineer has to know the heat of the reaction, whether it will require heating or cooling, and how much of either. Heating may be effected by a steam jacket, by electrical-resistance wire, by hot water or by a bath of a hot fluid such as Dowtherm, or, in the case of high-temperature inorganic reactions, by open flames. Cooling may be effected by a water jacket on the reactor, by cooling coils within the reactor, by the condensation of a refluxing liquid, or by designing the reactor so that only small amounts of material undergo reaction at any one time.

In all this the chemical engineer has to know the efficiency of heat transfer between the heating or cooling means and the surface of the reactor and between the surface of the reactor and the particular reaction involved. Unfortunately these factors do not always remain constant. For instance, the collection of scale on heating coils seriously impairs their efficiency. This means that safety factors must be incorporated in the deisgn of a reactor.

The chemical engineer likewise must run a complete energy balance on the entire process in order to minimize energy costs. Often the heat generated in one part of a process can be utilized by another part of the same process. Obviously the insulation of pipes and equipment becomes a whole field in itself.

D. Solid–Gas Separation and Contact

When a chemical reaction is finished, a wide variety of products may result. They have to be separated. This separation usually is based at least as often on their physical form as on their chemical nature. In order to orient the chemist, the problem will be approached primarily from the point of view of physical form.

Just to keep life simple for a moment, the first assumption is that a solid must be separated from a gas. This may mean a product must be collected from a pneumatic conveyor or dust from a dryer or smelter. Anyone who reads the newspapers knows this also is a factor in air-pollution control, which is not always completely obvious. It not only involves removing particulates from an exhaust stream, but it also involves providing clean air to a piece of process equipment such as a dryer.

What does the chemical engineer worry about particularly in solid–gas separation? Most important are the gas speed and the particle size and density of the solid, as well as the relationship between the solid and the gas volume.

How does the chemical engineer handle the problem? If the gas velocity is not large, a settling chamber can be very effective.

However, altogether too many chemical processes do not operate that slowly. Cyclones are used very commonly to separate solids from gases. They operate on the principle that the mixture of gas and dust strikes the side of the cyclone tangentially. The dust collects around the sides of the chamber, whereas the gas goes out the middle (Fig. 5). Suspended liquids can be separated from gases in this same way.

Bag filters are very effective for separating solids from gases. The materials

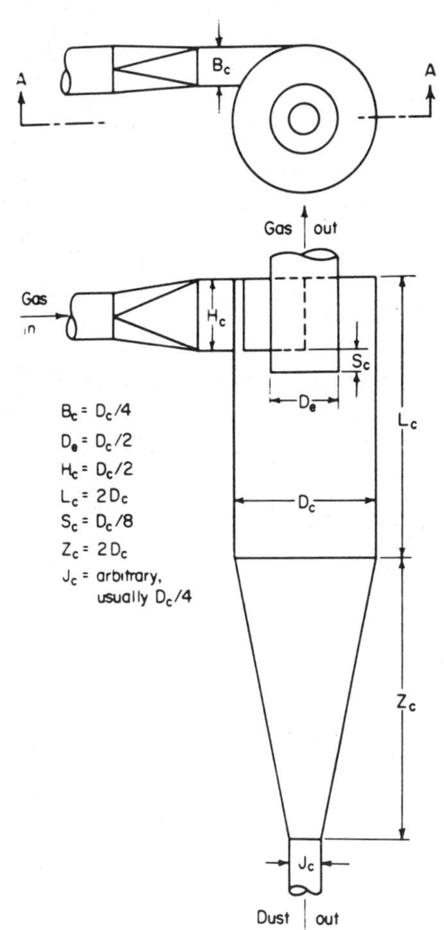

FIG. 5. Diagram of cyclone separator. *Chemical Engineer's Handbook,* p. 19-88, by R. H. Perry and C. H. Chilton. Copyright 1973 by McGraw-Hill, Inc. Used with permission of McGraw-Hill Book Company.

of filter construction and the filter weave depend on the speed of the gas, the amount of solid to be separated, its particle size, and its corrosiveness. Usually a bag-filter operation involves a number of bags in parallel. Thus while most bags are in operation, the dust can be shaken out of those that have been filled up, while they are taken off stream for regeneration.

Any gas stream can be run through some type of liquid (almost always water-based) scrubbing apparatus, of which the various designs are infinite. Scrubbers can remove solid particles both physically and chemically.

One of the most widely used methods for removing solid particles from a gas is the Cottrell and related electrostatic precipitators. They are based on the principle of ionizing the solid particles in the gas stream and then collecting them by means of the charge. The collecting electrodes may be in the form of plates, screens, rods, wires, or pipes.

In many chemical engineering operations the same type of equipment is used in contacting reactants that exist in two different physical states as is used in separating reaction products that occur in the same two different physical states.

For the purposes of this chapter there are three modes of gas–solid contact:

1. *Parallel Flow.* The gas flows parallel to the solid phase, although the solid bed often is static.

2. *Perpendicular Flow.* This is essentially the same as number 1 except that the gas flows across the solid bed, which may be fixed or may be moving across the gas stream.

3. *Through Circulation.* Here the gas flows through the solid bed, which may be fixed or moving as in a fluidized bed.

Obviously, any of these types of contact can be used to treat a solid with a gas, or can be used when the solid is catalyzing the reaction between two gases.

E. Drying

Probably the most extensive use of gas–solid contacting is in drying—separating a liquid as a vapor from solid particles.

Basically there are three types of dryers:

1. *Direct.* In this case the wet solid is contacted directly by the hot gas, and the vaporized liquid is carried away by the heating medium. The solid may be supported on trays (batch) or on a moving belt (continuous). Drying also may be effected in a pneumatic conveyor or in a fluid bed. In a circulating dryer, which may be batch or continuous, the solid is supported on a screen and the hot gas is blown through it.

2. *Indirect.* In this case heat is supplied through a retaining wall, and the vapors are removed independently of the heating medium. Tray driers can be used here also. The heat is supplied to the tray by means of a hot gas rather than directly to the material being dried. Drum dryers are used extensively for liquids, slurries, and flowable pastes. The material to be dried is dropped onto a hot, rotating drum, and the solid is scraped off continuously. In a rotary dryer

the solid material to be dried usually is introduced continuously at the high end of a slightly inclined rotating cylinder, and the dried product is removed from the low end. Heating may be either direct or indirect. All the indirectly heated dryers mentioned in this paragraph can be operated under vacuum if such is desirable.

3. *Radiant Heat.* In this case by means of a high-frequency electric field, heat is generated within the solid to be dried.

There is such an endless variety of dryers that there seems little point in drawing pictures here. It is hoped that the principles of construction and operation that are involved are reasonably obvious from these very brief descriptions. It should be evident that the whole field of heat transfer is complex.

F. Liquid–Gas Separation and Contact

As in all such operations, there must be maximum exposure of the surface of each phase to that of the other, in this case the surface of the gas to that of the liquid.

Distillation is the separation of these two phases, even though at the start of the operation the more volatile component may be in the liquid phase also. As the operation progresses, the lower-boiling component is volatilized and passes off as a gas from the top of a column, while the higher-boiling component is withdrawn as the liquid still bottom.

For many years one of the most common pieces of equipment used for both contacting and separation was a bubble cap column. Such a piece of equipment comprises a vertical pipe containing a series of horizontal plates (each equipped with a number of bubble caps) across which the liquid passes as it progresses from one plate to another down the column. The gas comes up through a series of central pipes, hits the top of each cap, reverses direction, and is dispersed through slots into the liquid (Fig. 6). In distillation the liquid mixture often is fed continuously onto a plate partway up the column, so that the more volatile

FIG. 6. Diagram of bubble cap column. Expanded from *Chemical Engineer's Handbook,* p. 18-1, by R. H. Perry and C. H. Chilton. Copyright 1973 by McGraw-Hill, Inc. Used with permission of McGraw-Hill Book Company.

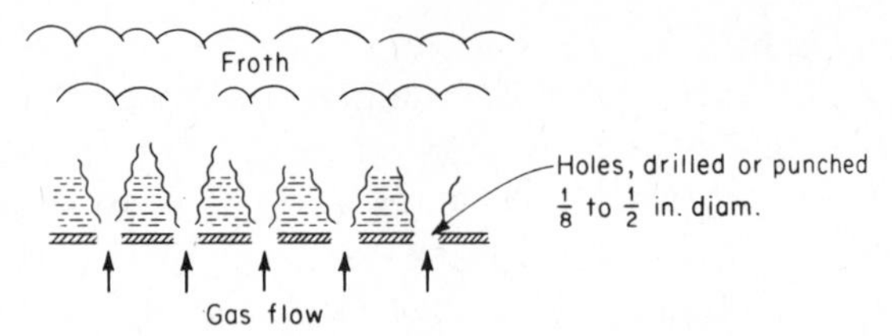

FIG. 7. Sieve-plate dispersers. From *Chemical Engineer's Handbook*, p. 18-4, by R. H. Perry and C. H. Chilton. Copyright 1973 by McGraw-Hill, Inc. Used with permission of McGraw-Hill Book Company.

constituent can be drawn off the top of the column and the less-volatile one, comprising the still bottom, can also be drawn off continuously.

More recently, bubble caps have been replaced extensively by sieve and valve plates. These can be operated in cross-flow fashion as bubble caps are, or in countercurrent fashion, in which both liquid and gas use the same openings (Fig. 7).

Packed columns also are used, although more for absorption than for distillation. They are nothing, but pipes full of packing. Liquid is distributed at the top of the packing, and the gas is introduced at the bottom. The flow is countercurrent (Fig. 8). Many types of packing have been employed, all the way from rocks to specialized shapes. Perhaps the two most widely used special shapes are Raschig rings and Berl saddles (Fig. 9).

The chemists' reaction may require intimate dispersion of a gas in a liquid, as in an air oxidation or in a hydrogenation. This means that some form of mechanical agitation of the liquid phase and some type of equipment for dispers-

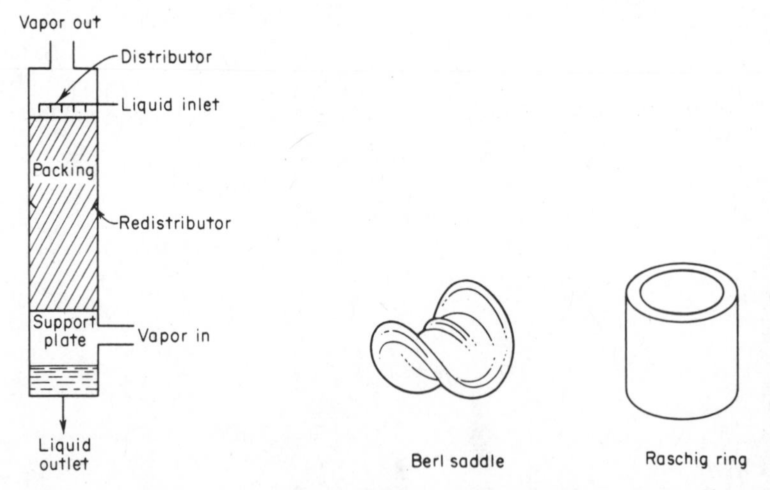

FIG. 8. Scheme of packed column. From *Chemical Engineer's Handbook*, p. 18-19, by R. H. Perry and C. H. Chilton. Copyright 1973 by McGraw-Hill, Inc. Used with permission of McGraw-Hill Book Company.

FIG. 9. Sample packings. From *Chemical Engineer's Handbook*, p. 18-20, by R. H. Perry and C. H. Chilton. Copyright 1973 by McGraw-Hill, Inc. Used with permission of McGraw-Hill Book Company.

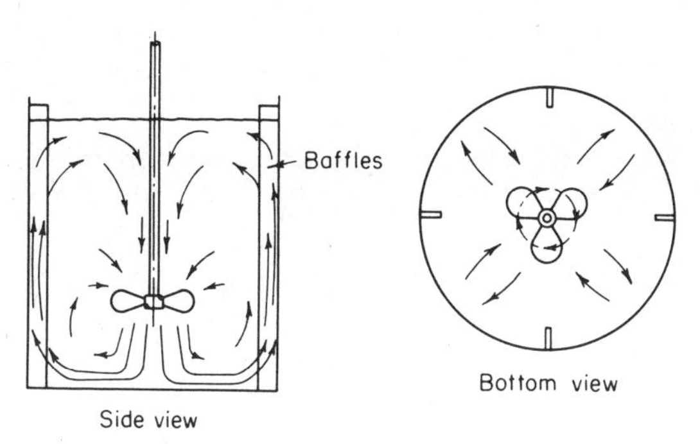

FIG. 10. Diagram of propeller with baffles. From *Chemical Engineer's Handbook,* p. 19-6, by R. H. Perry and C. H. Chilton. Copyright 1973 by McGraw-Hill, Inc. Used with permission of McGraw-Hill Book Company.

ing the gas will be necessary. The laboratory fritted glass plate is analogous to similarly designed metal dispersers used in a plant. All kinds of liquid agitators are available on a plant scale, all the way from simple propellers to more-complex devices such as turbines. To increase agitation, baffles may be affixed to the sides of the reaction vessel, of which the laboratory analogy is the Morton creased flask (Fig. 10).

G. Solid–Liquid Mixing and Separation

Pastes require special equipment, of which the charge can mixer is the most widely used. Essentially it is a scaled-up version of the old-fashioned ice cream freezer, except that in some models it is possible to lower the can away from the stirrer. In other models the stirrer is lifted out of the can, as in the case of the ice cream freezer. The stirrer may comprise smooth blades or a variety of intricately intermeshed paddles. This kind of solid–liquid mixing is particularly widely used in the paint and ink businesses.

Before discussing methods for separating solids from liquids, it should be noted that at the start of such an operation the solid may be dissolved in the liquid. This necessitates crystallization. There are many types of commercial crystallizers, which often are quite complicated in design, since the heat-exchange problems are complex. Usually the liquid is highly agitated. Crystallization may be induced by cooling alone or, more likely, by continuous evaporation of the liquid and continuous removal of the solid crystals as they are formed.

The simplest method of separating a solid from a liquid is by gravity settling and subsequent decantation of the liquid. The diagram of a typical settler is shown in Fig. 11. There may be no agitation or the suspension may be stirred slowly through zone *B*.

The principles for operating decantation equipment are obvious. In many such operations several decanters are set up in series, and the suspension is stirred and allowed to settle in each settling tank before decantation.

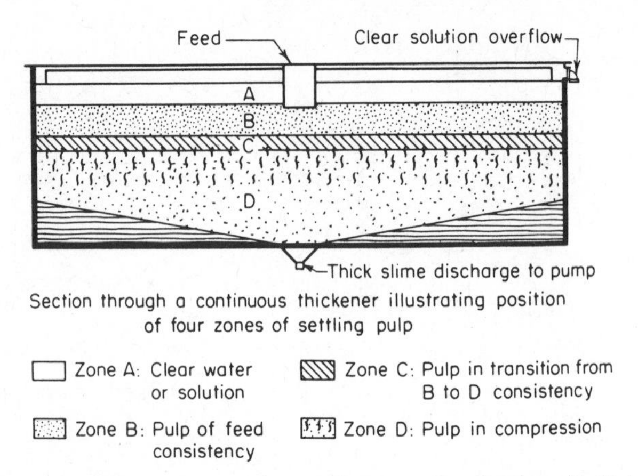

FIG. 11. Settling-tank diagram. From *Chemical Engineer's Handbook,* p. 19-44, by R. H. Perry and C. H. Chilton. Copyright 1973 by McGraw-Hill, Inc. Used with permission of McGraw-Hill Book Company.

Another method of separating a solid from a liquid is by filtration. The filter medium is most important. It must retain the solids it is supposed to and it must last. Following are some of its necessary characteristics:

1. It must be able to quickly bridge the solids across the pores so that there is no bleeding.
2. There must be low entrapment of solids in the interstices so that there is minimal binding.
3. There must be minimum resistance to filtrate flow so that there will be a high production rate.
4. The medium must resist attack by the chemicals involved.
5. The medium must be strong enough to support the pressure or vacuum that is applied.
6. The medium must last, which means it must withstand the mechanical wear and chemical attack to which it is subjected.
7. The solid cake must be easily removable.
8. The filter medium must conform to the equipment with which it is used.
9. Costs must be low.

A wide variety of filtration media are used—fibrous material, woven cloth, perforated solids, beds of particulates such as sand, and so forth. Whatever medium is chosen must be permeable. To achieve this a thin layer of a filter aid sometimes may be placed on top of the filtration medium, or the filter aid may even be incorporated in the slurry to prevent clogging.

In many filtration operations it is the solid cake that is the actual filtration medium. This is an engineering problem that the chemist probably never thinks of, but it explains why it may be advantageous to incorporate a filter aid in the slurry.

One type of filtration equipment is called a Nutsche filter. It is nothing but a scaled-up Büchner funnel. The solid may be separated by gravity, by pressure applied to the surface of the slurry, or by vacuum beneath the filter medium.

The most widely used type of filter press is the plate and frame. It is constituted of a series of plates that are perforated or grooved to permit drainage. Both faces of each plate are covered by a filtration medium, usually a cloth. Between each plate is a hollow frame in which the solid being separated by filtration collects. There are all kinds of feed and liquid-discharge arrangements. In order to remove the cake, the equipment must be disassembled and the cake removed by hand scraping. Filter presses can be used either as cake filters or as clarifying filters (those used to remove very small amounts of solids from a liquid).

A standard plate-and-frame press usually has a horizontal arrangement of rectangular plates held together in a vertical position with a solid half plate at each end. The whole business is held together by a capstan screw, gear and pinion, and so forth. There are many, many modifications of the principle involved. It would take several catalogues to describe all of them.

Another very widely used filtration apparatus is the centrifuge. An industrial centrifuge is shown in Fig. 12. The vertical sides are perforated and are covered inside by some kind of filtration medium. The bowl is rotated at high speed, so the contents are thrown horizontally against the walls by centrifugal force, and the liquid is expelled through the filtration medium and the perforations in the vertical sides. As in the case of plate-and-frame presses, there are many equipment models based on this same principle.

H. Solid–Solid Mixing and Separation

The mixing of solid–solid systems usually is involved in formulation rather than in preparation for, or in actually conducting a chemical reaction. The big exception is rubber vulcanization, in which the accelerators, antioxidants, and other reagents must be thoroughly mixed with the rubber latex before the final cure (chemical reaction).

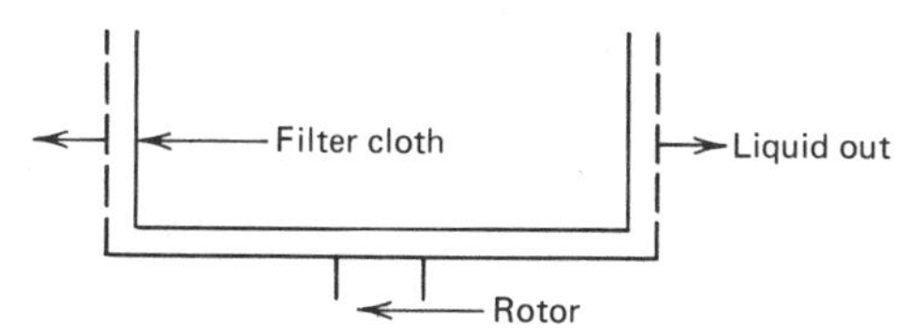

FIG. 12. Centrifuge diagram. From *Chemical Engineer's Handbook,* p. 20-82, by R. H. Perry and C. H. Chilton. Copyright 1973 by McGraw-Hill, Inc. Used with permission of McGraw-Hill Book Company.

For the mixing of solids and very heavy pastes, there are all kinds of kneaders. No particular one is in such general use that it must be described here.

For mixing small batches of solids, the Banbury mixer is preeminent and is used extensively in the plastics and rubber industries. Schematically it is shown in Fig. 13. Mixing is accomplished in the narrow clearances between the two rolls and the walls. The rolls can be operated at different speeds, thus enhancing mixing. A Banbury mixer requires considerable power, since the ram forces the solid ingredients into the mixer at what can be quite high pressures. With all the friction involved, heat buildup can also be a problem.

Similar in principle to the Banbury mixer is the roll mill. It can comprise two

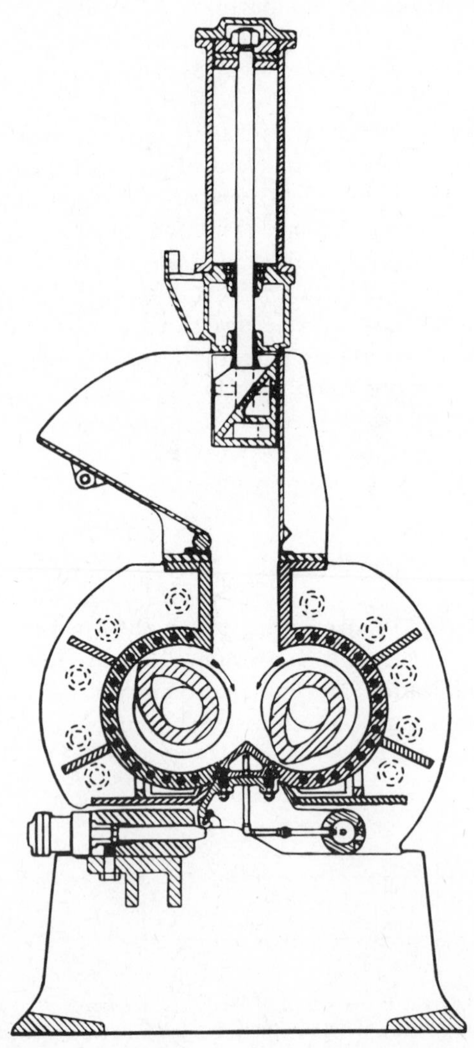

FIG. 13. Diagram of Banbury mixer. From Farrel Company (division of United Shoe Machinery Corporation). Used with their permission.

or three unencased rolls, usually operated (in the case of three rolls) at different distances from each other, often at different temperatures, and at different speeds. A roll mill provides very high localized shear. Since mixing is slow, roll mills usually are used as batch mixers. They find wide use in the rubber, ink, and paint industries. Even with precautions they can be dangerous to operate, particularly if the mix has to be removed from the rolls and restarted by hand. There used to be an old saying in the rubber industry that you could tell a veteran roll-mill operator because he had only one arm.

There are a variety of other mixers. A common one depends on a screw's forcing the solids through a pipe, with mixing being effected during the transfer process (see Section II.B). Again there are many models of this type of equipment.

The problem of the separation of solids often is solved in the laboratory by dissolving the entire mixture in some solvent and separating the materials by crystallization. The same technique is employed by chemical engineers in industry.

In the mining industry, froth flotation is used extensively in ore beneficiation. In 1960, 198 million tons of material was so treated to recover 20 million tons of ore worth $20 billion. In this process the entire ore body is ground finely, suspended in water, and one particular ore becomes coated selectively with a surface-active agent. The mixture is then agitated in a machine, and air is blown through it to create a froth, which is separated from the rest of the suspension. The desired product may be either in the froth or in the bottoms.

Another method for separating solids is by leaching, which includes the washing of a solid filter cake. The essential principle is to remove one component by treating the entire solid mixture with a solvent in which only that particular component is soluble. This may involve suspending the solids in a solvent, agitating the mixture, and then separating the liquid and solid phases by filtration or centrifugation. The process can be conducted on a Nutsche filter. An alternate approach is to carry the solid mixture on a moving belt and continuously contact the mixture with the solvent. Here the principle is similar to that upon which a Soxhlet operates. The choice of method and equipment depends upon the problem in hand.

I. Unit Operations Summary

This concludes a very brief discussion of unit processes. Process control, which is a whole field in itself, has not been discussed. All kinds of specialized instruments are involved in process control. Materials of plant construction have been mentioned here and in other chapters. Obviously the proper selection has a tremendous bearing on plant investment and plant maintenance.

Energy considerations, mentioned here and in other chapters, have always been extremely important and are becoming much more so. What is it going to cost to power processing machinery? How much heat will the various processing operations require? Can any of it be recovered? Again, this whole problem of heat generation and of heat transfer is reemphasized. It is extremely important that the chemist be aware of it.

III. ENGINEERING DEVELOPMENT OF A CHEMICAL PROCESS

Now that several unit operations and some of the equipment used therein have been examined, how the chemical engineer puts it all together will be illustrated. For this purpose it is assumed that only one engineer is involved; the engineer starts by assisting the chemist with the laboratory study in glass. The chemical engineer supervises the bench-scale work in metal, designs and operates whatever pilot planting is necessary, designs the final production plant, and coordinates its construction [2].

A. Laboratory Scale

The first thing the chemical engineer does is to draw a tentative flow sheet of the process. Examples of such flow sheets were given in Chapter 8. Each unit of equipment is shown, although its precise type and design are determined as the study progresses. All chemicals are shown—raw materials, intermediates, catalysts, products, by-products, solvents, and wastes. All temperatures and pressures are shown. Likewise the detailed compositions of all streams as well as conversions, yields, and residence times are indicated. Residence times are very important in determining the size of the equipment needed and therefore its cost.

The next step is to run a material balance on the entire process and on each individual piece of equipment. These balances include the moles, weights, and volumes of every component entering and leaving each vessel, and for every stream including recycles. From this it can be seen that the chemical engineer has to know the chemistry of the process in detail.

The next step is to run a similar set of energy balances. These comprise how much heat must be added to or removed from each unit. Presumably the heats of all reactions and of all phase changes either are known or have been measured accurately in the laboratory. All phase changes are indicated. These energy balances include all such equipment as furnaces, heat exchangers, reactors, compressors, and pumps, together with an estimate of their probable efficiency. Energy is extremely important. For instance, the largest operating expense of the Kerr-McGee Chemical Corp. plant at Searles Lake, California, is the cost of the fuel for concentrating the brine.

From these data it is possible to estimate equipment sizes—the weights or volumes of vessels, the areas of heat exchangers, the horsepower of pumps, and so forth. The next step is to estimate the capital cost of the plant. It is based on the cost of the major pieces of equipment. A chemical plant will cost roughly 4.5–6 times this figure [2]. This includes such items as site installation, piping, footings, foundations, and so forth.

The chemical engineer next estimates operating costs. These are discussed in detail in Chapter 12. From the estimated capital and operating costs the project is evaluated and the probable profitability and return on investment is calculated.

At this point the chemical engineer is prepared to go back to the laboratory study in glass and assist the chemist with it. Both the chemist and the chemical engineer can see where an improvement in yield would reduce costs. The chemical engineer is also very aware of places where an improvement in conversion would be equally helpful. Equipment for recycle and recovery may cost more than that for the reaction itself.

Another cost saving may be effected by a shorter residence time so that the step in question would require smaller equipment and a concomitantly lower capital investment. This may be brought about by a better ctalyst for the reaction or by a different crystalline form of a solid that makes its separation by filtration from a liquid quicker and easier. It may be desirable to eliminate a difficultly separable by-product altogether and thus eliminate a very expensive separation step. It is assumed that by this time the chemist has based the raw-material specifications on commercially available chemicals and has determined whether technical or reagent-grade raw materials are needed.

B. Bench Scale in Metal

The next step in developing a commercial process is to duplicate the laboratory results on a bench scale in metal. It is at this point that the chemical engineer may take over the direction of the project, with the chemist remaining a very active member of the team. Years ago it was standard practice to conduct testing in a total-process pilot plant. However, pilot plants are very expensive to build and operate. Therefore, the current trend is to check out as much as possible on a bench scale. Bench-scale equipment is comparatively inexpensive to buy, and the whole plant can be torn down and completely rebuilt in a short period of time, which is impossible in the case of a pilot plant [3]. Many reactor designs can be tested as well as a wide variety of catalysts and raw materials of varying specifications. Operation also is reasonable costwise. Even for 24-hour continuous operation, all that is needed is an operator for each shift plus the chemical engineer and the chemist during the day. Modern instrumentation makes stream and product analysis easy.

What then is tested on a bench scale in metal? The answer is just about everything:

1. The chemistry of the process as conducted in metal, including catalyst life and by-product and waste disposal.
2. The energy balances with particular emphasis on heat transfer.
3. The mechanics of the process in commercial-type equipment. This includes such things as reactor design; types and functioning of pumps;

methods of mass transfer; dispersion of gases or solids in liquids; methods of phase and chemical-component separation; flow meters; and the automatic feeding of gases, liquids, and solids.

4. The feasibility of continuous operation. This includes the ability to maintain and regulate flows, temperatures, and pressures.
5. Whether or not process control and operation is within the skill of the average operator.
6. Laboratory corrosion studies.

All of this checking leads to the design and selection of the best equipment (optimum performance for the price the process can stand) for the final manufacturing plant. These studies also are the basis for the development of control instrumentation.

C. Pilot Plants

It would be nice if all this information could be effected on a bench scale. Many times it can, but often it cannot. The scale may well be too small to measure the buildup of minor by-products sufficiently accurately. Temperature control may be so sensitive that checking on a larger scale is necessary. There can be other problems such as the evaluation of questionable equipment.

The only practical answer is to gather the necessary additional information in a pilot plant [4]. Pilot plants are expensive to build and expensive to operate. Usually they require a junior engineer plus several operators on each shift. Therefore, the tendency has been to pilot only those steps in which problems are envisioned, rather than the whole operation. The important thing to avoid is errors as the scale, and therefore the cost, of the operation escalates. Cost escalation was outlined in Chapter 7. It is better to make a mistake in a $1 million pilot plant than in a $50 million commercial plant.

It is important to remember that a pilot plant is not a scaled-up laboratory or bench-scale operation. It is a scaled-down plant. In a well-run company the plant manager helps in its design. (In this example, the chemical engineer has already been designated the plant design engineer as well.)

A pilot plant is not the place for the manufacture of sales-development samples. The cost is prohibitive. The scale of operation is anything but optimal, and there are far too many professionals involved in its operation. The experimental products produced are not uniform and probably do not conform to final specifications without extra refining. And key units in the final plant may have been omitted if no problems were encountered at the bench level. However, some companies do use a complete pilot plant to prepare sales-development samples (even though it greatly reduces the pilot plant's flexibility for experimental purposes—the reason it was built in the first place), although a flexible semiworks facility would be much preferable.

On the other hand, a pilot plant can be a good training ground for young chemical engineers who will be shift supervisors in the final plant, and for senior operators who will move in as foremen in the final plant. Also, a pilot plant is an excellent place to insure the adequacy of safety features that ultimately will be needed in the final manufacturing plant.

D. Plant Design

For a large plant a computer model of the plant design may be helpful. Likewise it may be desirable to construct a scale model of the projected plant [5]. Such a model is particularly valuable in seeing if equipment is arranged best and if floor levels are correct. Of great importance is whether or not enough space has been left for ladders, aisles, equipment operation, and equipment removal for maintenance. Are there enough safety exits and are they readily accessible? Can spilled liquids be cleaned up quickly and adequately? In case of fire can all parts of the plant be reached quickly and efficiently? Are safety showers and fire-hose racks properly located?

A model is perhaps the first place to check the adequacy of piping. Are there points where there may be line plugging due to solids buildup? Are line slopes adequate for drainage and sample collection? Is there adequate access to valves and is handwheel orientation correct? How about flange maintenance?

During all these studies—laboratory in glass, bench scale in metal, and pilot plant—probable costs are continually monitored and refined by the chemical engineer as illustrated in Chapter 7. Changes must be watched closely. For instance, a change in plant location may double raw-material costs because of changes in methods of delivery [6].

Although the chemist probably will never really be involved in plant construction, he or she may be concerned about delays in getting things done. For the chemist's own edification and "amusement" the various personal factors in this operation have been superbly described in detail [7].

This chapter has attempted to show the chemist how the chemical engineer thinks and operates. This is something the chemist in industry must understand. In fact the chemist should incorporate many of the considerations mentioned in this chapter into his or her own thinking.

REFERENCES AND NOTES

1. This information was gleaned and summarized from R. H. Perry and C. H. Chilton, *Chemical Engineers' Handbook,* McGraw-Hill, Fifth Edition, New York, 1973.
2. The approach is standard but very elegantly outlined by J. P. Clark, *Chemtech,* **5,** November, 1975, p. 664.
3. The advantages of bench-scale work in metal are discussed by V. J. Anhorn, *Chemtech,* **1,** September, 1971, p. 574.
4. F. R. Bradbury, *Chemtech,* **3,** September, 1973, p. 532, discusses the uses of pilot plants in some detail.
5. C. O. Utley, *Chemtech,* **6,** August, 1976, p. 488.
6. A. H. Schutte, private communication.
7. J. Mizrahi, *Chemtech,* **2,** August, 1972, p. 459.

PART TWO

MANAGEMENT AND MARKETING

11

CORPORATE TECHNICAL PLANNING

I. INTRODUCTION

Now that the technical side of the chemical industry has been glanced at, it is time to examine the financial and business side, which now can be done in perspective. Financial and business considerations in a corporation govern the allocation of funds to every facet of the entire technical program, including the research, development, and manufacturing departments. In this and succeeding chapters the methods by which corporations plan and evaluate their technical programs will be considered. And a look will be taken at various company organizational structures and techniques that corporate managements may use to implement their technical programs with maximum financial effectiveness.

The resources of even the largest companies are limited. Therefore, companies have to plan the allocation of their assets to obtain the maximum benefit from them.

How are these resources used? By far the largest part have to be employed defensively, to keep the company competitive in its established business or businesses. However, all companies wish to grow and become more profitable. Some of their resources have to be used for this purpose.

II. CORPORATE STRATEGIC PLANNING

The first step in corporate strategic planning is to set overall corporate objectives [1]. What are these? Needless to say, they are strongly financial:

1. What return on capital investment is desired?
2. What profitability is desired? Almost invariably this is higher than at present. It is measured by increased earnings per share of stock.

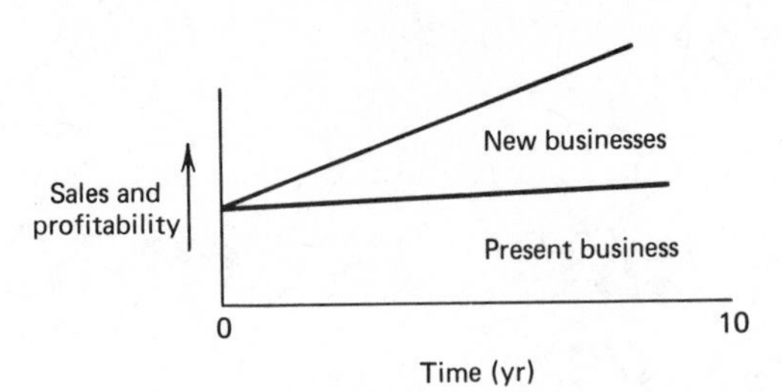

FIG. 1. Graph of 10-year goal I.

3. How fast does the company want to grow? All companies want to grow. Growth is measured by increased sales and increased return on stockholders' equity.
4. What kind of a company is desired 10 years from now? To say sales and profits must be much bigger is like an endorsement of motherhood. To meet the previous objectives the company may have to become quite different from the one now known.

The statement of these objectives leads immediately to other questions:

1. Will the present business achieve these objectives? Usually it cannot.
2. If it cannot, what other businesses can?
3. In which of these new businesses can the company expect to succeed with its present resources or with those it can reasonably expect to acquire?

Figure 1 is a very simple graph of the 10-year goal. This is a relatively simple goal for management to cope with. Even if the new businesses do not pan out as well as expected, the company has a solid basis for gradual growth in the present product lines. On the other hand the graph could look different (Fig. 2) and often does. This second graph probably is much more common for companies whose business, like the chemical business, is moving reasonably rapidly technologically. Obviously, this means that new products and new businesses are much more important to the 10-year plan than in the first instance. Although the author has no idea of the 10-year plan of the E. I. Du Pont de Nemours

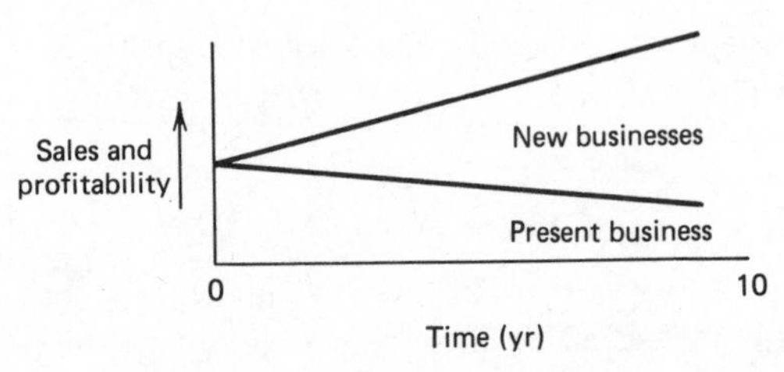

FIG. 2. Graph of 10-year goal II.

and Company, it is interesting to note that in 1972 new products introduced during the previous 10 years amounted to 15–20% of that company's sales [2].

III. CORPORATE TACTICAL PLANNING

Tactical planning is the selection of specific means to meet the objectives stated in Section II.

A. Maintenance of Present Business

In a sense the first route is defensive: expand the sales and earnings of the existing business. This is always indicated very strongly. However, there is a slight dichotomy in that certain phases of the present business may be becoming unprofitable. There is no sense in expanding their sales. This is why as a company grows it may divest itself of small-volume products that can no longer support the overhead. Large companies are always sloughing off such products to smaller companies that still find such products profitable. The author can remember when a new Monsanto Company president got rid of quite a number of such items. Thus divestiture does not increase sales, but it can increase profitability. On a large scale, E. I. Du Pont de Nemours and Company's discarding of Corfam undoubtedly must have done so. Likewise this probably is why both E. I. Du Pont de Nemours and Company and Celanese Corporation of America are closing cellulose acetate fiber plants [3].

Even more defensively, the present business may require a heavy capital expenditure just to maintain it. Sometime around the late 1960s, Swift and Company shut down many of its older meat-packing plants and replaced them with new, larger, and more-modern ones, just to stay in the meat business. Grocery chains are constantly shutting down small, unprofitable stores and replacing them with modern, efficient supermarkets. Chemical companies likewise shut down old, no-longer-profitable plants and replace them with new, modern ones, often employing greatly improved or completely new and more-efficient processes. Many examples of this were illustrated in Chapter 3.

Safety and environmental regulations constantly require big investments just to keep the plant running. It has been stated that by 1983 E. I. Du Pont de Nemours and Company will have to spend $500 million to install the "best available technology" for water-quality control [4]. Unfortunately two thirds of this expenditure is forecast to achieve no really significant improvements [4]. It will simply increase the price of the products to the consumer.

Process improvements and plant modernization are but one part of the technical effort to improve the present product line. Product improvement is equally important. Tire companies are always developing new belting materials and new tread configurations. The replacement of oil-based paints by latex paints in many uses, because of greater ease of application and the convenience of a water-based system, is another example. Oil additives contain a polymer that enables an automobile lubricating oil to maintain a more constant viscosity over a wide temperature range. The author can remember when alkylated polystyrene was replaced by long-chain polymethacrylates for this purpose.

Product lines constantly require new items to remain competitive. Dyestuff manufacturers are always introducing new dyes that impart a different shade or one more light fast. The author is familiar with a label-machinery-manufacturing company that has to add a new adhesive for its labels every time the chemical industry comes out with a wonderful new plastic film. Otherwise the labels do not stick. The author also is familiar with a manufacturer of food thickeners (for ice cream, desserts, etc.) that constantly has to add new items to its line to meet the changing needs of the food industry.

B. New Businesses

The second method for developing additional sales and earnings is through new businesses. There are several means available.

1. *Develop the New Business Internally.* In the chemical industry this means developing all the technology and marketing skills within one's own organization. The larger companies tend to follow this route because of antitrust pressure.

2. *License the Technology.* When Standard Oil Company (Ohio) came out with its process for preparing acrylonitrile from propylene, not only the then producers but also any new ones had to take out licenses for what obviously was the best process. Chemstrand Corporation entered the nylon 6,6 business through a E. I. Du Pont de Nemours and Company license. Several companies entered the high-density polyethylene business by licensing the basic Ziegler patents.

However, licensing does not mean that one buys a turnkey plant. When Monsanto Company first started manufacturing acrylonitrile, it used the acetylene–hydrogen cyanide route. Although the process was based on purchased German technology, the company had to conduct a very appreciable amount of development work to make the process commercially feasible.

3. *Go the Joint-Venture Route.* Perhaps a company has developed the technology and can acquire the marketing skills through a partner. Texaco, Inc. teamed up with American Cyanamid Company and formed Jefferson Chemical Company, Inc. in order to develop certain petrochemicals. In the days of the abortive boron-fuels development, American Potash and Chemical Corp., FMC Corp. and National Distillers and Chemical Corp. formed AFN for this purpose.

4. *Buy the Business.* E. I. Du Pont de Nemours and Company bought Endo Laboratories Inc. to help it get into pharmaceuticals. Monsanto Company bought Lion Oil Company to gain a source of hydrocarbon raw materials.

All four of these routes must be evaluated in relation to cost, lead time, and compatibility.

C. Tactical Options

The tactical options available to managements for increasing sales and profitability have been summarized by D. W. Collier together with recommendations on how best to handle each option [5]. His summary is as follows:

1. Present product
Present technology
Present market

This means product and process improvements. Companies do this best because they can use their normal operating organization and skills.

2. New product
Present technology
Present market

In this case the engineering and the technical service department must learn new things. Presumably the new product can be made by present manufacturing technology and the sales department will be contacting the same market.

3. Present product
Present technology
New market

In this case the sales department has to learn how to serve the new market.

4. New product
Present technology
New market

This case moves toward some kind of new-venture group, since the engineering, technical-service, and sales departments all have a lot to learn. If a new-venture group is set up, manufacture can be subcontracted to the company manufacturing department.

5. New product
New technology
Present market

In this case the sales department probably can handle its end of things. However, the engineering, technical-service, and manufacturing departments all have to learn new things. A new plant probably will be necessary.

6. New product
New technology
New market

Basically this is a new business for the company. It is in this area that new-venture groups (see Chapter 14) have proved most useful.

Although these classifications appear simple, many company managements have been fooled by them. Many petroleum companies have entered the chemical business, because the technology appears to be the same—chemistry and chemical engineering. It is similar but different. Petroleum-company engineers are experts at handling liquids and gases, but most of them have limited experience in handling solids. You do not use square corners on pipes carrying slurries. They plug.

More than that, the petroleum and chemical businesses are quite different. This brings up compatibility. A capable management in one kind of business may have trouble with a different one. A petroleum refinery is a turnkey plant, and its output usually is contracted for before plant start-up. A refinery starts up immediately and runs at full capacity at once. Chemical plants usually develop bugs, which may take months to iron out. Production seldom is sold out in advance. The plant has to be sized in relation to the expected market. It must be small enough that it can operate at least at break-even capacity within a reasonable time after start-up, and large enough that a duplicate plant does not have to be built beside it within a year or two.

IV. KINDS OF CHEMICAL BUSINESSES

From this brief discussion it can be seen that probably the first step both strategic and tactical planners have to take is to evaluate the company's resources. What does it have and what does it not have? What kind of business is it in?

The reader may feel that because all chemical companies deal with chemicals, they are in the same business. They are not. They can be in at least four different kinds of businesses, each with its own requirements and characteristics [6]. Perhaps the biggest contrast is between manufacturers of undifferentiated chemicals and those who make differentiated products.

Undifferentiated chemicals have a specific generic formula and carry a chemical name—sodium chloride, benzene, methanol, carbon tetrachloride. All producers make essentially the same product. Although there may be different grades (resin-grade and food-grade adipic acid), specifications are based on what the product contains. Undifferentiated chemicals find many applications.

In the case of differentiated products, there are real, or at least imputed, differences among products. Specifications are based on what the product does, not on what it contains. There often is only one and usually only a very few applications for each product. How does one differentiate?

1. One makes specific functional chemicals such as dyes or herbicides. One makes specific formulations such as fragrances or adhesives.

2. One dominates so small a market that there is no competition. The Loctite Company makes adhesives for such things as engine bolts. The polymerization of these adhesives is oxygen inhibited, so the adhesives are stable in air, but when the bolt is screwed in place and the air contact is eliminated, the adhesive sets up and locks the bolt in place. The total tonnage is very small but highly profitable, and the author does not believe anyone else is in the business yet.

3. One sells one's chemicals under a brand or code name rather than by generic formula. For instance, Monsanto Company's Santoflex 1 P, a rubber antidegradant, is N-isopropyl-N^1-phenyl-*p*-phenylenediamine [7].

4. One promotes the product directly to the final user and does not sell it on the merchant market.

5. One makes the chemical part of a system involving durable machinery and a disposable chemical such as a Xerox toner.

6. One patents new synthetic chemicals even though it may not be worthwhile to patent new formulations. This gives one composition-of-matter protection for the chemical (Chapter 9). It may not be worth the money to determine the scope of utility of a formulation (Chapter 9).

7. One continually innovates in products or packages or systems or in distribution and services offered.

Basically there are four different kinds of chemical products and therefore four different kinds of chemical businesses. These businesses blend into each other, and the larger chemical companies often are in all four of these businesses at the same time.

Following are the four categories:

1. *True Commodities.* These are undifferentiated chemicals. They are large-volume. Often they are based on captive raw materials. Their specifications are composition. They find many end uses. They are sold to many customers, who buy in large quantity, usually on a contract basis. True commodities are tonnage synthetics such as methanol and acetic acid, gases, gum and wood chemicals, fatty acids, and fertilizers.

2. *Fine Chemicals.* These too are undifferentiated chemicals. Although low-volume, they generally are unpatented and again are sold on the basis of generic name and composition. They have one or more end uses. Often there are a small number of customers. Fine chemicals are low-volume intermediates, medicinals (such as aspirin, which has to meet USP specifications), and aroma chemicals such as phenylethyl alcohol (not the final odoriferous formulation).

3. *Pseudocommodities.* These are differentiated products. They are large-volume and often are based on captive raw materials. They are widely used, al-

though as in the case of commodities, most of the sales may be to a few large customers. However, specifications are based on performance, not on chemical content. Pseudocommodities include tonnage resins and plastics, fibers, elastomers, carbon black, tonnage pigments, and tonnage surfactants.

4. *Specialty Chemicals.* These too are differentiated products. They are low-volume and are based on purchased raw materials (usually commodities). They are designed to solve specific customer problems and are sold to many different small-volume customers. Specialty chemicals include low-volume adhesives and surfactants, biocides, dyes, thickeners, formulated pesticides, and rubber chemicals.

Although the knowledge of what kind of business one is in is extremely important to the chemical-business executive, why should this interest the average chemist? The answer is simple. It tells him or her where various professional opportunities do and do not lie.

In the following discussion (highly pertinent to management's appraisal of its company's resources), companies specializing in commodities (category 1) will be contrasted with those specializing in specialty chemicals (category 4). Those producing fine chemicals and pseudocommodities lie in between these two extremes, but are more like those producing commodities. So their characteristics must be blended.

Technology will be examined first. In commodity companies, research and development emphasizes synthesis (new processes) and process engineering. The focus is on raw materials and on new or improved processes. The research and development facilities comprise synthesis laboratories engaged in research directed toward the discovery of these new processes, and include many process-development and pilot-plant facilities.

In specialty companies, the technical focus is on the customer and customer needs. The focus is on formulation and end-use knowledge, and on the practical know-how of customer-plant operations. Specialty-company research and development facilities comprise applications laboratories and field-testing facilities.

Now what about manufacturing? In commodity companies, one or only a few chemicals are made in a large, centralized plant. This plant runs 24 hours a day, 350 days of the year. The equipment is complex, automated, sophisticated, continuous, and inflexible. Production is for long periods of time on fixed schedules.

In a specialty company, many products are made in one plant, in small units. Usually the company has several plants in different parts of the country. Equipment is simple and flexible. There is no integration to raw materials; they are purchased on the merchant market as needed. Production is for short periods of time on flexible schedules.

Marketing differs just as much in the two types of companies. In a commodity company, sales are large-volume, usually to a few big customers on contract, plus a few other, smaller ones on the open market. The commodities usually are key raw materials for the customer in question (ethylene for a producer of polyethylene). Sales per sales representative are of the order of $10 million per year. The sales representatives really do not need to know much about a customer's needs. Although there is lots of market research, there is little technical service and test marketing.

Specialty companies sell to many small-volume customers. Sales per sales representative are of the order of $1 million per year and can be as low as $100,000 per year. The sales representatives must have an extensive knowledge of their customers' needs. The sales representatives are backed up by lots of technical service. There is little market research, but lots of test marketing.

Commodity prices follow the business cycle. Specialties hold steady.

Finally, here is the financial picture. Commodity companies have a high capital investment (big, automated plants), of the order of $1.25 per $1 of annual sales, with a concomitantly low return on the capital investment. Selling costs are low, and selling, general, and administrative expenses are about two to four times raw-material costs.

Specialty companies have a low capital investment, of the order of $0.25–0.50 per $1 of annual sales, with a concomitant high return on capital. Selling costs are high (see annual sales per sales representative as mentioned previously), and selling, general, and administrative expenses can run from $0.20 to $0.50 per $l of annual sales. Selling prices run around three to eight times raw-material costs.

V. APPRAISAL OF CORPORATE RESOURCES

What is the thinking of corporate planners in appraising the resources of a company.

1. *Financial.* How much money will be available each year during the next 10 years for additional support to the present business and for diversification? These monies come from present profits and from borrowing. The amount available from loans (both bank loans and bond issues) depends on the company's credit standing and on how far into debt it feels it is prudent to go. And these resources must be well managed.

Poor financial management can prevent the necessary reinforcement of the present business and stifle growth in desired directions. The stock market performance of a noninvesting company can look very good in the short term. On a long-term basis such a policy can be disastrous.

2. *Management.* What is the competence, depth, and aggressiveness of company management? What are their areas of business experience? What kinds of company policies are they used to?

Petroleum-company managements often encounter some of the problems just mentioned when they enter the chemical business. Altogether too often a large company buys a prosperous small company in a different business area and kills it with unfamiliar and more-elaborate administrative policies. This is particularly true if the two companies are in different industries.

Bad human management hurts immediately and, if allowed to continue, can wreck the company. With bad human management, capital is required to rebuild the company. If company management is thin, then when the company expands by acquisition, it is better to buy a company with strong management that can supplement that of the purchaser rather than buying a company with weak management that will require help from an already-too-small pool of competent executives.

3. *Technology.* What kinds of technical people does the company have? What can they do? What can they not do? If a strictly chemical company plans to expand into pesticides, agronomists and entomologists will be needed to conduct the testing of candidate chemicals. If a chemical company plans to go into fibers, then expertise in this area will have to be acquired. If a chemical company discovers polymers that appear to be of value in the paper industry, then professionals with a knowledge of paper technology will have to be hired. A commodity company that desires to manufacture specialty chemicals may find that its technical people have no idea of applications research (Chapter 7) and the need for a knowledge of consumer end usage.

4. *Marketing.* In what business areas can the sales department sell? A dyestuff sales representative would have to learn a lot before he or she could sell pesticides. The same would be true of a rubber-chemicals sales representative or an adhesives sales representative who had to sell paper chemicals.

How do the sales representatives sell? A person who sells tonnage chemicals on contract for a commodity company would probably have problems selling specialties on a door-to-door basis, and vice versa. Are the sales representatives accustomed to working with distributors, or do they sell to the customer directly? Do they need extensive technical-service backup? This latter is a concept that commodity-company managements may have a hard time understanding.

Is the sales department competent and aggressive? Will it be necessary to add sales management or more sales representatives?

5. *Manufacturing.* What is the condition of the plants and processing equipment? Will a lot of money be needed to replace worn-out equipment or an aged plant just to stay in business? Are the plants efficiently located in relation to markets both present and potential?

What are the skills of the manufacturing department? Can it handle explosives? Does it know how to process plastics? Can it make medicinals in conformance with FDA specifications? Can it run large, automated plants as well as small, batch operations? To a specialty company planning to manufacture a commodity, this is very important.

6. *Raw Materials.* Does the company have ready access to a continuous supply of the raw materials it requires? This can mean that several merchant producers are conveniently located. It also can mean that the company controls the source of its own raw materials. Monsanto Company and Stauffer Chemical

TABLE 1. Energy Content of Certain Products[a]

Product	Energy Content (%)	Product Price Rise (%) Necessary to Offset a Fourfold Increase in Energy Cost
Glass	15	45
Steel	20	60
Cement	30	90
Aluminum	40	120
Petrochemicals	60	180

[a] Reprinted with permission from M. J. Montet, *Chemtech*, May, 1974, p. 268. Copyright 1974 American Chemical Society.

Company each have their own phosphorus deposits, on which their phosphorus-chemicals businesses are based. The U.S. Borax and Chemical Corp. has a huge borax mine for its boron chemicals. Finally, is the company dependent on a raw material that comes from a foreign country where the government is unstable or prone to nationalize industry?

Since the costs of raw materials are constantly increasing, poor raw-material management can be very expensive. Poorer ores have to be processed. As mentioned in Chapter 3, most of the organic chemical industry many years ago shifted from coal to petroleum as its basic raw material. With the recent big increase in the price of crude oil and more increases to be expected, will it be desirable to shift back to coal in the not-too-distant future? Are there other sources of a raw material, such as sulfur from fuel oil?

7. *Energy.* Does the company have a continuous supply of the energy it needs? What changes are going to have to be made in the next 25 years? The supply of petroleum and natural gas is finite, and both will be greatly depleted, if not all gone, in 25 years.

Energy and raw materials are somewhat interchangeable. Ninety percent of all organic chemicals are made from petroleum, although they consume only 5–6% of the annual crude-oil production. Most of the copper for electrical purposes can be replaced by aluminum. However, the production of aluminum takes a lot more energy than the production of copper.

Poor energy management can be very expensive. Table 1 shows the energy content of several products and the price increase necessary to offset a fourfold increase in energy costs [9].

VI. APPORTIONMENT OF CORPORATE RESOURCES

There is only so much money available to any company, and that company has to decide where it is wisest to spend it.

1. *Personnel.* In support of the present business, does the sales department need reinforcement? Are more sales representatives needed? Is top man-

agement too thin? Should some younger people be brought in as backup? Does technical service need beefing up? Are research and development adequate?

In relation to new business, how much personnel is needed for the new plant? How many engineers, operators, and supporting personnel will have to be hired? A new business area is being entered. What will be needed in terms of sales and technical-service people in order to serve the new market? There are plans to enter the pesticide field. Besides the synthetic organic chemists already in the research and the development department, what biologists are going to be required?

2. *Facilities.* Must an old plant be replaced? Is some processing equipment so antiquated that it must be replaced? Is the installation of a new manufacturing process necessary in order to remain competitive? A good example of this necessity is the change from phthalic acid to toluene as the raw material in the manufacture of benzoic acid.

A successful new-product-development program has just been finished. How much money should be budgeted for the construction of the new plant? The company's research and development facilities are scattered and not too well located. Should they be combined in a new technical center? The central office is overcrowded, and some people are in rented space elsewhere. How much is the badly needed new office building going to cost?

3. *Raw Materials and Intermediates.* Is the raw-material position shaky? Should a hydrocarbon source be insured as Monsanto Company did when it bought Lion Oil Company? Is there a cheaper raw material? What will it cost to shift over? E. I. Du Pont de Nemours and Company made just such a shift when it changed from furfural to butadiene as the raw material for some of its nylon 6,6 intermediates.

R. H. Yocum has described an interesting situation with regard to intermediates [10]. In 1971, of Dow Chemical Company's some 700 products, about 100 were based on nine vinyl monomers, and 8 of the 10 U.S. product departments used these monomers. Thus there was a tremendous internal driving force to manufacture these monomers within the company. Vinylidene chloride was needed for Saran. Divinylbenzene was needed for ion-exchange resins. Styrene was needed for polystyrene and SBR latex. As a result, processes were developed, and a new monomer department was formed. Obviously, Dow Chemical Company saved money by manufacturing its own raw materials, and a new business was created based on supplying other companies with these monomers. Certainly this type of situation is not unique to Dow Chemical Company.

4. *Products.* Does the present product line need reinforcement? For instance, the rubber industry is served with a good line of accelerators and antioxidants, but an antiozone agent is needed (particularly a nondiscoloring one for whitewall tires). The company markets several grades of polystyrene. Now, to meet competition, a new, high-impact variety is needed. The company manufactures phthalic anhydride and could make a great deal more of it if it could be sold. This situation originally helped Monsanto Company bolster its plasticizer business with phthalate esters.

Every company that has a process that produces two products has a by-product problem. This requires the expenditure of research and development money to try to keep things in balance with the market. For a long time Monsanto Company accumulated appreciable quantities of *o*-chloronitrobenzene from the nitration of chlorobenzene, because products based on the *p*-isomer sold in larger volume. In the nitration the distribution between the two isomers then was about equal. The author does not know whether Monsanto Company has solved this problem, but the company certainly spent a lot of research and development effort on trying to do so. Monsanto Company was more fortunate in coping with the problem of the terphenyl by-product from biphenyl manufacture. After an appreciably sized lake of terphenyl had accumulated, it was found that partial hydrogenation of terphenyl leads to HB-40, an extender for polyvinyl chloride. The author believes the lake now has been pretty well exhausted.

What about forward integration? Can it be done without competing too seriously with one's customers? Monsanto Company moved into the consumer detergent business with All. However, All competed so successfully with the products of the company's phosphate customers (who used phosphates as builders in their own detergent formulations) that Monsanto Company finally had to choose. It decided to stick with phosphates and sold All to Lever Brothers. Other companies may have had better luck.

However, not all stories are that bad. Monsanto Company's position as a manufacturer of phthalic anhydride coupled with its experience with sulfonamide plasticizers led to a very extensive phthalate ester plasticizer business. Monsanto Company's position as a nitrator of chlorobenzene lead to its advantageous position in *p*-chloronitrobenzene chemistry. On the one hand are the *p*-nitrophenol insecticides, parathion and methyl parathion, and on the other hand the several *p*-phenylenediamine-based rubber chemicals shown in Chapter 6. Also, Monsanto Company's knowledge of phosphate chemistry led to a line of fire-resistant hydraulic fluids.

What about new technical areas to exploit? R. W. Gunder has mentioned several of these [11]. Some fifteen years ago the idea of a selective herbicide looked very appealing. Stauffer Chemical Company investigated the field and came up with the dithiocarbamates. The company received an added bonus, because these compounds were excellent environmentally as well. Although the market projection was poor, flame retardants (another field) still were felt to hold promise. This was indicated by such clues as tougher building codes, increased use of plastics in construction, and more government regulations in general. Stauffer Chemical Company now is cashing in. These are too good illustrations of how the market must be anticipated.

5. *Government Regulations and Inflation.* Any discussion of corporate needs has to include a consideration of government regulations and of inflation and their effect on the corporate business picture. Every so often, price controls are imposed on finished products. Usually this applies to a lesser extent to company raw materials and labor costs, so profit margins are reduced automati-

cally. Therefore, there is less cash available to support the present business, and particularly less cash with which to expand into other business areas.

Environmental regulations require financial outlays. Obviously these regulations reduce the amount of money available for expansion into new businesses and the creation of new, permanent jobs in such businesses.

Next comes implementation. Enough has been said about the support of the present businesses. It is hoped that the chemist and the chemical engineer have gained enough insight into the economic problems facing corporate managements that an understanding of the reasons for defensive financial technical outlays has been gained.

VII. EVALUATION OF NEW BUSINESS OPPORTUNITIES

How are new businesses and new technical areas evaluated? The evaluation can become very complicated. Following are some of the simpler criteria, not necessarily in order of importance:

1. What is the growth potential? Is this business area growing or is it stagnant? There is no sense in investing money in a stagnant business area.
2. What is the level and character of the technical effort required? In other words, does the company have the requisite people in its organization to support it, or does the company have to go outside and hire them?
3. What is the profitability of the possible new business area? Companies always want to increase their profitability, so this is very important.
4. How much will it cost to enter the field in terms of new plants, new research, development, and marketing department additions?
5. What is the return on investment?
6. How complex is the market? As a company used to dealing with a few large customers, can it exploit a market comprising many small customers?
7. How much technical service is required? This can be a big expense and can be totally unexpected to commodity-company managements.
8. Does the field have raw-material limitations? The classic example is the abortive boron-fuels program; very, very few airplanes could have been so fueled because of the total supply of boron in the world.
9. What is the patent situation? Do one or a very few companies control the business through patents.
10. What is the competition?

On the basis of these criteria, company managements can decide whether to enter the field in question through internally generated technology, licensed technology, a joint venture, or an acquisition. Often a combination is used.

Now comes the setting of the research and the development budget. In the chemically based industries, budgets vary in amount from about 0.3% of sales in food and beverages and about 0.5% in pulp and paper to about 3.5% in chemical and allied products [12]. Within the strictly chemical industry itself, variations are from about 1 to 4.5% [13].

Some companies set their research and development budgets at about the industry average. Although this is fashionable, it does not necessarily reflect the needs and aspirations of the company in question. A much better way is to tie the budget to the long-term plan for the company's growth.

For instance, if there are going to be heavy demands for capital to replace old plants and worn-out equipment or to meet environmental regulations during the next few years, there is no sense in having a large development program, because there will be no money available to build the new plants if the developments are successful. However, more research than usual may be called for, so that there will be plenty of new products and new processes ready for development when the money does become available in a few years.

The author is familiar with a company that wanted to expand in two areas of its present business that were relatively small in comparison with its main line of endeavor. The company was spending 15% of sales on research and development in each of these two areas in order to speed up expansion.

It must be remembered that a new plant or a new product line requires a lead time of 2–10 years. The $1 billion E. I. Du Pont de Nemours and Company planned to spend on plant construction in 1977 probably did not begin to become profitable for more than five years [14].

It is hoped that from this discussion the chemist has gained some idea of the financial problems a company faces and some of the ways business executives have to think. It is hoped that it has given the chemist and the chemical engineer an idea of how a company has to apportion its resources, only some of which are available for research and development. The financial, technical, and marketing inputs have to be balanced. All three are important.

REFERENCES AND NOTES

1. Some of this thinking is drawn from A. E. Albright, *Chemtech*, **3**, April, 1973, p. 197.
2. Anon., *Chem. Week*, **111**, November 29, 1972, p. 5.
3. Anon., *Chem. Eng. News*, **55**, February 7, 1977, p. 7.
4. I. S. Shapiro, *Chemtech*, **5**, December, 1975, p. 706.
5. D. W. Collier, *Chemtech*, **5**, February, 1975, p. 90.
6. C. Kline, *Chemtech*, **6**, February, 1976, p. 110.
7. *Monsanto Company Product Catalogue*, Thirty-Eighth Edition, Monsanto Company, St. Louis, Missouri, p. 58.
8. Some of this is taken from D. S. Davies, *Chemtech*, **4**, March, 1974, p. 135.
9. M. J. Montet, *Chemtech*, **4**, May, 1974, p. 268.
10. R. H. Yocum, *Chemtech*, **1**, June, 1971, p. 358.
11. R. W. Gunder, *Chemtech*, **1**, March 1971, p. 130.
12. Anon., *Chem. Week*, **118**, May 26, 1976, p. 20.
13. Anon., *Chem. Week*, **119**, July 14, 1976, p. 27.
14. I. S. Shapiro, *Chemtech*, **6**, November, 1976, p. 688.

12

EVALUATION OF INDUSTRIAL RESEARCH AND DEVELOPMENT

I. INTRODUCTION

Research and development cost money, and many people are involved. Obviously, companies do not invest money in research and development unless the effort will contribute more to the company than it costs.

II. COSTS OF TECHNICAL RESEARCH AND DEVELOPMENT

In Chapter 7 it was shown how the cost of even a successful development skyrockets as the development progresses. However, most developments never achieve commercial success. The unsuccessful ventures have to be supported until their abandonment is indicated. And the cost of research and development is constantly increasing.

J. E. Johnson and E. H. Blair have illustrated this situation very well in the case of Dow Chemical Company's pesticide research and development [1]. In Table 1 their data are used to show how the cost of commercializing a successful compound has increased over the years.

In 1972 a successful agricultural commercial development took 8–10 years. If we assume the patent (as a result of the research) issues about the time serious development efforts start, this leaves only 7–9 years of the patent's life to exploit the development. Much of this delay is due to what many chemical companies feel is inordinately long safety testing [2].

As in the example in Chapter 7, Johnson and Blair showed how costs escalate as a development progresses. Although they do not use the same terms as the author, the terms are easily equated (Table 2). In stage III of Dow Chemical Company's pesticide development, the efficacy, toxicity, and metabolism of the compound are determined. The process is developed. The ecology is examined,

TABLE 1. Cost of Commercializing a Successful Pesticide[a]

	1956	1964	1969	1972
Overall cost per compound	$1,200,000	$2,900,000	$4,100,000	>$10,000,000
Overall survival rate	1/1800	1/3600	1/5040	1/10,000

[a] Adapted with permission from J. E. Johnson and E. H. Blair, *Chemtech*, November, 1972, p. 667. Copyright 1972 American Chemical Society.

TABLE 2. Pesticide Development Costs[a]

Stage I, exploratory =	phase 1, research	$500,000
Stage II, defines the discovery =	phase 2, definitive development	$1,000,000 Total
Stage III =	phase 3, laboratory development	$1,900,000 Total
Stage IV, outside research =	phase 4, design development	$4,400,000 Total

[a] Adapted with permission from J. E. Johnson and E. H. Blair, *Chemtech*, November, 1972, p. 667. Copyright 1972 American Chemical Society.

particularly in regard to residues. In stage IV, cooperative outside laboratories confirm the data on efficacy and safety.

A large element in the year-to-year and stage-to-stage cost increases is the recent big growth in the need for more toxicological, metabolic, analytical, and ecological studies. The increase in these studies also lengthens the time of the development during which no return is being received on the money invested. One can get an estimate of the extent of these studies from the fact that the backup for Blazer, Rohm and Haas Company's new postemergence broad-leaved herbicide for soybeans, comprises 92 volumes of data weighing 408 pounds [3].

One loser's failure was caused by a residue in the environment, which resulted in long-term toxicity. At this point $2.1 million had been spent on its development. If this amount had been invested at 8% interest instead of in development, it would have increased to $3.4 million. Therefore, the total loss was $5.5 million, not $2.1 million.

Johnson and Blair also give some examples of losers and winners (Table 3). (A winner is a compound that lasts nine or more years with annual sales of $10

TABLE 3. Losers and Winners[a]

Loser	Cost	Time	Winner	Cost	Time
A	$ 600,000		F	$6,500,000	6 years
B	1,000,000	7 years	G	6,700,000	7 years
C	4,000,000	6 years	H	1,800,000	In progress
D	1,500,000	4 years	Total	$15,000,000	
E	500,000	4 years			
Total	$7,600,000				

[a] Adapted with permission from J. E. Johnson and E. H. Blair, *Chemtech*, November, 1972, p. 667. Copyright 1972 American Chemical Society.

TABLE 4. Financial Summary of Dow Chemical Company Pesticide Research 1960–1970[a]

A	Continuing research projects	$15,000,000	Stage I
B	New-product successes	$20,000,000	Stages II, III, and IV
C	Product failures	$10,000,000	
D	Existing-product maintenance	$25,000,000	
Total		$70,000,000	

[a] Adapted with permission from J. E. Johnson and E. H. Blair, *Chemtech*, November, 1972, p. 667. Copyright 1972 American Chemical Society.

million to $20 million and gives a pretax return on investment of 40%. Everything else is called a loser).

The winners not only have to pay for their own developmental costs, but they also have to pay for the money invested in the losers. And the winners have to be profitable to manufacture after all these costs have been paid off. Otherwise, why bother?

Johnson and Blair's 10-year summary of Dow Chemical Company pesticide research and development illustrates this very well (Table 4).

In order to produce new products, one must finance areas *A*, *B*, and *C*, in this case at a total cost of $45 million. Four winners were obtained. Therefore, they cost $45 million/4 = >$11 million apiece.

Human health products is another field in which the costs of new-product development are available. L. H. Sarett has written about Merck and Company's experience in developing human health products [4]. Here government regulations and the processing of information have contributed greatly to increased drug development costs. As in the case of pesticides, not only has the sheer volume of necessary laboratory work increased appreciably, but so has the amount of expensive development time.

Table 5 shows development times for new human health products, from the time the drug is selected to the time the new drug application (NDA) is filed. It includes the determination of safety in animals and the determination of safety and efficacy in up to 1000 human patients.

Table 6 shows another cost factor, namely, the time in months to obtain regulatory-agency approval of a new drug. The difference between the times necessary in Western European countries and the United States is the reason half of U.S. new-drug development is being conducted overseas [5], even though for ultimate U.S. FDA approval, all testing has to be repeated in this country.

TABLE 5. Development Times for Human Health Products[a]

1958–1962	1963–1967	1968–1972
2 years	4 years	5.5–8.0 years

[a] Reprinted with permission from L. H. Sarett, *Chemtech*, January, 1975, p. 20. Copyright 1975 American Chemical Society.

TABLE 6. Months for Regulatory-Agency Approval of a New Drug[a]

Country	1961	1962	1967	1969	1972
United States	—	6	—	40	>40
U.K., Holland, France, Sweden, Germany	6 (0–24)	—	9 (2–24)	—	16 (6–24)

[a] Reprinted with permission from L. H. Sarett, *Chemtech*, January, 1975, p. 20. Copyright 1975 American Chemical Society.

TABLE 7. Developmental Cost of a New Drug in the United States[a]

1962	1967	1972
$1.2 million	$3 million	$11.5 million

[a] Reprinted with permission from L. H. Sarett, *Chemtech*, January, 1975, p. 20. Copyright 1975 American Chemical Society.

In the United States the time for development and approval of a new drug in the early 1970s was 7.5–10 years [4].

What does all this mean? Table 7 shows Sarett's figures for the average developmental cost of a new drug in this country. It has now risen to at least $55 million [6,7]. The paperwork for a successful applicant can amount to 34 volumes [7]. The backup for a new antiarthritic drug amounted to 120,000 pages [8].

Obviously this enormous increase in drug development costs has had a big effect on drug research and development as well. In the 15-year period 1947–1962, 641 new drugs were introduced in this country. In the following 15-year period (1962–1976), the number shrank to 247 [9]. The new FDA regulations went into effect in 1962.

These delays and costs keep new drugs from patients in this country and benefit foreign patients first [6,7]. Thus of 27 important U.S. drugs, 14 were first marketed in the United Kingdom. [10]. The majority were discovered here. Table 8 shows where the United States ranks as a nation in the introduction of new medicinal drugs [11].

Another way of looking at costs is from the angle of the cost per technical individual in research and development. Table 9 shows some annual figures for

TABLE 8. U.S. Ranking in New-Drug Introduction[a]

Drug	Use	U.S. Rank
Metaprolerenol	Antiasthma	30
Adriamycin	Anticancer	32
Rifampin	Antituberculosis	51
Cromolyn	Antiallergenic	64
Co-trimazole	Antibacterial	166

[a] From Reference 11.

TABLE 9. Cost per Technical Research or Development Individual[a]

1957	1968	1971	1977[a]
$23,000	$40,000	$45,000	$50,000–$75,000

[a] From Reference 13.

a medium to large company [12]. As indicated, these costs vary with company size.

When the reader looks at starting salaries, he or she must wonder where on earth these numbers came from. For a research chemist these figures embrace not only the professional's salary, but also the pension plan; the company medical plan; technical help; secretarial help; the library; the laboratory building; the day-to-day use of utilities; the glassblower; the mechanical and electronic shops; the glassware washer (who probably is mechanical); the cost of all the chemicals and apparatus the chemist uses; the chemist's share of the cost of instrumentation; such laboratory overheads as supervision, personnel, purchasing, accounting, medical, guards, and janitors; and corporate overheads including company officers and legal and patent departments. Is it any wonder that the major cost item in a research and a development department is the time of the professionals involved?

For corporations there has been one mitigating factor—the development of modern instrumentation by means of which a professional's time can be much more productive than it used to be. Only 30 years ago in a process study involving liquid organics each change in a reaction variable meant a preparation on a sufficiently large scale, so often as much as 1 or 2 liters of product had to be very slowly and laboriously distilled through a highly efficient fractionating column (such as a spinning band column) in order to determine with reasonable accuracy the amounts of products and by-products produced. Now the same determination can be made with a very small sample from a much smaller scale preparation by means of a gas chromatograph in a very short period of time. Likewise, in the strictly research area, nuclear magnetic resonance and mass spectroscopy have become of tremendous help in elucidating the structures of chemical compounds.

Thus the new research chemist in industry can see that he or she and one laboratory partner probably will cost the company a quarter of a million dollars in the space of two years. They had better produce.

At this point it must be obvious why the age of accountability has been reached. Research and development must pay for themselves. They can no longer be conducted solely to give the company a favorable image with possible investors.

The obvious result is that research and development are being given increasingly close scrutiny by top management. The many checkpoints in a development program are just one symptom.

III. EVALUATION OF RESEARCH

How do business executives and, for that matter, technical managements evaluate and monitor research? The evaluation of research is particularly baffling to business executives. Since research is technical effort directed toward the discovery of new scientific facts, there is no way the investigator can prophesy, much less promise, what is going to be discovered. And yet research has paid off in company after company.

In the more sophisticated companies it is recognized that research produces new scientific knowledge, not commercially profitable innovations. It is up to the rest of the company to take these new scientific facts and develop them into new products and new processes.

Since it is so difficult for a business executive (the one who is putting up the company's money) to evaluate a research program, how does the executive control it so that the company has a reasonable expectation of getting its money's worth from it? Basically the executive is taking a calculated risk.

One answer is to support defensive research. If the company has a large investment in several plants that manufacture a given chemical and the business in highly profitable, then a business executive can see why it is very desirable to study the chemistry of the reaction and the engineering of the process in detail and to ascertain all the scientific facts involved. As a result, perhaps the yield of product can be increased. Perhaps the formation of an undesirable by-product can be minimized. Perhaps the engineering can be simplified so that the investment in new plants can be reduced. The business executive also can appreciate the problems to be faced if a competitor using the same process came up with answers to these questions and his or her company did not. The executive can balance the risk against the cost of the research.

In offensive research the advantages to be obtained from a new process can be balanced against its chances of technical success. If this could be done, how much would it be worth? In Chapter 6 a brief raw-material cost calculation was made to show how attractive the production of benzoic acid by the liquid-phase oxidation of toluene appeared to be in comparison with the decarboxylation of phthalic acid. Taking the several possible yields shown, the increased profit could be estimated by multiplying the raw-material cost savings by the Monsanto Company sales volume. Conversely, it is easy to see what could have happened to Monsanto Company's then benzoic acid business if a competitor had discovered how to oxidize toluene to benzoic acid before Monsanto Company did.

Even riskier from the business executive's point of view is the support of research that it is hoped will uncover new scientific facts from which new products might be developed. Some control is accomplished by confining the research to a specific scientific field.

Pharmaceutical and agricultural chemical companies are interested in new drugs and new pesticides. Therefore, their research effort is directed toward the discovery of new chemical compounds that possess physiological activity in living organisms. The charters are very broad, but both types of companies have been highly successful.

Petroleum and petrochemical companies are interested in the conversion of crude oil into useful hydrocarbons and these in turn into tonnage organic chemicals of commerce. Tremendous volumes of material are involved, and the price per pound must be low. Therefore, these companies conduct research in catalysis. This was illustrated in the case of petrochemicals in Chapter 4.

W. H. Carother's research on condensation polymers must have increased E. I. DuPont de Nemours and Company's interest in polymer research immensely. The company certainly has continued research in this area and continually introduces products based on new polymers to the marketplace.

Another approach for controlling research costs is to say "Let's have a good researcher look into the field for a year, and then we'll take a reading." In other words the initial risk is $50,000–$75,000 [13].

When all is said and done, it is extremely difficult to more than generally direct, much less really evaluate, a research program at any one time. Probably as good a method as any is based on the track record of the principal investigator, even though this is hard on the younger professionals. The author knows of a university professor whose proposals to the government agencies from which he derives his financial support simply amount to: "I am going to keep on doing what I have been doing." He gets the money because of an outstandingly successful research record. There is a lot more of this type of research evaluation in industry than the beginning chemist might suppose.

IV. EVALUATION OF DEVELOPMENT

Since development can be programmed and its result in large measure predicted, it is much easier to monitor and evaluate than research. This is all to the good, because, as has been seen, this is where the costs really escalate.

All kinds of rating systems have been devised, an excellent one being that of J. S. Harris [14]. All companies have such rating systems. They include mathematical models, checklists, and project-ranking schemes. On joining a company the chemist should learn the one the company employs. In either research or development the chemist should know when the project cutoff ax may fall and why. For this discussion a typical development-project-rating scheme based largely on that of R. Steckler [12] will be employed. It includes most of Harris's criteria.

Following are the factors that are evaluated at each checkpoint:

1. *Probability of Technical Success.* This is a matter of judgment on the part of the technical members of the development team. Is the separation of an undesired by-product proving more difficult than expected? Has it been impossible to raise the yield, or, conversely, are much better yields being obtained than originally expected? Can such marketplace-determined specifications as improved color be met?

2. *Probability of Commercial Success.* This is a matter of judgment on the part of the marketing members of the development team. Can the sales department sell the product? The author's process for air oxidation of toluene to benzoic acid was a commercial shoo-in because the company was already selling the product at that time. All the benzoic acid prepared by the new process had to do was meet the technical specifications then being met by that prepared by the old process.

Is the timing right? It has been seen how the commercialization of All had to wait for the need for the product to develop. Will the market be gone by the time the development is completed?

Is the price going to be right? If the company is becoming a me-too producer, will the present suppliers cut the price? (One does not become a me-too producer of a commodity chemical unless one has the cheapest process.) In the case of a new product, how are potential customers meeting this need now? Will they pay a bonus for the new product? Are projected manufacturing costs becoming impossible?

3. *Anticipated Profit During Product Life.* Will the product meet the standard of profitability set by the policy of the corporation? What will the useful life of the product be? In a field that is advancing very rapidly technologically, this time could be as little as five years. As soon as sales are up, the profit on a new product from a new and efficient plant usually is high. Later on, profit margins deteriorate because of an older and less-efficient plant, competition, and inflation.

4. *Development Cost.* As has been seen, the development cost can be appreciable. Will this cost plus the other necessary expenditures exceed the anticipated profit during the product's life?

5. *Fixed-Capital Requirements.* How much will the pilot plant and the final manufacturing plant cost? Does the company have this much capital available? Would the company be wiser to invest the money in some other product or process? Would the company be wiser simply to put the money in the bank and let it draw interest? Will too much capital be tied up with no return because of a lengthy development period?

6. *Working-Capital Requirements.* Are these going to be too large? Will big inventories have to be carried, as in the whiskey business (aged in wood)?

7. *Fit With Corporate Business Objectives and Skills.* This criterion should have been looked at very carefully before the development project ever was authorized. However, corporate objectives do change with time, so fit must be reevaluated at every checkpoint. There can hardly be a worse blow to developmental-team morale or a more fruitless expenditure of funds than to carry a project through phase 4 (see Chapter 7) and then have the proposal for phase 5, plant construction, turned down by top management because the company does not want to be in that business after all. The author knows of several cases in which this actually happened.

Does the manufacturing department have the technical skill and experience to make the product? If not, can these skills be bought or developed successfully? Does the proposed process involve the use of very high pressure equipment that the manufacturing department has had no experience using? Are peroxides involved? Has the manufacturing department ever handled them? Does the process involve hydrogenation? Has the manufacturing department ever handled large volumes of hydrogen?

How does the product fit with the sales department? Will the product be the only one in a sales representative's bag when he or she approaches a specific industry such as rubber? The competitors may have 30 products to offer that fill a variety of needs. If a sales representative has only one product, the per-product sales expense could be appalling. In serving a specialized industry such as rubber, does the sales department have competent technical-service backup with a thorough knowledge of the technical problems of the industry in question?

8. *Competition.* What is the competitive situation? How many suppliers presently are serving this market, and how strong are they? Do one or two dominate the market, and if so, how? How many potential customers are there? Are there so few customers that they can pretty well dictate prices by playing off one supplier against another? This situation was touched on in Chapter 6 in connection with continuing research on a terephthalic acid process.

9. *Patents.* What kind of patent protection can be obtained? Based on a knowledge (see Chapter 9) of competitors' and potential competitors' technical activities, is it likely that an adverse patent may issue and bring the entire development to a screaming halt when halfway completed? The author has seen this happen. Is the procurement of patent coverage going to be inordinately expensive (Chapter 9)? Will foreign patent coverage be needed, and if so, how much will it cost?

10. *Government Regulations and Environmental Concerns.* These are becoming more and more important for financial planning in the chemical industry. Is the product a possible carcinogen? If so, it may have to undergo such extensive testing that its development may not be worthwhile. Does the process emit odoriferous fumes that may be difficult to eliminate completely? The neighbors may complain. Will a special disposal system have to be constructed to handle the plant effluent? There are oil refineries in which 50% of the cost of the facility is devoted to obeying environmental laws [15].

Although this discussion has been focused on monitoring and evaluating ongoing developments, it is more broadly applicable. Projects that were shelved for good reason a few years ago may now be viable because of changed conditions.

In evaluating development projects the ultimate cost of maufacturing the new product has to be estimated. As was seen in Chapter 7, the precision of the estimate increases as the project moves from phase to phase and more and more information becomes available. At the start a ball-park estimate often can be made very simply. All companies know the relationship between raw-material and manufacturing costs for the products they produce. Although these vary with plant size (usually the larger the plant, the lower the manufacturing cost per pound of product), for a medium-sized plant there may be some simple formula such as doubling the estimated raw-material cost.

V. ESTIMATE OF MANUFACTURING COSTS

At the checkpoint at the close of phase 4, a very accurate estimate of manufacturing cost must be made before proceeding to phase 5. Following are some of the cost elements in the manufacture of a chemical [16]:

A. Raw Materials

B. Operating Costs

1. *Utilities and Fuel.* The cost of energy is going up very rapidly.
2. *Regular Labor.* Primarily these are plant operators.
3. *Plant Administration.* This includes plant management and the various manufacturing supervisors.
4. *Clerical Labor.* Clerks, secretaries, telephone operator, receptionist, and so forth.
5. *Clerical Supplies.*
6. *Contract Labor.* The plant may employ an outside contractor for cleaning rather than hire janitors. Major plant modifications may be contracted to an outside engineering firm.
7. *Maintenance Labor.* These are the plumbers, electricians, and so forth who keep the plant running.
8. *Maintenance Materials.*
9. *Plant Services.* These include mechanical and electrical shops and laboratories and the cost of operating them. Also included are such services as purchasing, accounting, and personnel.
10. *Taxes.* These are the local taxes that are paid because of where the plant is located.
11. *Depreciation.* This includes both plant and equipment.

C. Other Costs

1. *Sales.*
2. *Distribution.* You have to deliver the product to the customer.
3. *Corporate Overheads.* These include the plant's share of the cost of the company's main office, the plant's share of corporate taxes, and the plant's share of such corporate services as legal, financial, and patent.
4. *Research and Development.* Research and development has to be supported by the sale of products.

In closing, the author again emphasizes that all the criteria used in evaluating development projects are estimates. They are subjective estimates. They are nearly impossible to quantify by a meaningful formula.

Development projects are compared with each other. The decision to proceed with or to drop a specific project is made on the basis of such comparisons. A company has only so much money for product and process development and for plant construction at any one time. It has to choose the most promising projects to support.

In these comparisons one factor may far outweigh all the others. The factors are never equal. For instance, a project may have to be shelved for a while at least because all the company's capital resources must be used to modernize old plants. There may not be enough money available to build a new plant at this time. The patent situation may become impossible. The market simply may not be big enough to support the development effort. Petroleum companies do not develop watch oils. The pharmaceutical companies are directing their research and development efforts more and more toward the commoner diseases; in these areas, sales of a successful drug will repay the escalating development costs and provide additional income [4].

REFERENCES AND NOTES

1. J. E. Johnson and E. H. Blair, *Chemtech,* **2,** November, 1972, p. 666.
2. Anon. *Chem. Eng. News,* **58,** April 28, 1980, p. 10.
3. Anon. *Chemtech,* **9,** August, 1979, p. 464.
4. L. H. Sarett, *Chemtech,* **5,** January 1975, p. 20.
5. R. H. Levin, private communication.
6. Anon. *Chem. Eng. News,* **56,** June 26, 1978, p. 5.
7. Anon. *Chem. Week,* **128,** February 4, 1981, p. 3.
8. Anon. *Chem. Week,* **122,** May 10, 1978, p. 59.
9. Anon. *Chem. Eng. News,* **57,** April 30, 1979, p. 40.
10. Anon. *Chem. Eng. News,* **56,** July 10, 1978, p. 20.
11. S. Peltzman, *Chemtech,* **10,** May, 1980, p. 311.
12. R. Steckler, *Chemtech,* **1,** October, 1971, p. 579.
13. Author's estimate. See also Anon. *Chem. Week,* **117,** December 24, 1975, p. 24.
14. J. S. Harris, *Chemtech,* **6,** September, 1976, p. 554.
15. M. A. Wiley, *Chemtech,* **6,** February, 1976, p. 134.
16. These are adapted from A. Bisio and V. D. Herbert, Jr., *Chemtech,* **6,** July, 1976, p. 422.

13

ORGANIZATION OF COMPANIES AND OF TECHNICAL DEPARTMENTS

I. INTRODUCTION

Previous chapters have shown how company managements plan the use of research and development and how they evaluate these efforts. This chapter examines how research and development departments fit into corporate organizations and what such departments actually comprise.

Company organizational structures have evolved to fulfill a variety of needs. Initially an enterprise often started as one individual who developed the product, manufactured it, and then sold it to customers. Perhaps as good an example as any is that of a patent medicine, which a physician concocted originally, mixed up in the kitchen, and sold to those patients he or she thought would benefit from taking it. Certainly many chemical companies started in a similar way with the manufacture of one product.

As the business grew, it became more than what one individual could handle alone. Perhaps the founder chose to develop new products and manufacture them (in the garage) and hired a sales representative to peddle them. Or the founder may have devoted his or her energies to sales and business development and therefore hired a chemist to make the product (in the garage). In either case there originated a need to formally clarify the relationship between the two individuals and to provide a mechanism for deciding what should be made that could be sold. With the growth of the company and the addition of other people, the relationships became more and more complex.

The basic purpose of a corporate organization is to provide a chain of command through which the activities in the lower echelons can be planned, assigned, supervised, approved, and summarized for transmittal to those making the basic decisions.

In any company there is a need for a housekeeping function. Likewise, a person who is not self-employed needs a boss to go to for help with on-the-job problems. In many companies good bosses also help their assistants with personal problems as well.

In any company an employee also needs a regular appraisal of his or her work performance with guidance toward its improvement. There is a personal need for training and leadership toward increased responsibilities. The company needs this kind of professional development on the part of its employees. An employee needs a regular salary review. A chemist also has requirements for space and equipment—office, laboratory, and so forth—and for such services as stenographic, shop, and as mundane as janitorial. The chemist also needs help in accounting and purchasing.

II. GENERAL COMPANY ORGANIZATION

How then are companies organized to meet these needs? The place to start is with the owners. These can range from one individual or two or three partners in a small company to several thousand stockholders in a large company.

In every case the owners elect a board of directors to supervise the operation of the company. When the company is owned by two or three people, these same people may constitute the board, often with such outside help as a banker and a lawyer. In the case of a very large company, the board may number 12–20 individuals. Often about half these people are insiders (company officers who know the business thoroughly), and the other half are outsiders, officers of other companies, who not only provide needed additional skills such as legal and financial, but who also provide other views and serve as windows to the outside world. The distribution varies widely from corporation to corporation. Some very successful companies such as Dow Chemical Company have essentially inside boards, whereas others such as General Electric Company lean heavily on outsiders. Large boards usually are broken down into several committees such as executive, finance, and audit.

What then does a board of directors do? The following excellent summary is by I. S. Shapiro [1].

1. It selects company officers and provides for the succession of people in key spots.
2. It determines broad policies and establishes the general direction of effort of the enterprise.
3. It sets performance standards, ethical as well as commercial, against which management is judged.
4. It monitors management's performance in meeting overall strategy and operational goals as judged by these standards.
5. It keeps people informed about its policies and standards and about the steps that are being taken to keep the organization responsive to the needs of people whose lives it affects significantly.

The board of directors is presided over by a chairperson, who often is also chief executive officer of the company (the individual who runs the show), and occasionally may be the company president as well. The various corporate officers report to the board through the president.

In a technically based company there are three broad areas of responsibility plus various supporting services: (1) manufacturing, which makes as much specification-grade product as possible as cheaply as possible; (2) sales, which sells as much product as possible at as high a profit as possible; and (3) futures, which is concerned with the growth of the company. The author is prejudiced toward putting all futures activities in one organization, since it has been his observation that companies that do so are the ones that are most successful in expanding. However, there is no one best organization; it depends on the people involved. In large companies there is a tendency for a discipline-oriented organization to evolve—all the accountants in one department, all the chemists in another, all the lawyers in a third, and so forth.

In the futures effort, there is a need for planning and for decisions about the course of action to be followed. There is a need for project assignment and performance supervision. There is a need to obtain help from other company departments. There is a need for close coordination with the manufacturing and the sales departments, so that the results of the futures effort will be utilized. There is no sense in developing new products if the sales department will not sell them or in developing new processes if the manufacturing department will not use them.

III. ORGANIZATION OF A SMALL COMPANY

In this section some typical company organizations are examined [2]. First, in a very small company, the organization chart probably resembles the one shown in Fig. 1.

In a small company like this one, the functions tend to become blurred, with everyone wearing several hats. The manufacturing department not only comprises plant operation and facilities maintenance, but may also include some engineering and process improvement. The sales department may embrace not only direct sales, but also market research, marketing research, and commercial development (these functions will be defined and discussed in detail in Chapter

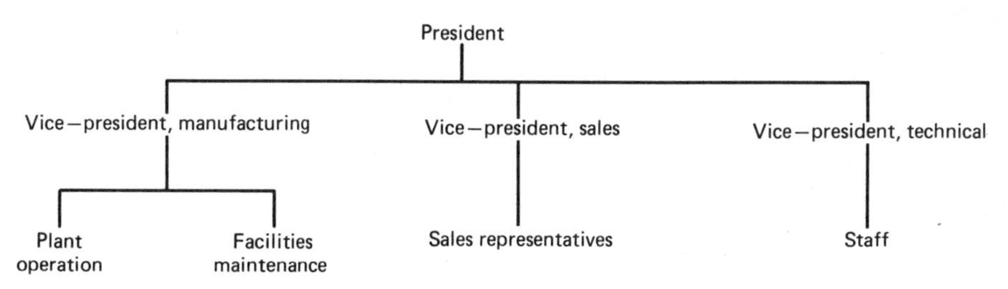

FIG. 1. Organization chart of a small company.

16). Under the vice-president, technical may well be found research, product and process development, applications research, engineering, quality control, technical service, pilot plants, and commercial development. As a company grows these functions become more clearly defined and more compartmentalized.

Another characteristic of a small company is informality and attendant ease of communication. All three of the vice-presidents shown in Fig. 1 supervise their entire staffs directly, and the subordinates in question all know each other and work together informally. Liaison between the three vice-presidents and the president is equally easy. For a plant trial, the vice-president, technical arranges things directly with the vice-president, manufacturing. When the plant has a problem, the vice-president, manufacturing goes directly to the vice-president, technical for asistance. When a customer has a complaint, the vice-president, sales goes to the vice-president, technical for help in resolving the problem. In fact all three vice-presidents probably discuss these problems together and with the president over lunch.

In small companies reaction to situations is speedy. Since all the key people have ready and immediate access to the president, the latter can authorize immediately the expenditure of funds, in addition to those budgeted, for developing a new product to meet a competitive market threat, or for a suddenly necessary remodeling of the manufacturing plant. This flexibility is what enables a small company to survive and grow. It enables it to quickly develop and market a new product or move rapidly in or out of a manufacturing area (as in the specialty-dyestuff business) as competition dictates.

Note that no service functions are shown in Fig. 1. The three vice-presidents probably do their own purchasing, and the accounts and personnel records may be maintained by the president's secretary. Legal assistance is hired outside when needed.

IV. ORGANIZATION OF A MEDIUM-SIZED COMPANY

In a medium-sized company the functions become more clearly defined. Staff activities have begun to be differentiated. However, there still may be only one manufacturing facility involved (Fig. 2).

Initially, responsibility for staff functions is focused on one individual, perhaps with the title of assistant to the president. These functions include accounting, finance, legal, patents, personnel, and purchasing. As the medium-sized company grows, one individual with whatever assistants are needed tends to take over responsibility for each function, still in staff relationship to the president and to company operations.

Plant engineering, although technical, is basically a manufacturing department responsibility and begins to appear there. However, the engineers in the futures area certainly continue to be involved in short-term plant problems, even though their basic responsibility has begun to be major process improvements and new processes.

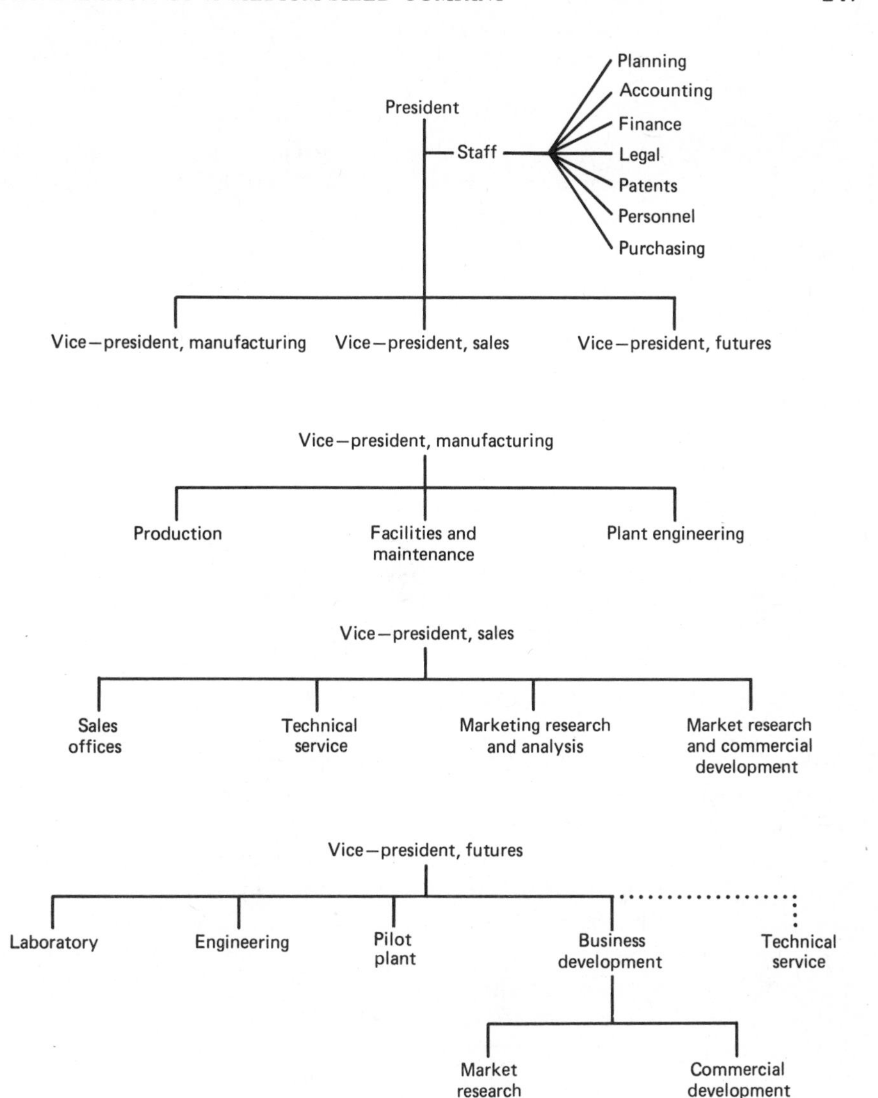

FIG. 2. Organization chart of a medium-sized company.

Technical service has probably moved under the sales department budget, where it belongs. However, technical supervision has to remain under the technical director, who is part of the futures area. Marketing research and sales analysis have begun to be differentiated and made a part of the selling responsibility. Even in a medium-sized company some market research, unrelated to present product sales, and some commercial development probably remain in the sales department.

The futures area has consolidated its technical activities into laboratory, engineering, and pilot plant. The staffing of the engineering department always has been a problem, not only in medium-sized companies, but even in very large companies. Corporations have fluctuated between having a very large engineering group and having a bare skeleton force. Once a modest average task load has been established, the tendency has been to maintain a staff of a size suitable to handle this day-to-day load and with enough senior people to monitor the work of the contract engineering companies that are hired to handle such peak loads as major plant expansions or new plant construction.

By the time a company has reached medium size, major aspects of new business development have at least begun to be focused in the futures area. These include market research on new products and the commercial development of the same, particularly to new customers, with the sales department continuing to help with old customers.

Although the various functions have become clearer and more specialized, communications have become more difficult. The various professionals tend to be housed in discipline-oriented groups. They no longer talk to each other informally every day at lunch. This means that messages have to go up through channels and back down again. There are many more copies of every memorandum. Reporting is more formal. There are more procedures and red tape. On the positive side, there is more delegation of responsibility and supervision.

V. ORGANIZATION OF A LARGE COMPANY

Last comes a very large company, in which the major responsibilities have probably been completely clarified. The manufacturing department has an engineering group, responsible for plant troubleshooting and minor process improvements. The manufacturing department no longer needs to keep running to the technical department for help with minor problems. Plant engineering is part of the cost of manufacturing.

The sales department also is self-sufficient, with its own market research and marketing research on present products. Technical service is a selling expense, and its budget is supported by the sales department, although its laboratory activities are supervised by the technical director.

The organization of the futures area now may look something like Fig. 3.

All the staff functions have been shown as reporting directly to the vice-president, futures, although most of the detailed responsibilities are delegated. Each technical director probably has his or her own stable of consultants, for whom the director is directly responsible. Facility maintenance probably is the responsibility of the director of engineering. Information services, including the main library and the storage and retrieval of information, may well be the responsibility of the director of research and development, although both the engineering and the business-development department probably have their own libraries as well.

There probably is a business manager who is responsible for purchasing and accounting. These areas have been separated from the corresponding corporate departments because the approach to life of the futures area is different. In a

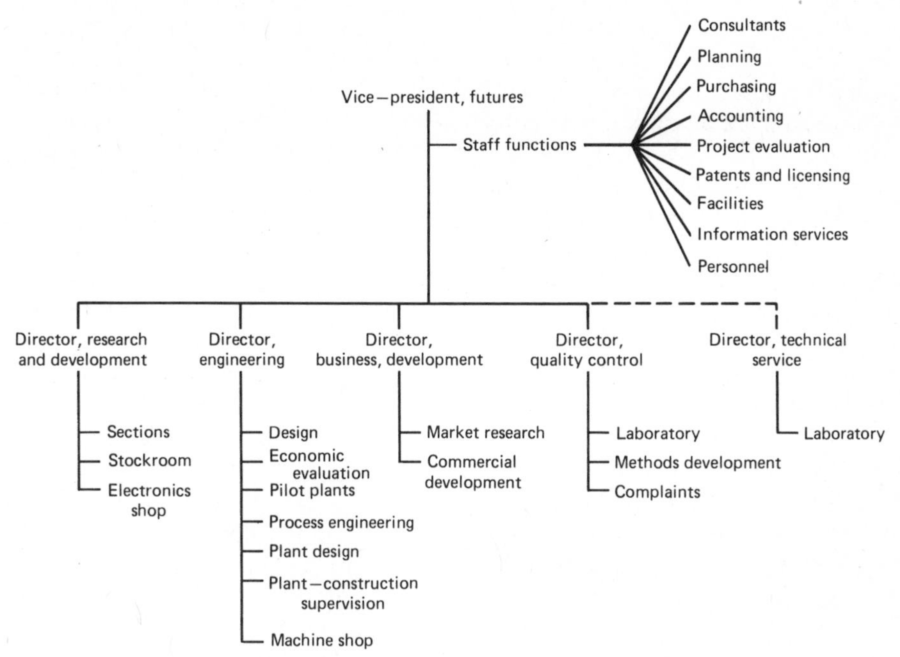

FIG. 3. Organization chart of the futures effort of a large company.

manufacturing operation the cost of raw materials, supplies, and services is an appreciable part of the final production cost and must be kept to a minimum through skillful purchasing. In a futures operation the major cost is the time of the professionals involved, so the object is to obtain raw materials, supplies and services of the quality required as quickly as possible. Futures inventories are small and constantly changing, particularly in the case of research chemicals and supplies. Corporate executives are interested in accounts of such expense items as rent, telephone, taxes, travel, utilities, and so forth. In a laboratory operation with any background of experience, these items can be estimated to within 5% simply by counting noses. What interests the vice-president, futures is the ongoing expense of each project.

Many large companies place their patent attorneys in the legal department simply because they are lawyers. (In smaller companies that employ outside counsel, liaison with the patent attorneys usually is conducted by the research and development people, and the corporate lawyers have nothing to do with patents.) A corporate legal department is concerned with the day-to-day problems of running the business—contracts, government regulations, and so forth. The patent department is the legal arm of the futures area and belongs there. It is concerned with the future technical position of the company, both as internally generated and as acquired.

The futures personnel department also has been separated out, mostly for convenience. The problems it faces are not tremendously different from those of the corporate personnel department, although the corporate people have a much larger number of union-labor problems. In the futures area a much higher percentage of the problems have to do with professionals. Corporate personnel directors exercises functional supervision over the futures personnel people to maintain uniform personnel policies throughout the company. They should.

The planning and project-evaluation department usually comprise a small staff group that works directly with the vice-president, futures and his several directors.

The quality control department is placed here, reporting to the vice-president, futures as the top technical officer of the company. In some pharmaceutical firms the quality control department is headed by a vice-president, who reports directly to the president. It never should be part of the manufacturing department, which it polices.

Technical service is shown as a dotted-line function to indicate technical supervision by the vice-president, futures. Budgeting, financial administration, and priorities are set by the sales department.

Although the stockroom, the machine shop, and the electronics shop serve all the futures areas, they have been placed under the supervision of the individual whose area probably is their best customer. The electronics shop could just as well report to the director of engineering.

The organization of the research and development sections varies widely from company to company, depending on the nature of the business and the managerial style in vogue at the time. Years ago the chemical sections tended to be discipline oriented—physical chemistry, organic chemistry, polymer chemistry, inorganic chemistry, analytical services, applications research, and so forth. In each section there were exploratory groups, product-development groups, process-development groups, and service groups. By service groups are meant technical laboratories that prepare large, although laboratory-scale, quantities of research raw materials; physical chemical groups whose primary function is to run thermodynamic and kinetic studies for the chemists and the engineers; and organic and inorganic chemical groups that prepare ultrapure samples of chemicals for physical chemical groups that are attempting to elucidate the last details of some manufacturing process of vital importance to the company.

More recently, particularly in technical areas in support of the present business, there has been a tendency to organize research and development sections in relation to product lines and markets—textile chemicals, paper chemicals, rubber chemicals, detergents, and so forth. In this case each section contains exploratory, product-development, process-development, and applications-research groups.

Obviously, in a very large company communications become much more difficult. By necessity they become more formal and standardized. The number of reports and people to be kept informed multiplies continually. There also is much more specialization and compartmentalization. Although good managements try to minimize their number as far as administrative efficacy permits, the number of intermediate management echelons increases too.

One solution to the communications problem, although very time-consuming, is a big increase in the number of committees and concomitantly in the number of meetings each person has to attend. Even though time-consuming, well-run committee meetings can be used to quickly impart a great deal of information to a large number of people and to garner a lot of advice from many people with different viewpoints.

Very large companies tend to break up into operating divisions, each with its own business. Originally, these too often were technologically oriented-organic chemicals, plastics, films, fertilizers, and so forth. The more recent trend has been to focus them around specific markets and employ whatever technology is needed in each case—retail products, agricultural chemicals, intermediates, rubber chemicals, and so forth.

Such divisions become profit centers. Although they have their own technical departments, the technical effort becomes mainly short-term in immediate support of the present business. The exploration of new technical and business areas and long-term, fundamental technical studies fall into the corporate area. Also, the vice-president, futures, if a technical individual, often takes over responsibility for coordinating the entire technological effort of the company, both corporate and divisional.

In conclusion, there is no one best company organizational structure. The ones shown in this chapter are reasonably typical and should give the beginning chemist some idea of what might be expected in industry.

REFERENCES AND NOTES

1. I. S. Shapiro, *Chemtech,* **6,** November, 1976, p. 688.
2. Every book of this type contains "typical" company organization charts. The ones shown here are all based on fact and are designed to give the reader a rough idea of company organizations and why they got that way. Other equally valid examples can be found in J. H. Saunders, *Careers in Industrial Research and Development,* Dekker, New York, 1974, pp. 35–45 and L. W. Bass, *The Management of Technical Programs,* Praeger, New York, 1965, pp. 44–52.

14

ALTERNATE ORGANIZATIONAL STRUCTURES

I. INTRODUCTION

In the last chapter the development of the typical pyramidal, discipline-oriented type of company organization was examined. As such pyramids grow bigger and bigger, they become more and more ponderous and cumbersome. They can no longer respond quickly to emergencies. Administration becomes very involved. Professional advancement and promotion are limited except through growth in the actual size of the organization. Innovation becomes stifled [1].

II. ACCOUNT EXECUTIVES AND PRODUCT MANAGERS

In order to cope with these problems, companies have devised a variety of organizational structures outside the pyramid. Usually large sales departments are organized on the basis of geography. Branch sales offices are located in key cities, and each sales representative has a specific territory or list of customers and potential customers, which he or she contacts on a regular basis. However, one customer may be so important to the company that a senior sales representative is made an account executive to coordinate all approaches to this one customer and to contact the customer at all locations and at several management levels (a number of plants, for instance) regardless of geography.

In some markets, particularly those very close to the consumer, such as cosmetics, it not only is necessary to respond very quickly to competition, but it also is necessary to coordinate the efforts of outside agencies with those of the company. To meet this need the product manager concept was evolved. For instance, the product manager for lipstick coordinates the technical effort directed toward new formulations, packaging development aimed toward new and more-attractive packages, and the overall sales effort. The manager obtains constant feedback from the market to guide the formulators and package designers. The manager works with the consumer panels to evaluate the company's products against those of competitors. The manager plans sales campaigns and works with the advertising agencies to support those campaigns. Thus the product manager carries a great deal of responsibility and yet is outside the pyramidal organization.

It is interesting to note that in the consumer-products business a new product can often mean only a new package. This puts no great strain on the chemist who does the formulating. However, the packaging engineer may be frantic trying to combine materials specifications and required mechanical properties with the artist's new and very attractive package design and achieve it at an acceptable cost.

III. TASK FORCES

In business an emergency may suddenly arise that necessitates throwing together a task force to meet the particular crisis in question. A senior individual is placed in charge, and whatever talents are needed are recruited from anywhere in the company. A friend of the author's used to handle real estate for a large oil company [2]. The company's distributor in a neighboring state sold out to a competitor. This necessitated the immediate building of at least one new bulk plant and the acquisition of sufficient real estate to again adequately serve the area with filling stations.

At that time the oil company sales department in question comprised 5 territorial offices, 15 division offices, 50 state manager's offices, and about 250 district offices. The task force which was formed was headed by a division assistant manager and included two state managers, four division real estate representatives, and eight salesmen, who did the paperwork and legwork for the real estate representatives. There also were four clerks to type all the reports and papers in support of the real estate proposals to company management.

At one point in his career the author was a member of several task forces (successively) besides performing his regular administrative duties of research management. These task forces were not so much of an emergency nature as the one just described. They involved merger and acquisition appraisals. They varied in size, but they usually included representatives from research, engineering, marketing, and finance.

IV. PROJECT TEAMS

In Chapter 7 it was shown that as a new product is developed more and more people progressively become involved in the effort. These people are trained in different disciplines and come from different parts of the company. Many contribute only a small part of their time. Even in phase 2, definitive development, direct administration through pyramidal channels has become impractical. The program hardly can be run on a day-to-day basis by a committee comprising the research and development director, the chief engineer, the chief patent attorney, and the director of marketing.

Based on task-force experience, the obvious solution is to set up a project and place one individual in charge of it. This type of administration, completely outside the pyramidal housekeeping structure, is used widely in industry, particularly in the development of new products and new processes in large companies. Since whole books have been written on the subject [3], only enough space will be devoted to it here to familiarize the readers with the project-task-force concept. They may well be assigned to such a task force on reporting to work.

A. Characteristics

Before considering one or two project examples, some of the characteristics of a project system need to be reviewed:

1. The project team is multidisciplinary. The example in Chapter 7 will be analyzed in detail, but it is clear that even as early as phase 2, a variety of research and development people, an engineer, a patent attorney, and a marketing individual were involved.

2. Obviously project administration crosses pyramidal administrative lines.

3. There is both full- and part-time participation. The amount of participation is determined by the project leader in relation to the needs of the assignment and to the budget. In general, most chemical analysts work on several projects at any one time.

4. The core team meets often, so that information and skills are fed in at the optimum moment and do not have to pass through pyramidal channels. The Monsanto Company core development team, (of which the author was a member) that was developing a low-density-polyethylene process met regularly every Monday morning. This project also will be analyzed in more detail shortly.

5. Acceptance and participation by team members is achieved by early involvement. For instance, the engineer helps even in the definitive development stage of a project. Thus a totally unfamiliar process in which the engineer has made no input, will not be dumped on him or her to pilot plant.

6. Members of the project team maintain their residence in their own discipline-oriented group, and in work outside the project continue to report to their immediate line-organization supervisor.

7. Project-team membership is a temporary assignment. When the project is completed, the individual team members are assigned to other tasks.

8. Project administration ensures continuity of personnel and administration. People heavily involved at the start may devote only a small amount of

their time as the project progresses, but at least they continue to be involved, if only on a consulting basis. And the actual managership of the project may change as the project progresses. A chemist may manage the laboratory phase, and an engineer (who has been involved all along) may take over when pilot-plant work starts. On the other hand, the chemist most certainly does not leave the team.

9. The project system leads to a flatter technical organization with administrative responsibilities spread more widely than in the pyramid itself. A senior project manager need not be a research director or an assistant research director. If a project is well run, everyone's line supervisor is kept informed through reports, but the line supervisors are not members of the team, although they may contribute some consulting help on occasion. The extreme of this situation is encountered in the consulting firm of Arthur D. Little, Inc., where all assignments are run on a project (case) basis, just as in a law firm. People who have only lived in a pyramidal organization have difficulty understanding how professional A in Arthur D. Little, Inc. can have professional B as a member of his or her case team (where professional A is the boss) and at the same time serve on one of professional B's teams (where professional B is the boss). Likewise the project (case) leader can have an administrative superior or even a company officer as a member of his or her project team, and in that respect subordinate to the project leader. The author did so many times.

10. Project teams are not used very widely in research. However, they are almost essential in development, where people of many disciplines pursue well-defined and carefully budgeted objectives.

11. With a project director and a core team free of other administrative supervision, the program and course of action can be changed quickly. New personnel can be added and old ones deleted as circumstances dictate.

B. Examples

The author was a member of a project task force for developing a low-density-polyethylene process. Table 1 shows a hypothetical research team for developing such a process. The exact team composition and the percentages of professional time expended are representative, based on the author's experience. The percentages of time put in by those on a part-time basis were not constant. They varied as the need arose.

Another example is the surface-coating project described in Chapter 7 (Table 2). Individual participation in each phase has been estimated, so that expenses can be quantified roughly. How such a project is coordinated and how responsibility changes from phase to phase is shown in the table.

C. Responsibility

Obviously the surface-coating project was a very large development project, typical of what would be encountered in a big company. How was it adminis-

TABLE 1. Polyethylene-Process Research Team

Member	Percentage of Time
Associate research director	10
Assistant research director	15
Research group leader	100
Senior research chemist	100
Junior research chemist	100
3 Technicians	100 each
Senior mechanical engineer (special high-pressure equipment)	100
Junior engineer	50
Draftsman	50
2 Mechanics	25 each
Analytical chemical group leader (new methods)	10
Analytical chemist	50
Physical Chemist	30
Polymer-evaluation group leader (new methods)	20
Polymer-evaluation technician	100
Polymer-evaluation technician	30
Consultant	—

tered? Figure 1 is a chart that shows the several responsibilities in each phase. The numbers on the chart correspond to the following responsibilities:

1. *Supervisory Responsibility.* These people planned the program, organized the team, supervised the work, and reported the results.
2. *Decision Responsibility.* These people decided whether or not to continue the project. They approved the proposals for future work and the budgets contained therein.
3. *Professional Responsibility.* These were the team members who did the work.
4. *Information Responsibility.* These people kept in touch with the project through reports and were consulted on specific points from time to time. In many cases the information they received early on gave them the background to actively participate in a later phase at a subsequent date.

In assembling this chart certain assumptions were made. Marketing and engineering as well as research and development were in the futures area. Market research and commercial development both were part of the marketing department. Engineering development, economic evaluation, pilot plants, plant design, and supervision of plant construction were all in the engineering department. Accounting was not included in the chart because these people knew what was going on anyway.

Phase 1 was run by research and development management (the chemical group leader) and there was no need to involve or inform anyone else, except insofar as they received copies of the regular research progress reports. The

TABLE 2. Surface-Coating Project Teams

Phase	Percentage of Time
I. Research	
Chemical group leader	5
Senior organic chemist	100
Surface-coatings expert	25
II. Definitive Development	
Chemical group leader	10
Senior organic chemist	100
Surface-coatings expert	50
Patent attorney	10
Market-research individual	10
III. Laboratory Development	
Chemical group leader (project director)	25
Senior organic chemist	100
Junior organic chemist	50
Surface-coatings expert	50
Patent attorney	10
Senior chemical engineer	10
Market-research individual	25
IV. Design Development	
Chemical group leader	5
Senior organic chemist	20
Surface-coatings expert	20
Patent attorney	10
Senior chemical engineer (project director)	50
3 Junior chemical engineers	50 each
6 Pilot-plant operators	50 each
Analytical chemist	50
Chemical analyst	50
Market research individual	25

In this phase the two-year time period was required for the outdoor exposure tests. The pilot plant only ran for one year.

V. Final Development

It is not possible to estimate at all closely just how many people were involved in this phase, since much of the expense represented the cost of the plant and the fees of the outside construction firm that helped with the plant design and did the actual construction work.

	Reasearch and Development	Reasearch and Development Management	Engineering	Engineering Management	Marketing	Futures Management	Patents	Purchasing	Quality Control	Sales	Manufacturing	Finance	Company Management
Phase 1, Research		1											
Product identification	3												
Preliminary exploration	3												
Definitive development proposal	3	2		4									
Phase 2, Definitive Development		1		4									
Product definition	3		4		4		3	4					
Process definition	3		3		4			4					
Applications research	3				3		3						
Economics	3		3		4								
Laboratory development proposal		3		4	3	2							
Phase 3, Laboratory Development		1		4									
Product evaluation	3		4		4		3	4	4				
Process evaluation	3		3		4		3	4	4				
Applications research	3		4		3		3		4				
Economics	3		3		4								
Design development proposal		3		3	3	2		4	4	4	4	4	4

FIG. 1. Management responsibilities in phases of a large development project.

chemist, with the help of the group leader, prepared the proposal for phase 2, and it was approved by research and development management.

Phase 2 also was run by the chemical group leader. However, besides the chemists involved, help was drawn from the engineering, marketing, and patent departments. The purchasing department was informed. The proposal for phase 3 was prepared by research and development management with help from the marketing department and was approved by futures management.

Phase 3 was run by the same group leader, now in the official role of project director. The engineering, marketing, and patent departments continued to

	Reasearch and Development	Research and Development Management	Engineering	Engineering Management	Marketing	Futures Management	Patents	Purchasing	Quality Control	Sales	Manufacturing	Finance	Company Management
Phase 4, Design Development		4		1		4							4
Pilot–plant design	3		3								4		
Pilot–plant operation	3		3		4		3	3	3		4		
Field evaluation of product	3		4		3		3			4			
Economics	3		3		4			3		4	4	4	
Final development proposal		3		3	3	3		4	3	4	4	4	2
Phase 5, Final Development		4		4	4	4							1
Plant design and construction	4	4	3	1	4	4		3	4	4	3	4	
Initial plant operation	4	4	3	4	4	4		3	3	4	1	4	
Market tests	4	4	4	4	3	4		4	4	1	4	4	
Initial sales		4		4	4	4		4	4	1	4	4	

FIG. 1. *(Continued)*

contribute. The proposal for phase 4 was prepared by research and development, engineering, and marketing managements and was approved by futures management. For the first time many other parts of the company received detailed information about the scope and progress of the development.

In phase 4, project leadership changed, and the senior engineer, who had been involved all along, took over as project director, with the big thrust then in the pilot-plant area. The research and development, marketing, and patent departments continued to contribute. In addition, both the purchasing and the quality-control department became actively involved. The proposal for phase 5 was prepared by all parts of futures management with help from the quality-control department. It was approved by company top management, and that approval meant that the decision was made to build the plant and enter the business.

Phase 5 was coordinated by a top company officer. Engineering management had the responsibility for designing and supervising the construction of the plant. The manufacturing department took over when the plant was ready to start up. The sales department was responsible for market tests, and, of course, for subsequent sales.

V. OTHER NEW-PROJECT APPROACHES

No matter how well administered, development projects like the two just described are slow and ponderous in their progress. Large projects such as these with their many decision points and large numbers of personnel are cumbersome. Worse, altogether too often, with so many different departments involved, there is a tendency among individuals to cover up and protect themselves and avoid taking risks no matter how dedicated the core project task force may be.

Another problem is that developmental efforts of this size cost money. In order for the project to pay off, the early sales revenue from the new product or process has to be large. The big-project approach does not produce small-volume new products whose sales eventually can grow to be very large.

Also, large companies have big overhead charges, which have to be supported by big individual-product sales volumes. A few years ago the author was in touch with a big chemical company whose excellent research and development department continually came up with potentially interesting new products. However, the company could not afford to carry these new possibilities in the red while they gradually grew, and the potential products might never have been able to profitably support their share of company overheads. The company sold these developments to other, smaller companies, where the overhead picture was not so demanding. Needless to say, this procedure was anything but good for morale in the big company's research and development department. Another example of this kind of large-company problem is the fact that at one point E. I. Du Pont de Nemours and Company reported Qiana (a new fiber) to be in the red, even though sales volume already had reached $35 million per year [4].

Yet new products are essential for a company's growth and continued profitability. Sooner or later every product either is displaced or finds itself in nearly fruitless price competition. For instance, in 1972 E. I. Du Pont de Nemours and Company reported that 15–20% of its $4 billion sales was represented by products that had been introduced to the market in the last 10 years [4]. And the company had many more products coming along in spite of the Corfam disaster ($100 million loss).

The basic problem for most companies, particularly large ones, in coming up with new products is organization [5]. How have such companies tried to solve this problem?

One approach has been to organize new-product task forces [5]. These do their job, or at least try to, and then disappear. Their weakness is that new-product development is a full-time, continuing job.

Another approach has been that of new-product committees [5]. These at least are reasonably permanent, but they have all the weaknesses of any committee in reaching decisions and implementing them. Also, the members of the committee are involved in present company business as well, which takes precedence over new-product development.

A commercial-development department has been mentioned; it can be reasonably effective if tied in closely with technical research and development in a futures operation. A product manager also has been mentioned. However, in both cases the people are in staff positions and have no muscle. They can recommend, but they cannot act or compel others to act. Altogether too often they just try to sell what research and development discovers and develops [5]. They are most effective in introducing new items to the present product line, not in innovating completely new products and product lines.

VI. VENTURE MANAGEMENT

Ideally innovation should be conducted by one individual (like the founder of the company) who knows the market segment, who can determine exactly what the significant needs are, who can invent the products to fill those needs, who can do the required development work, and who can plan and execute an effective selling program [6]. There are not too many of these people with the required background to determine what action should be taken and with the broad ability to accomplish it.

How, then, have large corporations tried to capitalize on the all-too-rare initiative and leadership skills, on the enthusiasm and energy, on the perspective and common sense, and on the feeling for commercial realities of the entrepreneur? One attempt, basically to recapture their corporate youth, has been venture management, usually through a new-enterprise division. This approach has been tried by Monsanto Company, Hercules Inc., E. I. Du Pont de Nemours and Company, M & T Chemicals, Inc., and Minnesota Mining & Manufacturing Company, to name a few examples.

Following are the requirements for a successful new-product-innovation effort:

1. It must be a continuous activity.
2. It must be a full-time activity.
3. It must have muscle and commitment.
4. It must be entrepreneurial.
5. It must be fast on its feet. The broad futures-area concept is fine for overall guidance and support, but specific projects have to move along very rapidly, particularly if initial sales and therefore the corresponding development effort are expected to be relatively small. Speed is essential. Not only must a company be in the right business, but it must be there at the right time.

Innovative needs in large corporations have often been met by product champions. These are people, often the inventor of the product, who identify themselves with the invention at an early date. They put themselves on the line and take many personal risks before all the facts are in. Admiral Hyman G. Rickover's identification with the atomic submarine is a good example. These product champions build both the technical and the marketing people into a team. They usually have a tremendous internal selling job to do. They are entrepreneurs within a company.

Basically what venture management does is institutionalize the product champion. The entire new venture is placed in the hands of no more than two or three people, as in a very small company, and these people are backed with money.

What are the characteristics of these new-venture groups?

1. Each venture is single minded with its own written charter.
2. Each venture is a separate profit center just as a new business is.
3. All the necessary skills—technical, marketing, and finance—reside in the one small group.
4. The group is entrepreneurial. There is no not-invented-here (NIH) factor or buck passing. The group members are all in the same boat, and their future success in the company is tied to the success of the venture.
5. The risks are big, as is the responsibility, but this is where the action is.
6. The operation is very flexible. It is dedicated to change.
7. There is no bureaucracy. Communications are just as easy as they are in a very small company.
8. New ventures are a proving ground for top management.

These ventures often are housed in a separate new-enterprise division, which may be part of the futures area or may be on a level with the operating divisions.

REFERENCES

1. C. G. Arcand, *Chemtech,* **5,** December, 1975, p. 710.
2. F. S. Tinthoff, private communication.
3. L. W. Bass, *Management by Task Forces,* Lomond Books, Mount Airy, Md., 1975.
4. Anon., *Chemical Week, 111,* November 29, 1972, p. 5.
5. D. I. Orenbuch, *Chemtech,* **1,** October, 1971, p. 584.
6. P. R. Lantos, *Chemtech,* **3,** October, 1973, p. 588.

15

REPORTS

I. INTRODUCTION

Now that a variety of corporate organizational structures have been examined, it should be evident to the reader that there is a big need for mechanisms of communication within every company. Meetings simply do not solve the problem, even in small companies. The standard mechanism, which is essentially universally employed in one form or another in every company, is the written report.

Therefore, very early in an industrial career, the technical professional is faced with the problem of writing reports. These vary in number and format from company to company and from department to department within a company. However, all companies require some reports in connection with nearly every activity. Research and development people write progress reports, sales representatives write sales-call reports, and those in the manufacturing department write periodic summaries of operations.

II. REPORT OBJECTIVES

Following are some of the overall objectives of reports [1]:

1. To provide the facts and conclusions necessary to enable management to decide on action and to take such action in as short a period of time as possible. This may mean beefing up or terminating a research and development program, coping with new competition in the marketplace, or changing a production schedule.

2. To enable management to coordinate the responsibility and accountability of people in different programs. Reports are a mechanism by which management guides the growth and advancement of its subordinates and evaluates those subordinates.

3. To provide clear communications between members of a project team, particularly when some members are not in day-to-day contact with each other.

In a large process-development program, this could include people in the departments of research and development, engineering, production, patents, marketing, and personnel.

4. To provide a permanent, well-organized, and easily retrievable record of information.

5. To provide the writer of the report with his or her fair share of recognition within the company.

In addition to these general objectives, there are several of particular concern to the individual writer:

1. A report enables the writer to organize the information and to evaluate it. Research and development people can detect leads that were not obvious until placed in perspective and can see gaps where additional information is needed. It enables them to draw relevant conclusions and thus is an excellent tool for planning future work.

2. A report encourages suggestions from its readers. Thus on a large development project, an engineer on the team may see where some additional laboratory data are needed, or a patent attorney may suggest a few experiments that will provide the basis for better patent protection.

3. A report informs others of the writer's contribution.

From these objectives it can be seen that a report to a very real extent is a personal sales document. It is a means by which an individual becomes known and establishes a reputation in a company. A reader of a well-written report may think, "I do not understand exactly what is being done, but there seem to be excellent reasons for doing it, and the results appear to be of real value to the business. It looks as though this could be a very useful contribution." Conversely, a reader of a poorly written report may think, "I do not think this individual knows what he or she is doing. To me the effort appears to be a waste of money."

Since a report is a sales document, it must provide its readers with the information they need in clear and understandable form and demand a minimum of effort on their part to obtain it. There is a real premium on brevity and conciseness.

III. REPORT RECIPIENTS

Who, then, are the readers of a report? For a research and development report they fall roughly into three groups.

1. *Top Management and People in Other Departments of the Company.* Many of these individuals have no technical training or have been trained in disciplines other than that of the report writer. They are not interested in technical findings as such. They want to know the reasons for the work (why the

writer is spending the company's hard-earned money in this fashion) and the significance to the business of the results already obtained and those that are hoped for in the future.

2. *Technical Management.* These people are interested in the same things as the previous group. They also are very much interested in the technical approach to the work, the technical reasons for the actions taken, the technical results, the significance of these results, and the recommended course of future action. This group uses reports not only to guide the professionals and the course of their work, but also to appraise them for salary increases and promotion to larger responsibilities.

3. *Other Technical People Who Need to Use the Detailed Information.* For example, an engineer needs process data in order to design, build, and run a pilot plant. A patent attorney needs examples for patent applications. A fellow chemist may need detailed directions for preparing a specific compound.

From the needs of these readers it can be seen that company reports are directed toward action and decision making. They are reader oriented. They emphasize the information that is of strategic importance to the company's business and to the specific program in question. Reports must be complete and contain all the information that needs to be known. They must also be concise, objective, and timely. A report that provides information long after it is needed is of limited value. A pilot-plant project can be delayed for an appreciable period of time if the research and development process report is not available. It can be disastrous if the sales department and top management are not informed immediately of the appearance in the marketplace of a new, competitive product line. To be effective a report must be clearly understood by those responsible for planning and action.

IV. REPORT TYPES

What then are the types of reports a professional is likely to be expected to write? These vary from company to company, but certain types are generally encountered. Since the author's experience has been in research and development, reports of this type will be discussed here, but the general reasoning is applicable to other company operations.

A. Requests for Investigations

Many companies require such reports as a means of formalizing the authorization and funding of research and development projects. The requests usually are short, often limited to one typewritten page. They are directed primarily to technical management, but those involving large expenditures of money may require top-management approval as seen in the previous chapter.

Since these reports are requests for money, the objectives in relation to the company's business must be clearly stated. In the case of a process-improvement study, it may be possible to estimate the likely annual savings or increased annual profit that will be generated if the project is successful. In the case of a product-development study, it may be possible to show how the product will fill a specific need and round out the company's product line. Annual sales for the new product often can be estimated. In the case of the exploratory investigation of a broad field, such as flame-resistant polymers, the justification is more general, but the technical and business opportunities are clearly delineated.

A brief description of the technical and business background as well as the patent situation usually follows. The writer hopes to say, "There presently is no product that fills this need. Our technology indicates how such a product can be developed. There are no adverse patents." This is the ideal case in a product-development project.

The approach usually is described. Technical management needs to appraise the technical feasibility of the project even though the business opportunity may be obvious. There is no room for the description of detailed experiments, although in discussing the proposed project orally with technical management, the hopeful investigator probably will have to describe in some detail the first experiments that are planned.

Finally, the cost of the project and its duration are estimated. In the case of development projects, both the cost and duration, at least to the first checkpoint, can be estimated with reasonable accuracy. In the case of research projects, a level of effort for a certain period of time usually is indicated, with the understanding that at the end of the specified period the project will be reviewed and reappraised.

B. Progress Reports

Almost all companies require regular research and development progress reports, often on a monthly basis. The exact format varies, but the basic elements are pretty much the same from company to company. These reports are directed primarily to technical management (which usually summarizes them for top-management consumption and review), even though top management may have a strong interest in some particular project. Obviously the report-distribution list includes people in other departments who are or may be involved with the project in one way or another, such as engineers, commercial-development people, and patent attorneys.

What are the essential ingredients of a progress report? It is desirable, if the company's format allows it, to include a very brief statement of background and objectives (perhaps three or four sentences) to put the work in perspective. Unfortunately in many companies this is done by title only.

There follows a very brief summary of the work accomplished, or not accomplished, during the reporting period. Since a report is a sales document, the following approach is worth considering. Create a favorable impression by reporting some positive results at the start. Then include all the bad news. Finally, report something favorable, so that the reader ends up with a good taste in his mouth. Needless to say, the reporting of the results must be accurate and not misleading. Improved color does not mean that you can now see a welding arc through a quart bottle of the stuff.

Finally come conclusions and recommendations. The latter must be spelled out and justified particularly clearly, since they are a request for more money. It is assumed that the project is sufficiently attractive technically that the investigator is reluctant to terminate it.

C. Interim and Final Reports

These are written at a key checkpoint in the investigation or as a final summation of the results achieved at the conclusion of a project. They include the same items as progress reports plus some others.

Although brief and concise, the introduction gives the complete justification for the project and the background of earlier work in the field. If patentable results have been discovered, the background involves a sufficiently complete literature survey that the patent attorney does not have to repeat the search. It is assumed that the investigator has worked sufficiently closely with the patent department that the possibly relevant patents in the field have been identified.

In some projects the objective and scope may have changed during the course of the investigation. The introduction delineates these changes, the reasons therefore, and the additional benefits that were expected from the changes thus made, both technical and businesswise.

The summary is the most important part of the report of an extended investigation. In some companies this is merely a summary of the technical results. In others it is a summary of the entire investigation and is placed at the front of the report, even before the introduction. In this case it summarizes the introductory material, the experimental results, the conclusions, and the recommendations for future action.

The conclusions are just that. To be effective they must be objective and critical. They constitute the basis for future action and are an assessment of the project.

The recommendations for future action are just that. Perhaps the project should be terminated. If so, that would be the recommendation. Perhaps more laboratory work is indicated. This would be particularly true in an interim report covering the work up to a specific, previously agreed upon checkpoint. In this case the recommendations for future work take on much the same form as a request for investigation.

Usually and almost always at the completion of the laboratory phase of a program, future plans call for action by other groups in the company. This may mean that the patent department should consider filing patent applications. More often the decision needed is whether to proceed with additional work such as bench-scale process-development and pilot-plant studies in which the engineering department and ultimately the production department will be involved. Usually the justification for this additional step (except in the case of a new process for a present product) involves a survey of potential markets and appreciable commercial development by marketing people.

In interim and final reports there usually is a fairly extensive discussion. Just where this is placed in the format depends on company policy. It must be brief and objective but at the same time comprehensive. The diary, or "when I went to the bathroom" type of narrative, is strictly frowned on. The discussion covers strategically important, completed work, not totally complete reporting.

Interim and final project reports contain an extensive experimental section. This records the experimental results for the permanent record. The presentation is directed toward those who will use the technical details. Therefore, the highest standards of scientific reporting obtain. If the report format allows, reference is made to the specific laboratory-notebook pages involved.

For those in exploratory work, the description of scientific experiments also is necessary for patent applications and, it is hoped, for publication in scientific journals. It is easy to write up experiments with this in mind and therefore avoid duplication of effort. If a specific chemical preparation has been tested under 8 or 10 different experimental conditions, the best experiment can be described in detail for the purposes of the report, the patent application, and the publication. Then the other experiments usually can be summarized in tabular form.

In interim and final reports there often is an appendix. It usually includes factual information not found elsewhere in the report, such as experimental results that proved irrelevant to the resulting main course of the investigation, extensive tables and graphs, and extraneous literature coverage. Preliminary experiments and unsuccessful trials often are included.

All such reports include a bibliography, which is self-explanatory.

D. Annual Reports

Some companies require annual reports, which are written at the end of the calendar year. They are a sort of interim report on active projects. They hardly assist management in budgeting [2], since they are written at the end of the calendar year and company budgets usually are prepared late in the previous summer or early in the previous fall.

Their major usefulness is to the writer. They enable him or her, particularly in a continuing project in which an interim report is not called for, to assemble the results over the year, examine them in critical perspective, and plan the future program thereby. It is essential both to the individual investigator and to management that this be done regularly and effectively.

In summary then, reports are an extremely important and effective means of communication within a company. They must be reader oriented and are a necessary vehicle for effecting timely decision making and action. They are one of the best sales documents available to individuals in a particular company. To write a good report the writer must have a clear understanding of company objectives, both technical and business, as well as an understanding of the objectives and significance of his or her own individual assignment.

REFERENCES AND NOTES

1. H. Skolnik and L. F. McBurney, *Chemtech,* **1,** February, 1971, p. 82. This is an excellent article, and many ideas contained therein have been added to the author's own experience in preparing this chapter.
2. B. J. Luberoff, *Chemtech,* **1,** December, 1971, p. 705.

16

MARKETING

I. INTRODUCTION

Part Two has concentrated on the business side of company affairs and has shown how the technical side (research and development) fits into the overall picture. In concluding this look at the business side, there is one area, in large part technically based, that requires more-detailed examination. This area is marketing. As they pursue their industrial careers, many technically trained people move into marketing, which has a very important corporate function.

In previous chapters industrial research and development were discussed, with strong emphasis on the technical side of innovation. Innovation is the translation of an invention or a technical improvement into the economy [1]. Marketing is the business side of innovation. It is not the same as sales. It mobilizes all the resources at an organization to create, stimulate, and satisfy the customer at a profit. Marketing research, market research, and commercial development are three of its tools.

Marketing, like research and development, is an area in which problems of semantics can arise quickly. In this book, the following definitions are used.

Sales. Like manufacturing, sales is a line (operating) function. The sales department's job is to sell as much of the company's products as possible at as big a profit as possible. The actual sale is the consummation of a transaction [2].

Marketing. Marketing is a staff function devoted to the expansion of the company's business. Not only does it embrace all the analytical, promotion, and service functions that lead to profitable selling of the present product line [2], but it also embraces the business strategy and business planning for entering new areas with new product lines.

Market Research. Market research is research of the market, either for a present product or for a possible new product or product line. It characterizes the market as to size, growth potential, business and technical opportunities, competition, and such other factors as government regulations, which contribute to the picture.

Marketing Research. Marketing research is research on methods of selling. It embraces such things as sales strategy, pricing, advertising, and selling methods (direct or through distributors, and so forth).

Commercial Development. Commercial development is the mechanism of introducing new products to the trade.

Like research and development, marketing can be both defensive (in support of the present business) and offensive (in opening up new business areas).

II. DEFENSIVE MARKETING

Defensive marketing embraces several functions.

A. Commercial Intelligence

The first function is commercial intelligence, which necessitates continual contact with the trade—executives, research, development, marketing, and purchasing people in other companies. Certain questions are of continual concern. Has a competitor developed a new product that may displace one or more of the company's? Has a competitor developed a new process that will enable it to seriously undercut the price of the company's product, or even make the company's product completely unprofitable to manufacture? Is a new competitor about to enter the field, perhaps by cutting price?

If a new product in a competitor's product line is involved, then samples must be obtained if possible, and product development must be alerted, so that it will be able to come up with new products to meet the competition. If a new process is involved, sufficient information must be gathered so that the research, development and engineering people can evaluate it.

B. Marketing Research

Besides commercial intelligence, defensive marketing has the major responsibility of generating greater sales of present products in existing markets.

This involves marketing research, some market research, and the various service functions in support of direct sales. Under marketing research the optimum terms and conditions of sale must be determined. These include such things as pricing, credit, discounts, consignment, lease/sale, scheduling, specifications, warranties, packaging, service, and returns. Market segments have to be established. These involve customer class and location; warehousing; and physical distribution. Market channels have to be determined—direct sales, through dealers, or private label. This leads to the establishment of sales territories.

Following are a couple of examples. A company wants to take business away from a competitor. If the company offers the competitor's customer a lower price, the customer often may allow the competitor and present supplier to meet the lower price, and the net result usually is less profit for everyone in the business. It may be much better policy to offer improved service or specifications that better meet that customer's needs. If a product is used in such totally different industries as food and electroplating (vanillin is), it may be desirable to have two different sales representatives handling the product, one for each market segment. Then each one can become an expert in the particular industry served.

C. Market Research

Although an appreciable number of factors have to be examined in a market-research study, a reasonable degree of accuracy may be expected most of the time in a field in which a company already is doing business. As will be seen later, life gets much more complicated in completely new fields.

What does market research embrace? It is a systematic analysis of the market, both present and potential. What are the underlying trends? Where is the industry that consumes the product going? How much more product will be needed in the next few years (sales projection)?

Market research involves a complete characterization, both business and technical, of the field or industry in question—its past, its present, and its future [3]. The market in this industry must be characterized qualitatively and quantitatively. Its segments and product lines must be defined, as well as their interrelation and profitability. This involves a determination of the volumes and prices for existing products, from which predictions may be made for the future. Other factors must also be weighed. Who presently is in the business (competition)? Who is likely to enter? What will those presently in the business do if there is a new entry? (Will the new entry be given a hard time through price reductions, and, if so, what reductions?) How will this affect profitability? Specifically who are the customers and on what basis do they buy (price, performance, technical service, long-term contracts, and so forth)?

How does a market-research individual come up with answers to all these questions? Just as there is a technical literature, so is there a market literature that can be consulted—compendia of statistics, journals, government reports, company annual reports, and so forth. When the researcher has a general knowledge of the consuming industry, trends can be graphed. A company's own sales and technical-service people can be helpful. Outside contacts are essential—other market-research and commercial-development people, technical people in other companies (particularly customer companies), purchasing agents, and executives. These can be reached through questionnaires and telephone calls, but most effectively through personal interviews.

Although studies of present products in a known field are likely to be reasonably accurate, the existing assumptions must always be challenged. It does not happen often, but something dramatically new can change the whole picture. The commercialization of the automobile had a revolutionary, devastating, and completely unanticipated effect on the markets for barns, horses, harnesses, carriages, sleighs, hay, and oats.

Service functions in support of direct sales are fairly obvious and need only be mentioned here in order to give an idea of their scope. They comprise the acquisition, training, and deployment of staff; incentives for staff; advertising and sales promotion in general; public relations; agency liaison; brochures and price lists; participation in merchandising shows; product management; product and package improvements (which means close liaison with product development); and behavioral research.

III. OFFENSIVE MARKETING

A. Present Products in New Markets

Generating greater sales of present products in new markets also requires marketing research, which is very similar to that required for old markets, except that members of the marketing staff have limited familiarity with the new fields to be served. This means a lot of extra groundwork rather than largely updating previous studies. Also, marketing channels may be different, through dealers rather than by direct sale. New conditions of sale may be indicated and new packaging may be needed.

Generating greater sales of present products in new markets also requires market research and plenty of it. Although the research is similar to that for old markets, the emphasis obviously has to be on new uses. It also means the adaptation of present products for these new uses with attendant new specifications, as for instance in the case of such formulated products as adhesives and surface coatings. These may really constitute new items in the present product line. This means close liaison with and the guiding of applications research, but not its direction. Applications research sets its own priorities on the basis of the information it receives. Care must always be taken that the new application and therefore the new market justify the necessary laboratory work. Is the market worth the cost?

The basis for these new markets has to be a real need that can be met. One has to have a complete picture of the industry in question in order to see these needs and to estimate reasonably quantitatively whether each need is worth filling. The need may be met technically by a present product, with perhaps slight modification, or perhaps by a new business approach.

It is important to remember that market research and sales are two entirely different and conflicting occupations. A sales representative's objective is to make a sale. A market-research individual's objective is to acquire information. It has been recommended that one not use line sales representatives for market research, because a good sales representative often cannot differentiate between products to be sold and those on which information is needed [4].

A friend of the author's has recounted a classic story that shows how differently market-research and sales people approach a problem [5]. A market-research man in a company that manufactured pumps called on a customer with the local sales representative. The customer was having real problems handling a viscous and corrosive liquid. Just as the market-research man was getting a really good fix on the viscosity and corrosiveness of the liquid and the inadequacy of presently available equipment, the sales representative chimed in: "Here on page 32 of our catalogue is just the pump you need to solve your problem." End of market research.

Generating greater sales of present products in new markets also entails additional service functions. New brochures must be prepared, the sales representatives require additional training, new advertising in new media is required, and different merchandising shows must be attended and probably exhibited in.

B. Innovation

Innovation is the introduction of new products into new fields. Organizationally, the marketing effort for increased sales for present products usually is very close to current operating functions. In a divisionalized company, this means it is located in the appropriate operating division. Conversely, innovation usually is a corporate function because of the time factor. Divisions are responsible for annual profits. Products or systems that are innovative usually require education and nursing to survive until profitable. Divisions are not likely to devote either the manpower or the money for the necessary length of time. Thus corporate research and development are part of the strategy of innovation.

Marketing-research and service functions are very much the same for innovation as they are for increasing sales of present products. They are obvious corollaries to the innovative process. The main marketing emphasis is on market research. Needless to say, the market research is far more complicated, much more likely to be inexact, and fraught with far greater risks than in the previous two cases. In fact it can be completely unreliable. In the case of a totally new product, people have no idea whether they need it or want it. Neoprene, polyethylene, nylon, Teflon, transistors, xerography, and the Polaroid camera are seven of many examples [6]. The failure of a project is more often due to business than to technical reasons.

In Chapters 4 and 6 the chemistry of two technically based fields was shown. These fields were developed by starting with very simple raw materials and synthesizing a wide variety of chemicals from them. Uses were found and needs filled by applications work on the part of the producers, or, more often, on the part of their customers.

How might a market-based field—furniture finishes—be developed [3]? What products are worth making and why? Twenty years ago the market survey would have been conducted entirely from the producer's point of view. It would have embraced a 10-year history of the size of the total market, from which a straight-line projection could be made for the next 3–5 years. It would have been broken into such specifics as sealers, stains, finish varnishes, and flat varnishes. The key users would have been identified. Typical container sizes and typical order and shipment sizes, as well as pricing and discount practices, would have been identified.

Now particular emphasis would be placed on such trends in the user industry as (1) the increasing scarcity and cost of wood, (2) the increasing use of plastics in furniture construction, (3) the use of more chipboard and particle board, and (4) the use of overlay papers over composition board. Major problems in the user industry would be identified such as (1) the large amount of floor space necessary for the conveyors and dryers and other finishing equipment, (2) the high cost of waste incurred in trimming fine wood, (3) the large amount of hand labor in furniture manufacture, and (4) the pollution, union, OSHA, and community problems stemming from the use of finishes containing large amounts of volatile solvents.

From these trends potential needs and opportunities would be identified such as (1) finishes for plastics, (2) finishes that give a woodlike appearance to plastic composition board, (3) rapidly drying systems, and (4) solventfree or nonpolluting finishes. Possible new product lines might be a system of graining inks or a system of solventless finishes with the equipment for curing them. In developing such lines, the developer keeps in mind questions such as the following. How will the customer use these products? How can the customer be helped? What is the sensitivity in the price–performance relationship? (The better one always costs more. This is a corollary of the "butter-side-down law" [7].)

In spite of all the care that may be used in studying the market, unforeseen factors may be overlooked completely and ruin a venture. Conversely, these factors may so influence the market that it becomes far larger than anyone would have dreamed. Scotch tape originally was developed to mend books. The huge market with its multitude of uses was hardly foreseen. Nylon originally was developed for hosiery and parachutes. No one thought of tire cord, sweaters, carpets, machine parts, and so forth. The competition for nylon hosiery was silk at $4.50 per pound, the competition for sweaters or carpets was wool at about $2.00 per pound, and the competition for plastics was $0.82–1.20 per pound [8].

The success of new products in new markets may require considerable faith on the part of their backers in the face of adverse indications [9]. The early pioneers of the automobile industry must have had such faith. Fifteen years ago it was felt that there might be a market for 50 computers. International Business Machines Corp. thought differently and now has antitrust problems on its hands. Two consulting firms used by Xerox Corp. said there was no market for the Xerox Corp. 914 copier. In fact, Chester Carlson, the inventor of Xerox, went to 22 of the best U.S. companies and was rejected 22 times. He was told that they saw no future for copying machines [8]. A third consulting firm said a few thousand might be sold over a period of several years. As of 1983 sales have been at least a quarter of a million items. Needless to say, many of the 22 companies later regretted their decision when Haloid Corporation, the predecessor of Xerox, became so profitable. It has been stated that 50–60% of our economic growth has been due to technological innovation [10]. There were no computers in 1945, nor was there any television, nor was there any jet-airplane travel. In 1965 these three industries represented $13 billion of the gross national product and employed 900,000 people [10].

In contrast to these success stories there have been any number of failures [11]. Some companies fail because their products fail to fill a need. Other factors can also cause failure, for example, cross-cultural communication problems. A pharmaceutical company conducted a worldwide market survey in which the Orient appeared to be a most promising market. The company brought out a whole new line of products packed in attractive white packages with multilanguage labels. The whole thing fell flat because to many Orientals white is the color of death. When frozen orange juice first came out, the word *concentrated* could not be used because people thought it had to do with bishops and popes. A chemical-equipment manufacturer put on a major promotional campaign in England for his scrubber. It was a failure because in England a scrubber is a lady of the evening. There are many similar instances [11].

These examples give some idea of the reasons for the success or failure of an innovation. The following section summarizes these reasons.

C. Requirements for Innovation

To be successful, innovation must have the support of management [12,13,14], which must in turn provide the proper atmosphere. This must have been true at International Business Machines Corp. and at Xerox Corp. in the success stories cited just now. In addition, the innovation must have a champion [12]—an entrepreneur, who organizes, manages, and assumes the risk for the change [14]. Entrepreneurs are committed to opportunity and often combine in themselves both the technical and marketing functions.

What then are some of the elements of a successful innovation [1,15]?

1. A thorough understanding of the user's requirements.
2. Good marketing. By this is meant a good market-research study, thorough education of the potential user, and an anticipation of possible customer problems.
3. A sound technical effort so that the new product has no technical defects.
4. Often the use of outside technology and advice in the particular area of concern.
5. Authority. Senior people in charge of the project.

Interestingly, successful innovations often come from companies outside the industry in question. Synthetic fibers and dyes have been developed by the chemical companies, not by the textile industry. Much of paper chemical technology also has come from the chemical industry and not from the paper companies themselves. Likewise many of the active ingredients that form the bases for detergents have come from the chemical companies and not from the soap companies.

Also, large companies tend to be poor innovators [1]. Thus the Polaroid process was developed by Polaroid Corp. and not by Eastman Kodak Company or Ansco. Transistors were developed by Texas Instruments Inc. and not by Radio Corporation of America or General Electric Company. Xerography came from Xerox Corp. and not from Eastman Kodak Company or Addressograph-Multigraph Corp.

Why is this? For one thing, operating units reject innovation [1]. Why do they do so?

1. Discounted cash flow (a present method of accounting) favors small increments of improvement in the present business, not major innovations.
2. Since engineering designs must work, most engineers are innately conservative.
3. Manufacturing is most efficient when it is routine. This will be discussed in more detail in Chapter 18.
4. Labor unions are very conservative. They are not in favor of labor-saving improvements and the possible use of new products that might revise standards.
5. Sales usually are rewarded by commissions, so time spent educating sales representatives and customers on new products is largely wasted from the point of view of the sales department.
6. An operating manager's pay is based on return on investment. A new product may lower that return temporarily, just when the manager is being considered for advancement.

A small firm innovates; a large firm develops [16]. As a company grows, its management changes into a bureaucracy, great for efficiently running the present business, in which future activity is defined by past experience, but stifling to innovation. This leads to a production orientation, with the attendant need for measurable controls. Since innovation is very risky for a middle-management bureaucrat because of the penalties for failure, particularly in connection with the large sums of money that may be involved, bureaucratic managements do not innovate. An individual who makes a mistake tries to place the blame elsewhere.

Thus even a large company such as E. I. Du Pont de Nemours and Company, which has a splendid record for innovation, can fail occasionally. Although the failure of Corfam has been attributed to the use of extensive computer and mathematical models that got out of touch with the market and with the business [16], the author has heard of other reasons. The company never had a really good technical process. The author has heard that the company never was able to use a cheaper urethane as a component of Corfam, nor did it ever achieve the economy of scale-up in production that was projected. A more serious flaw was in pigmentation. The light color shades so essential to women's shoe styles were irregular. Color matching from batch to batch was nearly impossible. Hides (the competition) were expected to become in short supply. They did not. Corfam will not stretch; leather will. Thus a woman could not buy a Corfam shoe that was basically too small for her foot (vanity) and "break it in." Also many women's shoes are for limited use with a specific outfit, for which the much cheaper polyvinyl chloride is perfectly suitable, even though it does not breathe. Market research did not estimate the competition of vinyls in shoes because E. I. Du Pont de Nemours and Company was not a producer of polyvinyl chloride.

D. Commercial Development

Commercial development, another tool of marketing, is the process of actually introducing a new product to the trade. It is the sales arm of the corporate futures activity, and involves putting corporate research and development to work. It involves taking samples of the new product to prospective customers and showing them how to use it to help their businesses. In the case of a formulation, or even of a commercial chemical, it involves establishing specifications and prices with the help of the prospective customer.

The following specific example illustrates how the process should work. (There are plenty of examples in which it did not.) A chemist discovers a new polymer, which a few preliminary tests indicate might find application as a surface-coating resin. The chemist contacts the surface-coating specialist in commercial development. The latter spends as much as a whole afternoon on the telephone talking to three or four friends in the coatings industry. The specialist describes the known properties of the new polymer—color, hardness, adhesion, solvent solubility, emulsifiability, light stability, and possible cost. On the basis of these telephone conversations, the development specialist advises the chemist that the possible cost is too high (it always is) and the solubility in hydrocarbon solvents should be improved (to a specific level also). On the basis of this information the chemist (supported by research and development management) decides that further laboratory work is justified. The possible cost can be reduced, structural modifications probably can be made that will increase hydrocarbon-solvent solubility without damaging hardness, adhesion, and weatherability, and color certainly can be improved. After three or four months' more work, the chemist goes back to the commercial-development individual with the new results. The chemist is asked to prepare a 5-pound sample of the new polymer. Then the commercial-development individual spends a week or so calling personally on the people contacted previously by telephone, giving them samples of the polymer and helping them formulate coatings based on it. The development specialist follows this outside evaluation work on the telephone and through further personal visits. The chemist may be included in one or two calls to help the potential customer with the evaluation, and so that the chemist can personally see the actual problems the potential customer is encountering—what the advantages the new polymer has over those presently in use and what disadvantages it has. This first field evaluation usually leads to the need for further product improvement and further potential-cost reduction. As these improvements are made, the commercial-development individual oversees more-extensive field trials.

The essence of this example is that the technical-development and the commercial-development departments have to work together very closely for the innovation to be successful. Each one puts in more time (and therefore spends more money) as the efforts of the other justify doing so. The sure way for this kind of teamwork to fail completely is for the commercial-development individual to insist that it is impossible to get a fix on selling price, size of market, and commercial utility until the research and development individual has given all the technical details. The research and development chemist cannot provide the answers to these questions without pilot-plant production of what may well be wrong-specification material. The chemist cannot request his or her superiors to authorize such pilot-plant expense without some idea of the potential market. End of this possible innovation.

Dr. Donald Schon [17] has described a classic case of how the technical-development and the commercial-development department can ostensibly be working together, but have their efforts collapse through faulty communications.

In this case the marketing department felt that there was a need for a new cleaning powder that could handle a wider range of stains, including cigarette and rust. Obviously, it should cost no more than those powders on the market and it should be just as efficient in all other respects.

Of some 10 formulations that were tried against cigarette and rust stains, one looked pretty good. However, the new ingredient that made it effective was oxidized too readily. An antioxidant was needed. One worked fine except that it reduced the effectiveness of the powder and provided a bad odor. Stronger abrasives were added to complement the chemical effectiveness, and new odorants were added to cover the bad odor the antioxidant had generated.

The formulation was now ineffective against rust after one week's standing. Substitutions had to be made for two of the components. Thus after six months' work, the formulation was effective against cigarette and rust stains and as effective as existing products against all other stains, and it had a good odor and a good shelf life. Therefore, the market-research department conducted a consumer test with 40 women over a two-week period. The product failed totally because it had a purple color. Color had never been revealed as a relevant variable. The product-development department had no way of realizing this, and the market-research department never told them (if they had ever found out).

In a very elementary way this chapter has summarized the various functions of marketing and how it interrelates with the technical functions of a corporation. Obviously, marketing is an area of possible interest to a technically trained individual. The beginning chemist entering industry should be aware of it.

REFERENCES AND NOTES

1. D. W. Collier, *Chemtech,* **5,** February, 1975, p. 90.
2. Much of the material on marketing is based on I. D. Canton, *Chemtech,* **4,** October, 1974, p. 581.
3. Much of the material on market research is based on P. R. Lantos, *Chemtech,* **3,** October, 1973, p. 588.
4. N. M. Draper, *Chemtech,* **5,** August, 1975, p. 464.
5. Frederick H. Greene, Jr., private communication.
6. J. H. Berg, *Chemtech,* **4,** May, 1974, p. 278.
7. G. J. Hahn, *Chemtech,* **4,** January, 1974, p. 16.
8. H. G. Johnson, private communication.
9. G. B. Zornow, *Chemtech,* **2,** October, 1972, p. 594.
10. A. E. Brown, *Chemtech,* **3,** December, 1973, p. 709.
11. W. A. Jordan, *Chemtech,* **6,** November, 1976, p. 671, gives many examples, a few of which are mentioned here.
12. M. A. Glaser, *Chemtech,* **6,** March, 1976, p. 182.
13. J. J. Schwartz, J. D. Goldhar, and T. J. Gambino, *Chemtech,* **6,** July, 1976, p. 418.
14. R. C. Springborn, *Chemtech,* **4,** May, 1974, p. 274.
15. Anon., *Chemical Week,* **110,** March 1, 1972, p. 41.
16. C. G. Arcand, *Chemtech,* **5,** December, 1975, p. 710.
17. D. A. Schon, *Technology and Change,* Delacorte Press, New York, 1967, p. 104.

PART THREE

CAREERS

17

PLANNING AN INDUSTRIAL CAREER

I. INTRODUCTION

In planning an industrial career the chemist or chemical engineer, even while still in school, is in a position to begin. In this chapter some guidelines will be laid down for this planning and an attempt will be made to acquaint the beginning chemist with the kinds of career opportunities that are available in industry. Training, both professional and nonprofessional, that would be helpful to an incipient industrial chemist or chemical engineer will be suggested.

The first job is the start of an individual's career. Formal training is now over. A job is a means for providing income and a way to pass the time pleasantly. Both are important. Everyone wants sufficient income to provide a decent living for his or her family and enough money to enable the family to do the things they want to do. Pleasure also is important, since after leaving school, professionals will be spending 40 hours a week minimum on the job for the next 40 years. Finally, all professionals, want the job to be a means for contributing to human progress and welfare.

In planning a career three important questions must constantly be asked and answered [1]: What have I done well? What do I do well? What do I like to do?

There are many career opportunities for chemists and chemical engineers in industry, and somewhere between 50 and 85% of them are not in research and development. It has been stated that 56% of American chief corporate executives have degrees in science or engineering [2].

For the purposes of this discussion the author has classified petroleum and pharmaceutical companies as part of the chemical industry. But there are many opportunities for chemists and chemical engineers in such nonchemical companies as General Electric Company, International Business Machines Corp., and Eastman Kodak Company. Likewise, the paper, textile, rubber, metallurgical, automotive, and building-materials industries have chemical problems and rely heavily on chemists and chemical engineers for their solution.

II. EDUCATION

It is assumed that the chemist or chemical engineer has or will have obtained a degree from a good undergraduate school with a major in chemistry or chemical engineering. At this level the major courses are pretty much prescribed. In graduate work at a good university it is desirable to keep course work at a minimum, particularly after the first year, since it is the thesis that is most important. The aim of education is understanding; so one should not get bogged down in details.

Besides a chemistry major, an industrial chemist needs a general understanding of chemical engineering, such as can be obtained in a standard unit-operations course if one is available at the institution in question. The chemist needs to appreciate material balances, heat balances, what metal equipment looks like, and how it is used. Although much of chemical engineering is applied physical chemistry, the chemical engineer in industry needs a reasonable knowledge of inorganic and organic chemistry. Knowledge is needed of what is going on in a reactor and of the physical and chemical properties of the chemicals being dealt with. The chemical hazards involved in the operations must be understood and possible environmental problems appreciated.

Although the technical aspects of a chemist's education are more or less obvious, there are several nontechnical areas that must be emphasized very strongly. Most important is the ability to communicate, both orally and in writing. Actually this is as important in academic life as it is in industry. In industry the professional has to write proposals for such things as research programs and plant modifications. Reports must be written covering research projects, engineering evaluations, and sales calls. Inability to communicate is one of the biggest complaints industry has with technical people. At one point in his career the author supervised a section comprising some 50 research chemists. He had to spend two days every month polishing up their monthly progress reports. Courses in writing and perhaps one in public speaking could be helpful to both chemists and chemical engineers.

Closely related to the ability to communicate is the ability to sell. All through life one has to sell oneself, one's ideas and one's programs. A technical professional does not need to act like the high-pressure life-insurance salesman who terrorizes his prospects with the possibility of imminent demise, but the professional does need to sell himself or herself as an individual and present ideas in clear and concise fashion. To a nontechnical individual the image of a technical professional is that of someone who can present facts in just such a way.

Every citizen needs to understand what goes on in the world and to appreciate the forces that shape events. In particular, a technical professional in industry needs to know enough economics to understand how the industrial system works. Such a professional also needs at least the rudiments of accounting, to understand budgets and the control and allocation of expenditures. For a broader understanding of the political and social forces at work in the world, some knowledge of history, political science, and sociology is desirable. Much of this can be acquired by outside reading in school and reading subsequent to graduation from school.

Many industrial operations are conducted by teams and task forces, so it is of primary importance to the technical professional in industry to be able to work with others and to understand how nontechnical people think. Basically this concerns political skill in the best sense of the word—the ability to compromise and work toward goals, the ability to motivate and lead others, and, ultimately, the ability to accomplish one's objectives with the help of others. A certain amount of back scratching may be involved. An organic chemist who takes the time to explain some chemistry to an instrumental analyst may find that an analysis is obtainable in a hurry when badly needed. The author has no set prescription for acquiring this political ability. Knocking around with a variety of people in fraternities, sororities, student government, athletics, and other extracurricular campus activities while still in college often is helpful. Along these lines the author once knew a chemical manufacturing executive, who, when hiring people for plant supervision, wanted to inverview C+ students who had been active on campus, and not the very top members of the class scholastically. Community activities are a logical extension of this form of education after graduation.

Finally, summer jobs while still in school contribute to an individual's education. Although the monetary consideration is obvious, it is here that an individual learns to work hard in the outside world and to get to know different kinds of people. Also, it is a chance for a student who may be thinking of a career in research to get an idea of what sales or manufacturing is like, and vice versa.

III. GENERAL RESOURCES OF THE TECHNICAL GRADUATE

Following is a list of some of the general resources of the technical graduate that the chemist or chemical engineer presumably has or will have acquired as a student [3]. The technical graduate

1. Has professional competence.
2. Knows how to learn.
3. Can apply knowledge.
4. Can solve problems with the logical use of data.
5. Knows the value of mathematics.
6. Knows that the frontiers of knowledge are never still.
7. Can continue to learn. (This is very important. One big reason that people with many years' service with a corporation may be terminated in middle age is that they have allowed themselves to become technologically obsolete.)
8. Is able (one hopes) to communicate with other experts and use their help.
9. Is qualified to cope with change. (Strangely, many people in industry are

not. For instance, plant managers often do not welcome the introduction of a new process in their plant. Everything is going smoothly, all the operators are well trained and experienced, and the plant engineers have a good preventive-maintenance and process-improvement program underway. The introduction of a new process means transferring some of the experienced operators to it and having to train new ones for the present processes. Preventive-maintenance and process-improvement programs are interrupted while the plant engineers get the bugs out of the new process. All this leads to lower efficiency and less profit and therefore a poorer image for the plant manager.)

10. Understands the importance of concentration.
11. Can use concepts to explain the inexplicable.
12. Has intellectual quality and intellectual integrity.
13. Is capable of independent effort. (A completed thesis is a good example.)

IV. CAREER OPPORTUNITIES IN THE CHEMICAL INDUSTRY

Although not all possible careers are included, the following sections should give the beginning chemist some idea of the breadth of opportunity available in the chemical industry, as well as some idea of the kinds of things in which he or she would be involved in each case [4].

A. Research

As defined at the start of Chapter 6, research is technical effort directed toward the discovery of new scientific facts. It obviously has a strong science orientation and, unfortunately, altogether too often can be very narrow. In industry its product is technology, not dollars. However, the technology must be in a field of company interest where its application will bring in revenue. For oil companies this means catalysis and hydrocarbon chemistry, for pharmaceutical and agricultural chemical companies it means compounds with physiological activity, and for petrochemical companies it means new reactor designs. Both chemists and chemical engineers are engaged in research in industry.

Involvement: *The science or technology in which the professional has been trained.*

B. Product Development

Product development is the use of known scientific knowledge to develop new products. Often the new product may be a new formulation such as a slow-release drug, a lubricating-oil additive, a surface coating (paint), an electroplating bath plus the electrodes, or a textile-finishing agent. It also can be a system such as an office copying machine, which includes the machine, the paper sheet, and the various sensitizing and developing chemicals; or photography, which includes the camera, the film, and the developer. Product development is much broader and can be shallower than research. It includes product testing and usually involves development on a larger-than-usual laboratory scale including pilot planting. Both chemists and chemical engineers do product-development work.

Involvement: *Technical, usually several different fields of chemical and engineering technology; market needs.* (*There is no sense in developing a product if no one needs it or wants to buy it.*)

C. Process Development

Process development is the use of known scientific knowledge to develop new processes. Typical examples are the hydrocracking of heavy oils, in which both catalytic cracking and hydrogenation are involved; the development of the present commercial phenol process from the initial preparation of cumene hydroperoxide and its scission to phenol and acetone in an academic laboratory; and several syntheses of adiponitrile from butadiene. All these processes involved known chemistry when their development was undertaken by industry. Also there is a lot of process-improvement work, including pilot planting, in process development. This phase of industrial chemical technical activity needs both chemists and chemical engineers. Again, it is much broader than research and draws on several branches of chemistry and engineering.

Involvement: *Technical, embracing several fields of chemical technology; economic.* (*Competition between different chemical processes for the manufacture of a specific product depends on costs.*)

D. Quality Control

Quality control is the assurance of product quality, and the function is vested in a specific group in a company's technical organization. Such groups are important everywhere in the chemical industry, but find their greatest strength in pharmaceutical companies and their suppliers. Quality control depends heavily on chemical analysis and to some extent on statistics. There is a particularly big opportunity for analytical chemists in methods-development groups, which convert research analytical methods into procedures that can be conducted quickly and accurately by semiskilled technicians. Usually only chemists find quality control a rewarding career.

Involvement: *Technical* (*the methods both chemical and statistical had better be right*), *federal and state regulations* (*those of the FDA in particular*), *customer relations.* (*Diplomacy is needed since most product complaints are referred to the quality-control department.*)

E. Engineering as Such

The engineering functions in a chemical company were described in detail in Chapter 10 but are listed here again for reference: research, product development, process development, pilot planting, plant design, plant site selection, project engineering, plant construction, and technoeconomic evaluation. The personnel involved usually are chemical engineers (plus mechanical, electrical, and civil engineers), although chemists quite often get into pilot-plant work and technoeconomic evaluation. Also one or more of the chemists who helped develop a given product or process quite often follows it through all subsequent stages until the commercial plant is in production. Such chemists may even be responsible indefinitely for all chemical problems encountered throughout the life of the plant.

Involvement: *Chemical engineering; economics; safety regulations; pollution control; federal, state, and local regulations* (*including zoning*); *labor relations; transportation* (*one has to bring raw materials to the plant and ship product to the customer*); *taxes.*

F. Manufacturing, Plant Management, Maintenance

This concerns the actual day-to-day operation of a chemical plant. In many companies manufacturing is part of the training for top management. Both chemists and chemical engineers do this kind of work, although it probably is more true of the latter.

Involvement: *Technical* (*one has to know the chemistry and the mechanical operation of the plant*), *trade-union practices, work organization, rates of pay, safety regulations, federal and state legislation, pollution control, method studies, efficiency* (*industrial engineering*), *trouble shooting, and minor process improvements* (*plant engineering*).

In plant management there is the obvious necessity for the ability to deal with people. Also, the plant manager cannot be an expert at everything, but must rely on the help of a variety of experts.

G. Commercial Development and Technical Service

These are grouped together because their requirements and much of their methodology are very similar. Commercial development is the introduction of new products to the marketplace. In technical service a chemist solves a customer's problems mostly by the use of present company products. In both activities a chemist needs a thorough knowledge of the company's product line and a knowledge of how each product performs in the various applications in which it finds use. In both activities a chemist needs to know the customer's requirements. Both people work very closely with the research and technical-development departments, and these departments often are a good source of new ideas. Here again the chemists and chemical engineers engaged in this kind of work must be able to deal with people.

Involvement: *Technical* (*as described above*), *economics, customer relations.*

H. Sales

Sales representatives persuade potential customers to buy the company's products, as well as obtain continuing orders from present customers. Sales and manufacturing are the two principal line functions in any chemical (or for that matter any manufacturing) company, and sales as well as manufacturing often is used as part of the training for top management.

Most companies have specific sales-training programs. Sales representatives must have the ability to persuade and negotiate (in the chemical business, chemicals are not always sold at list prices, and there are long-term contracts besides spot sales). A chemist or chemical engineer should not consider a career in sales if personally untidy or if easily discouraged. (One does not make a sale on every call.)

Involvement: *Customer relations, scheduling* (*a record of prompt deliveries helps make sales*), *distribution, transportation, packaging, economics, advertising.*

I. Planning and Marketing

As discussed in Chapter 11, the corporate planning and marketing group is the one that lays out the details of company strategy and often is staff to the board of directors. It is concerned with the financial resources of the company and broadly guides the technical and marketing efforts. It recommends strategies for growth and diversification, and locates and evaluates possible acquisitions. The membership of such a group usually includes both chemists and chemical engineers, most of whom have broad business experience and know the company well. Market research is one of the group's tools. An individual who cannot stand being wrong should avoid market research because he or she will be wrong.

Involvement: *Technical very broadly; mathematics; economics and finance; marketing, marketing methods, and markets very broadly.*

J. Patent Law

As discussed in Chapter 9, patent lawyers prepare patent applications and prosecute them in the patent office. They guide research in that they often ask for additional examples in order to obtain the best possible patent coverage.

Both chemists and chemical engineers become patent lawyers. They must have very broad technical backgrounds, since their work embraces a wide variety of subjects covering the whole area of the company's technological interests. Patent law is an ideal place for the technical professional who loves the technical field but dislikes laboratory work. A patent writer does not need a law degree, but such a degree is necessary for an individual to argue in the courts. In many cases the necessary legal training to enable an individual to become a patent lawyer is acquired in night school. A patent lawyer has to write well and argue effectively. (Patent examiners usually expect the attorneys to prove the validity of their claims. They do not do it for them.)

Involvement: *Technical very broadly, legal.*

K. Purchasing

Obviously purchasing agents buy a company's raw materials and supplies. Both chemists and chemical engineers become purchasing agents. They must be able to negotiate. (Again, many chemicals are not always sold at list prices.) For effective negotiation a knowledge of a supplier's processes and probable costs can be most helpful.

Involvement: *Technical, economics, scheduling, transportation, inventory control* (*inventories of raw materials have to be kept as low as possible, but they had better not run out when needed*), *quality control* (*it helps to understand the quality-control methods of a supplier who has delivered off-specification material*).

L. Personnel and Technical Recruiting

Personnel departments are concerned with organizational and people problems. These problems involve internal counseling, personnel training and development, personnel appraisal, salary administration (in a large company, salary policy must be administered uniformly from one location to another), personnel records (when a plant manager needs a new assistant manager and does not have a qualified candidate on deck, the personnel department provides a list of possible candidates available elsewhere in the company), labor relations (most companies provide personnel expert in labor relations to assist a plant manager with labor negotiations), hiring, and recruiting. The last mentioned can be of particular interest to chemists and chemical engineers. The companies most successful in field interviewing (at colleges and universities) are those that send interviewers who can talk the same language as the interviewees and thereby sell the company [5]. Both chemists and chemical engineers get into personnel work. In both cases they must like people.

Involvement: *Relations with and between people, psychology* (*everyone uses it, but personnel people usually get extra training*), *sales* (*in all interviewing of prospective employees the interviewer has to sell the company*).

M. Technical Information

Time was when technologists wanted specific technical information, they went to the company librarian for directions to the appropriate source. Now technical information has become almost a science in itself, involving computerized storage and retrieval of information, particularly in pharmaceutical companies, in which research people have to have ready access to a mountain of chemical and biological data generated both internally and externally. In many large companies the technical information people also send out to the technologists some sort of bulletin that calls attention to recent literature that might be of interest to the professional in question. The chemists and chemical engineers working in this area have to have broad technical backgrounds, a knowledge of company interests both present and potential, and a reading knowledge of foreign languages.

Involvement: *Technical, computerization, foreign languages (both in abstracting primary sources and in translating specific articles for company technologists).*

N. Export and Foreign Business

This field has been growing very rapidly and offers a big opportunity. The chemists and chemical engineers working in this area must have a sound knowledge of the company's domestic business and a working knowledge of foreign languages. (One only becomes fluent in a language by living in a country. In foreign business one has to be able to get around and deal with people whose knowledge of English may be sketchy at best, so even some knowledge of the language is helpful.)

Involvement: *Economics, shipping, agency arrangements (in many countries one does business through an agent), currency control (currency fluctuations can be very disconcerting), laws on contracts (they are different in every country), debts, credit responsibility, transportation, distribution.*

O. Management and Administration

One becomes a manager almost before one realizes it. Anyone who has two laboratory assistants or two pilot-plant operators reporting to him or her, is a manager and is no longer dependent only on his or her own individual efforts for achievement. Managers use specialists, but do not need to match their knowledge. (One does not have to be an expert mechanic to tell if a good repair job has been done on one's car or an expert carpenter to tell if the new cabinets in the kitchen are well made and properly installed.) However, a manager in industry does need enough professional competence not to be bullied or snowed. A manager must have the ability to delegate and to develop subordinates. A manager encourages order, cohesion, and purpose in the group and defends the group against external hostility. The group achieves external success in which all share.

The chemists and chemical engineers who become managers must have a thorough knowledge of at least their phase of the company's business. For self-defense they need some knowledge of accounting, since the accountants are the watchdogs of industry. A manager must be sure the accountants are providing the information that is really needed, not what they think is needed. For instance, a quality-control manager needs to know how much it is costing to control each product, not such items as chemicals, laboratory apparatus, rent, and utilities, which can be estimated within 5% on the basis of experience, simply by counting noses. Members of top management need to understand finance in order to cope with the bankers and other financial people. Financial people are strong in industry; the cuts of technical people in the early 1970s may be an indication of this.

Involvement: *First the manager's own professional work and closely related activities, ultimately all phases of the company's business.*

This summary has shown many of the career opportunities available in the chemical and related industries. In most cases much of the involvement has to be learned on the job, although specific outside study often is indicated. The author cannot emphasize too strongly the need for the ability to work with people and the need for the ability to communicate clearly with others both orally and in writing. Finally, he would like to emphasize again the desirability of enjoying one's career. Both people and organizations are most successful when the work is fun.

V. CAREER CHANGES

Before leaving the subject of specific career selection, it is desirable to see why many people change their careers during the course of their life. Perhaps the commonest example is the move into management—a research chemist becomes a research section supervisor, a plant engineer becomes an assistant plant manager, a sales representative takes over a sales office. Besides a move into management (promotion) there are many other reasons for career changes such as the following:

1. An individual may grow tired of the original career, and other opportunities may have greater appeal.
2. Someone in the research and development department may develop a product or process and decide to follow his or her "baby" into the sales or the manufacturing department.
3. The company may decide someone is better suited to another line of endeavor.

4. There may be better opportunities for advancement elsewhere as the company's needs change.

5. The company may feel someone has real potential for top-management responsibility and may wish to broaden his or her experience by exposure to other aspects of the company's business.

6. There may be a change in the company's business that compels a change for the personnel involved. For instance, the company may have to close a plant, and the supervisory engineers may be moved from manufacturing into a central engineering group, perhaps only temporarily.

The author recommends that a chemist or chemical engineer joining a company get a good look at other career opportunities as early as possible, so that a move can be made to advantage if and when the opportunity offers, and, at the worst, the professional will not get stuck in a blind-alley job.

Finally, there is one big change a beginning chemist or chemical engineer makes when moving from academia into industry, a move that can often be very traumatic. It involves an attitude of mind—the change from being a scholar to being a professional. The following section discusses the characteristics of each category [6].

VI. THE SCHOLAR VS. THE PROFESSIONAL

A scholar seeks knowledge and understanding for their own sake. Besides the mental satisfaction acquired, scholarship is measured by publications, by articles in learned journals, and by books. This applies to all fields of scholarship besides chemistry and chemical engineering. A biographer goes through a great deal of background material, letters, public statements, and so forth, in order to understand the individual about whom he or she is writing and thereby to publish a definitive biography. An ornithologist who discovers a new species of bird studies its habits, territorial display, diet, and so forth, and collects specimens in order to describe it in detail and classify it appropriately.

On the other hand, a professional uses his or her knowledge and training for a specific purpose. Professional achievement is measured by the results produced. A civil engineer builds structures, bridges, and roads, and their utility and durability measure the achievement. A physician ministers to the sick and injured, and this achievement is measured by the return to health of the patients. The physician uses new knowledge, such as a new operation or a new drug, for the benefit of the sick or injured. A lawyer assists clients with legal problems, and this achievement is measured by good documents—wills, deeds, contracts, and so forth—and by success in litigation.

In school a chemistry or chemical engineering student is a scholar, acquiring knowledge and understanding. Thesis results are an addition to the world's knowledge.

In industry a chemist or chemical engineer is a professional. Knowledge and training are used to provide goods and services—new drugs and dyes (research); better and cheaper plastics (product and process development); efficiently run plants (manufacturing); companies growing rapidly and profitably and providing more goods, services, and jobs (corporate planning); and a cleaner environment (pollution control or product modification, as in the case of biodegradable detergents). In each case the professional's achievement is measured by the results produced.

The main business of industry is to make and sell goods, to satisfy people's material needs at a satisfactory return on investment [7]. Business must operate at a profit; very simply, its income must be greater than its expenditures. (If a family does not operate its own personal affairs at a profit, it ends up in bankruptcy court, and its personal property is sold to satisfy its creditors). A business's profit is proportional to its efficiency in satisfying people's needs. Businesses that are not as efficient as their competition as a rule disappear, and the owners (stockholders) receive a small return of their investment (whatever can be gotten from the disposal of physical assets in a bankruptcy sale), and all the employees lose their jobs. A successful business executive does things and makes them work efficiently.

The professional in industry must recognize that business executives often have to make very costly decisions on the basis of inadequate data [8]. This hardly bothers a scholar; a theory can always be revised or replaced, and the only damage is to the ego of the original proponent. If a newly built plant needs major modification in order to operate, it can be a financial disaster to the company and a professional disaster to the technologists responsible for its construction. In business one must estimate the probability of success with the resources available and do the best job for the money [8]; that job usually is not a finished piece of work. For instance, consulting assignments are conducted on the basis of proposals that become contracts. The proposed job has to meet the client's needs satisfactorily, and the price has to be low enough that the client can afford it. Overruns come out of the consultant's hide.

As emphasized in Chapter 12, technical effort in industry must pay for itself and give the corporation what it needs. It solves problems in support of the present business and provides the technology that can be exploited as new business opportunities. Costs are not a restriction to the technical professional; they are a challenge. The professional in industry has to be familiar with such business tools as forecasts, objectives, budgets, and operating plans, and use them as sources of needed knowledge and guidance.

VII. CHARACTERISTICS OF A PROFESSIONAL IN INDUSTRY

Following are some of the characteristics of a professional in industry [9]:

The professional has a strong sense of individual responsibility. The professional reevaluates assignments continuously and constantly asks, "What business am I in?" [10] One of the toughest decisions to make is when to stop a project. But it is worse still to finish one and find a good technical job to be of no value.

The professional answers first to himself or herself and second to the employer for all professional actions.

The professional adheres to the tenets of integrity of the profession.

The professional gives the employer what the employer ought to have, whether or not the professional likes to give it and whether or not the employer wants it or likes to receive it. This may require considerable tact and diplomacy. However, no physician likes to tell a patient that the diagnosis is terminal cancer, and no lawyer likes to tell a client that the contract brought in for examination is worthless.

The professional has flexibility in approaching problems.

The professional has a tendency to assume he or she has tacit authority to make decisions the professional feels are in the employer's interest.

The professional has the ability to recognize the need for more knowledge and the ambition to get it. Constant self-improvement is very characteristic of a professional.

The professional has a desire to serve that is not necessarily related to personal regard for the person being served.

Financial return is not the sole measure of success in the professional's own eyes or in that of others.

A good idea of the professional's role in industry can be gained from the following brief exchange [11]:

Professional: "But my problem is. . . ."

Boss: "I don't give a damn what your problem is. Your job is to solve mine."

REFERENCES AND NOTES

1. These questions evolved in a discussion with Professor Zeno W. Wicks of North Dakota State University.
2. G. P. Armstrong, *Chemtech,* **1,** December, 1971, p. 711.
3. Adapted from G. P. Armstrong, *Chemtech,* **1,** December, 1971, p. 711.
4. Some of the ideas in this section came from G. P. Armstrong, *Chemtech,* **1,** December, 1971, p. 711, and from G. H. Beeby, *Chemtech,* **1,** April, 1971, p. 199.
5. J. C. Bailar, private communication.
6. Some of the ideas in this section came from B. F. Gordon and I. C. Ross, *Chemtech,* **1,** March, 1971, p. 136, and from J. A. Young, *Chemtech,* **2,** February, 1972, p. 72. Both articles are well worth reading for their descriptions of a professional.
7. C. Pacifico, *Chemtech,* **1,** January, 1971, p. 15. A good elementary description of industry is given in Chapter 2 of J. H. Saunders, *Careers in Industrial Research & Development,* Dekker, New York, 1974.
8. C. Pacifico, *Chemtech,* **1,** May, 1971, p. 270.
9. J. A. Young, *Chemtech,* **2,** February, 1972, p. 72.
10. W. E. Hanford. See B. J. Luberoff, *Chemtech,* **2,** August, 1972, p. 449.
11. B. J. Luberoff, *Chemtech,* **2,** July, 1972, p. 385.

18

GETTING A JOB

I. INTRODUCTION

At least by the last year in school most students begin to think about who is going to provide the funds to pay their bills. Likewise young chemists and chemical engineers may have found their first jobs completely unrewarding and may desire a change. Chemists and chemical engineers of all ages in industry may have found themselves unemployed and therefore with no source of income. Presumably all of them have done some thinking and planning about the kind of work that will be desirable, along the lines outlined in the last chapter. The present chapter provides more-specific information.

II. COMPANY SELECTION

The first step is to select companies for whom it would be desirable to work. Following are some of the considerations that it would be helpful to keep in mind.

As a starting point, *Chemical Week* publishes monthly a list of the 300 largest companies in the chemical and related industries. Most libraries have Moody's and Standard & Poor's indexes, which give some idea of the nature of a company's business as well as a list of its principal officers. Companies can be checked out there. For those interested in research and development, Bowker Associates publishes a directory of the industrial research laboratories of the United States. Although not complete, it gives similar information about many of them. Chemical engineers also may be interested in the chemical construction companies such as C. F. Braun & Co., The Lummus Co., Bechtel Corp., and so forth.

A. Company Product Lines

In Chapter 11 company product lines were discussed in relation to corporate planning. These will be reviewed briefly as one way to characterize companies [1].

Basically there are two types of product lines. The specifications for the first type, undifferentiated chemicals, are based on what the chemicals contain, a specific formula. All producers make essentially the identical product, and there are many applications for each chemical. On the other hand, the specifications for the second type, differentiated products, are based on performance. Here there are real differences in products from different suppliers, and there is only one or a very few applications for each one. Undifferentiated chemicals comprise true commodities, such as tonnage synthetics, gases, gum and wood chemicals, fatty acids, and fertilizers, which are large volume; and fine chemicals, such as low-volume intermediates, medicinals, and aroma chemicals, which are low volume. Differentiated products comprise pseudocommodities, such as tonnate resins and plastics, fibers, elastomers, carbon black, tonnage pigments, and tonnage surfactants, which are large volume; and specialty chemicals such as low-volume adhesives and surfactants, biocides, dyes, thickeners, formulated pesticides, and rubber chemicals. Companies dealing with fine chemicals and pseudocommodities are more like those dealing with commodities in the nature of their business, but still lie between them and companies dealing with specialty chemicals.

Therefore, it is most effective to take the two extremes and contrast them from the point of view of the technical employee. Commodity companies emphasize processes—synthesis and process engineering. They focus on raw materials and improved or new processes, in synthetic laboratories and pilot and semiworks plants. Manufacturing is in large, centralized plants using complex, sophisticated, automated, continuous, and inflexible equipment. There is lots of market research, but little test marketing or technical service. The sales representatives do not need to know much about a customer's needs.

Specialty companies emphasize the customer and its uses. The focus is on formulation, end-use knowledge, and the practical know-how of customers' plant operations in applications- and field-testing laboratories. Manufacturing is in several plants in different parts of the country, with many products made in each plant. Equipment is simple and flexible. There is lots of technical service and test marketing, but not much market research. The sales representatives must have extensive knowledge of customers' needs.

In appraising a company in this way from the point of view of product line, it must be remembered that many of the big companies are in all four types of businesses, probably centered in different divisions of the company, but not necessarily so. For instance, Monsanto Company sells ammonia (commodity), aspirin (fine chemical), polystyrene (pseudocommodity), and herbicides (specialty chemicals). There are many other examples of each category in the Monsanto Company product catalogue, as well as in the product catalogues of most of the other large chemical companies.

B. Company Size and Atmosphere

Another way of looking at companies is from the point of view of size, the simplest criterion being volume of annual sales in dollars. Large companies tend to be stable, whereas small companies may make it big or go broke. The larger companies tend to be in several different businesses, so at least on paper they would appear to offer a greater choice of opportunities. However, large corporations tend to become bureaucratic [2], which can mean limited chances to move from one type of career activity to another. Smaller companies are more flexible, so it often is easy to move. Large, bureaucratic companies tend to become compartmentalized, so an employee may have little contact with other company activities and types of people. In a small company everyone has lunch together and everyone knows everyone else—research, manufacturing, sales, accounting, purchasing, and so forth—and knows what each one does. In a large company competition is from inside, one divisional budget versus another, whereas in a small company the competition is from outside [3].

Companies can be conservative or innovative. In a conservative company, such as a bank, one does one's job, and progress and promotion are on a service basis and may even depend on the demise or retirement of one's superior. In general, large companies tend to be conservative and bureaucratic and resist innovation [2]. Smaller companies are more likely to take risks and innovate [4] and therefore may provide bigger opportunities for their more ambitious employees. However, even here there can be a mixture, E. I. Du Pont de Nemours and Company has been characterized as financially conservative and technologically audacious [5].

Some of a company's flavor can be gleaned from a look at its history, its growth rate and the duration thereof, and its ability to use research and development results. When visiting a company, one can learn a great deal from the atmosphere. Some years ago the author called on two paper companies on the same day. They were totally different. One was extremely progressive, with dynamic and aggressive employees working in a modern, superbly equipped plant. The other company was totally hidebound. The author does not remember whether or not the office force actually worked at rolltop desks, wore green visors, and held their shirt-sleeves up with rubber bands; but that was the atmosphere of the place. Strange as it may sound, divisions in the same company can differ almost equally widely, although perhaps not quite as much as the two paper companies just mentioned.

C. Company Ownership

Another very important factor to consider is company ownership and management. If a company is family owned and controlled, there may be limited opportunities for advancement, in that the top positions are always occupied by members of the family, and younger family members are always being groomed to succeed them. Likewise, in family-managed companies, communications between functions in the company tend to go up to the top and back down again. Therefore, although all the employees may know each other well socially (at the company picnic for instance), each one knows very little about what the other one does or how that individual's activities might relate to his or her own. There is little opportunity for an employee to help determine policy in a family-owned company. (In order to avoid any trace of nepotism, some non-family-owned companies refuse to employ relatives of anyone holding a managerial position.) Theoretically at least, there is no limit to advancement in a professionally managed and publicly owned and controlled company. Likewise there usually can be plenty of opportunity to become acquainted with company functions other than one's own, since people in different departments are expected to work together on a team basis, and individuals usually are given considerable responsibility.

D. Other Company Characteristics

Employee-relations policies are important. These include medical plans, retirement benefits, vacations, and so forth. Although these policies are anything but a major consideration in choosing between companies with standard plans, there are disadvantages in working for a company whose plans are grievously deficient. Of much greater importance are a company's termination practices. It is worrisome to work for a company that discharges a percentage of its employees arbitrarily every time the economy gets a little tight. Also, such a company probably is not building as effective and productive teams as one that constantly weeds out those employees who do not fit and who have no future with the company. The American Chemical Society keeps records of discharges and periodically publishes records of mass firings.

All successful companies must meet legal standards, respond to social pressures, meet the technological facts of life, and respond to competition [5].

Following are some characteristics of the healthiest companies [6]:

1. Everyone's job is secure as long as it is done well. No one is terminated during a reorganization. There is a spirit of cooperation and trust.
2. There is fairness in pay differentials, profit sharing, and promotions. (Unfairness leads to envy and resentment.)
3. Each individual professional has autonomy and a chance for development. No one is replaceable.
4. People participate in those decisions affecting them, such as how to do the work. They have the right to speak out without penalty, and they work with others and with customers.

Geography is also important—where do the employee and spouse want to live? One can lead a very enjoyable life almost anywhere in the United States, although it is difficult to convince many Californians and New Englanders of the fact. When he ran a research laboratory in southern California, the author had little trouble in hiring capable people from first-class Eastern companies—Californians who wanted to come home. Everyone has preferences. People raised in the South dread a cold Northern winter, and Northerners dislike the intense summer heat of the South. Some people like to live in a city with all its urban conveniences, whereas others prefer a small town or a rural location. Long commuting rides can be both time consuming and boring. To many it is important to be near relatives and friends. The author knows people who have left good companies for all these geographical reasons.

To summarize then, chemists and chemical engineers will want to pick the company that best answers their present career choices and will give them the best opportunity for job satisfaction as professionals. They will want to pick a company where they like the people and will enjoy working with them. In particular they will want a good boss whose performance standards are similar to their own.

III. JOB HUNTING

The chemist or chemical engineer now is ready to start an active job-hunting campaign [7]. This has, one hopes, involved some research along the lines just mentioned. Solid leads may have been gained from professors. Faculty contacts with industry are a particularly valuable way to be apprised of job openings. Other leads may be obtained from fellow chemists or engineers in industry with whom the chemist or engineer is acquainted. Interviews with industrial recruiters on campus of course are very valuable for students.

Other resources include the national American Chemical Society clearinghouse in Washington, D.C., and particularly the employment clearinghouses at national American Chemical Society meetings. Advertisements in *Chemical and Engineering News* and in such nationally circulated newspapers as *The New York Times* can be a very valuable source of employment leads. Such advertisements should be answered diligently, even though the yield per letter can be low. Every so often the right opportunity will crop up.

Chemists and chemical engineers also should use college or university placement offices insofar as they can be of help. Although usually not of great assistance to the professionally trained, local employment agencies occasionally turn up possible opportunities. The national "headhunters" are a great help to senior people.

A. The Résumé

The first task of the individual seeking a job is to prepare a résumé. The object of a résumé is to obtain an interview. It is a sales document, not a personal history. What does the reader want? The reader wants to get a good picture of the applicant with a minimum of effort. Therefore, the résumé should be short.

What does a résumé contain? First, the career objective is stated as clearly as possible. If one wants to get into administration, it is best to say that one hopes to eventually. No one is hired for such responsibilities at a tender age.

Next comes education—colleges and degrees with dates. Thesis titles are listed with the names of the advisers. Since the résumé will be read by a professional, a sentence clarifying the subject of the thesis often is desirable. If grades are complimentary, they are given. On the other hand, if the applicant is a "late bloomer," one who really did not catch on until late in college, it is best to omit grades. Their nature can be explained subsequently in the interview.

Experience is important, both professional and otherwise. This includes previous jobs and, if still in school, fellowships and teaching assistantships.

Honors and professional-society memberships, such as American Chemical Society and American Institute of Chemical Engineers, are helpful.

Many companies like to hear about outside interests—campus activities, hobbies, and community work. Their interest in a well-rounded individual was mentioned in the last chapter.

Military service still is important, although becoming less so.

Finally one lists such personal information as address, telephone number, age, and marital and family status.

Usually about three references are given, although in some cases it is stated that these will be furnished on request.

If one has publications or patents, these are listed on a separate sheet with complete titles and the names of all authors and/or inventors. Articles in press and in preparation (if reasonably well in hand) are included.

The résumé is introduced by a brief covering letter. This letter states why the chemist or engineer is contacting the company and especially why this applicant might make a contribution to the company's business. The letter and résumé are sent (the letter addressed personally) to the head of the department for which the applicant wishes to work (research, development, manufacturing, engineering, sales) to insure action. The individual so contacted may decide the applicant is too good a prospect to pass up, even though there may not be an immediate opening. One never applies to a personnel department. They act only on specific requests from other groups in the company. Almost invariably the reply will be something like the following:

"Thank you very much for thinking of X Company. You certainly have an excellent record. However, at the moment we do not have an opening for someone with your specific qualifications. Therefore, we would like to keep your résumé on file and will contact you when we do."

The applicant almost certainly will never hear. The only difference between most company personnel-department files and a circular file is that in the former case the document is not destroyed immediately.

B. Interviews

If the résumé has made a favorable impression, the first interview usually is a screening interview, conducted by one company representative. If the impression in this interview is favorable, the chemist or engineer usually is invited to visit the company location—laboratory, plant, or office—to be interviewed by several people and shown something of the premises.

What does one wish to accomplish in the interview, remembering that it will be run by the company representative? Most important is to sell oneself and get the job one wants.

Presumably the chemist or engineer is familiar with the company through the research outlined in Section II, and perhaps through a copy of the latest annual report (available at any company sales office) and a product list. In the interview the information so gathered should be verified. An effort must be made to determine as much as possible about the nature of the company and the career opportunities available in it. Can someone move from one activity to another? The good interviewer will impart most of this information.

The applicant will want to ascertain specifically the nature of the work in view, and the types of problems to be solved. If he or she is applying for a job in research and development or in engineering, this may be difficult, since much of this work is proprietary. The applicant should try anyway. It is easier for the interviewer to be more specific about manufacturing, sales, or quality control (current operations, not futures).

Finally, the applicant will want to find out what kind of people work for and run the company. Will the boss be a good one? Will most of the colleagues be likable (one cannot possibly like every one of them) and enjoyable to work with?

An interview is a two-way street. Besides giving the applicant a picture of the company, the interviewer expects to accomplish several other things in 20–30 minutes. First, the interviewer expects to get some idea of the applicant's technical competence. The author knows from his own experience and that of others who have conducted screening interviews of this type for other companies that this can be accomplished without resorting to trick questions in most cases. It is particularly easy with PhD candidates simply by discussing their theses.

Secondly, the interviewer is just as much interested in personality as is the interviewee. Do I like this individual? Could I work with him or her? Would this individual fit in my company? Are there any obvious personality problems?

This business of fitting in a company is highly subjective, like belonging to a fraternity or a sorority. There are no absolute standards like grades in a course. When visiting a school not too long ago, the author was asked to interview a particular student, whom the faculty rated very highly. Much to the school's amazement, a company representative had said the student was nowhere nearly aggressive enough. The author interviewed the student and liked him very much. Had he been interviewing for a company, the author would have recommended that this student be invited to the company laboratory for further interviews. The student would have fitted in the author's company, but evidently not in the other man's. More than that, the author probably would not have fitted in the other man's company either, nor the other man in his.

Finally, in the interview, the company representative wants to get some idea of the applicant's motivation and his long-term interests. Personally, the author has felt rather dubious about this type of information in the case of students, although it is extremely important with chemists in industry when the individual wishes to move from one company to another.

C. Company Interviewers' Checks

Besides interviewing the applicant, the company representative will check the applicant's references. What does the company representative expect to find out?

First, the representative will want to check his or her impressions of the applicant's technical ability. Is the applicant imaginative and resourceful? Can the applicant work manually in the laboratory, pilot plant, or plant? Inability to work with one's hands is one of industry's big complaints about beginning chemists.

Secondly, the interviewer will want to check his or her impressions of the applicant's personality. After all, the applicant was on best behavior in the interview. Can the applicant work with others? Will the applicant fit into an organizational team?

What about the applicant's character, honesty, integrity, and loyalty? Can this individual be trusted with company secrets? The author has placed this fairly low on the list, since most professionals meet these criteria. He has never seen a letter that said even very indirectly, "You do not want to hire this crook."

Finally, is the applicant flexible and able to learn? For instance, a professional in a new job will not be working on the same subject as in the old job nor on the same subject as in academia. Does the applicant work hard? Companies do not want lazy people.

IV. COMPANY ENTRY POINTS

Suppose all this has lead to a job offer and the applicant has accepted. What are the entry points into the company, depending on one's choice or that of the company? For a chemist the likely entry points are the following:

1. A research group.
2. A product-development group.
3. A process-development group.
4. A quality-control laboratory (perhaps as part of the company's training program).
5. A pilot-plant group.
6. A sales office.
7. A manufacturing plant.

For a chemical engineer the likely entry points are the following:

1. A research group.
2. A process-development group.

3. A product-development group.
4. A pilot-plant group.
5. A manufacturing plant.
6. A sales office.
7. A technoeconomic-evaluation group.
8. A design engineering group.
9. A project engineering group.

V. THE FIRST DAY

When the new employee reports for work, what will the first day be like? The following description is reasonably typical.

The first port of call is the personnel department. Here the employee's personal history is recorded in detail. General company policies and rules are explained. An employment contract is signed, and, if in research and development, a patent-assignment contract also. Usually, in the employment contract, one agrees that if one subsequently joins another company, one will not work in exactly the same technological area for a specified period of time, such as two years. This is to prevent some company from acquiring detailed know-how by hiring a competitor's employee. It is standard for a professional to assign all patent rights to the employer. When the author worked for Monsanto Company, it was in return for $1 and other valuable considerations (salary raises, bonuses, and promotions).

The next stop on the first day is the new employee's boss. The boss explains the new employee's duties and responsibilities and provides the necessary background for the job and the first assignment. As a potential sales representative, the new employee may start with a formal sales training course. The new employee is assigned home base—a research laboratory or a desk in a sales office, a plant office, or an engineering room. Specific company regulations that relate to the job are explained.

Most companies require protective clothing in laboratories and plants. In the laboratory, this usually means a laboratory coat and safety glasses. Chemical plants usually require hard hats and safety shoes as well as protective clothing and safety glasses. Safety shoes can be helpful in the laboratory also.

Laboratory results are recorded in bound notebooks, often with a tear-out carbon copy of each page. In this way, the original, permanent record can be kept in a safe, and the carbon copy can be torn out and used by the chemist or chemical engineer for immediate reference before later storage in some convenient access area such as the library. Pilot-plant records usually are kept similarly. Plant records often are kept on forms especially designed for the purpose. Laboratory-notebook pages have to be signed individually by the chemist or chemical engineer and witnessed by someone else. The new employee is introduced to these.

Research and development projects right through pilot planting usually are assigned some sort of code number, so that the charges to them can be monitored easily. This means that all the personnel involved have to keep time cards, perhaps to the nearest hour. The most obvious need for this kind of accounting is in the case of an analytical chemist, who may work on 20 different projects during the course of a month. Keeping time cards (not punching a time clock) should not bother a chemist or chemical engineer. Other professionals, such as lawyers and consultants, also keep time cards in order to charge each client fairly for the time devoted to that case. Doctors and dentists behave similarly, although they tend to charge a flat fee for each item, such as an office call, a physical examination, an appendectomy, tooth cleaning, a routine extraction, and so forth.

As discussed in Chapter 15, every professional in a company writes some kind of a report. The new employee is told what these are and shown their format, whether they comprise sales calls, production summaries, or research results.

Likewise, on the first day the new employee learns the location of information sources, supplies, and services and whom to contact therefor. These include the library, the shop, the stockroom, stenographic help, instrument repair, the glassblower, and even the location of the lunchroom and the lavatory.

Fairly early on the company physician gives the new employee a physical examination. These usually are repeated annually.

VI. ADVICE FOR SUCCESS

It almost goes without saying that the chemist or chemical engineer has received all kinds of advice about how to succeed. The author nevertheless feels it would be helpful to provide a checklist here [8]:

1. Obtain a clear definition of each problem.
2. Know the economic value of each problem.
3. Follow a work plan.
4. Plan shortcuts.
5. Strive for simplicity.
6. Balance precision with economics.
7. Keep busy.
8. Have several company problems if possible.
9. Work safely.
10. Finish what you start.
11. Keep your boss informed.
12. Be neat and pleasant.
13. Cooperate.

14. Be reliable.
15. Write good reports.
16. Follow up.
17. Make suggestions.
18. Grow with the company.

VII. EVALUATION OF TECHNICAL PEOPLE IN INDUSTRY

How are technical people judged in industry? Obviously this judgment is based on such things as hard work, loyalty, cooperation, productivity, imagination, and the other items in the list in Section VI.

Another criterion is technical competence. In industry, unlike in school, the technical competence of a technologist is judged by professionals in other professions and by nonprofessionals. Admittedly a single job must be judged by someone in the same profession. However, over time a nonprofessional can judge a professional by such criteria as (1) the extent to which the professional meets needs, (2) comparison with others in the same profession, and (3) the professional's reputation among other professionals [9].

In the previous chapter it was emphasized that business must operate at a profit. Therefore, there is an economic factor in a professional's evaluation. Perhaps this can best be visualized graphically [10]. Figure 1 shows the technical potential of three professionals as vectors.

Figure 2 shows a plot of them where H = point of hire, P = business objectives, and K = technical knowledge. Thus from the technical point of view $A > B > C$, but from a business point of view $B > C > A$. Only in research does the technical point of view far outweigh the business point of view, but even here the research results must be developable into profitable products and processes.

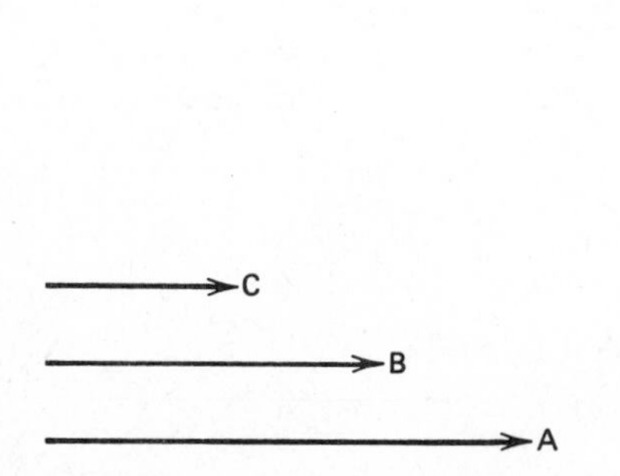

FIG. 1. Technical potential of professionals as vectors. Reprinted with permission from C. Pacifico, *Chemtech,* May, 1971, p. 270. Copyright 1971 by American Chemical Society.

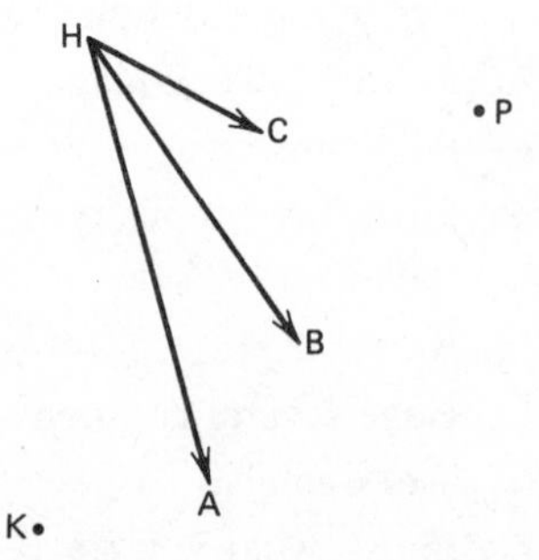

FIG. 2. Vector comparison of three professionals. Reprinted with permission from C. Pacifico, *Chemtech,* May, 1971, p. 270. Copyright 1971 by American Chemical Society.

REFERENCES AND NOTES

1. C. Kline, *Chemtech*, **6,** February, 1976, p. 10.
2. C. G. Arcand, *Chemtech*, **5,** December, 1975, p. 710.
3. H. Bishop, *Chemtech*, **5,** February, 1975, p. 107.
4. A provocative article along these lines is R. D. Clark, *Chemtech*, **2,** November, 1972, p. 656.
5. W. A. Franta, *Chemtech*, **3,** November, 1973, p. 650.
6. M. Maccoby, *Chemtech*, **5,** June, 1975, p. 336.
7. An excellent article along these lines is J. P. Corcoran, *Chemtech*, **3,** April, 1973, p. 200. The example of a good and a bad résumé is particularly helpful.
8. J. O. Percival, *Chemtech*, **5,** May, 1975, p. 300.
9. B. F. Gordon and I. C. Ross, *Chemtech*, **1,** May, 1971, p. 270.
10. C. Pacifico, *Chemtech*, **1,** May, 1971, p. 270.

INDEX